I.C. Parmee (Ed.)

Adaptive Computing in Design and Manufacture

The Integration of Evolutionary and Adaptive Computing Technologies with Product/System Design and Realisation

Springer

Dr Ian C. Parmee
Engineering Design Centre, University of Plymouth, Drake Circus, Plymouth, Devon, PL4 8AA, UK

ISBN 3-540-76254-X Springer-Verlag Berlin Heidelberg New York

British Library Cataloguing in Publication Data
Adaptive computing in design and manufacture : the
    integration of evolutionary and adaptive computing
    technologies with product/system design and realisation
    1.Engineering design - Data processing 2.Manufacturing
    processes - Data processing
    I.Parmee, I. C.
    620'.0042
    ISBN 354076254X

Library of Congress Cataloging-in-Publication Data
Adaptive computing in design and manufacture : the integration of
    evolutionary and adaptive computing technologies with product/system
    design and realisation / Ian Parmee, ed.
        p.   cm.
    Includes bibliographic references.
    ISBN 3-540-76254-X (alk. paper)
    1. Computer-aided design. 2. Adaptive computing. 3. Production
    management--Data processing.   I. Parmee, Ian, 1954-
    TA174.A32 1998
    620'.0042'0285--dc21   98-6382

Typesetting:Camera ready by editor
Printed and bound at the Athenæum Press Ltd., Gateshead, Tyne and Wear
69/3830-543210 Printed on acid-free paper

Preface

The third evolutionary / adaptive computing conference organised by the Plymouth Engineering Design Centre (PEDC) at the University of Plymouth again explores the utility of various adaptive search algorithms and complementary computational intelligence techniques within the engineering design and manufacturing domains. The intention is to investigate strategies and techniques that are of benefit not only as component / system optimisers but also as exploratory design tools capable of supporting the differing requirements of conceptual, embodiment and detailed design whilst taking into account the many manufacturing criteria influencing design direction.

Interest in the integration of adaptive computing technologies with engineering has been rapidly increasing in recent years as practical examples illustrating their potential relating to system performance and design process efficiency have become more apparent. This is in addition to the realisation of significant commercial benefits from the application of evolutionary planning and scheduling strategies. The development of this conference series from annual PEDC one day workshops to the biennial 'Adaptive Computing in Engineering Design and Control' conference and this year's event reflects this growth in both academic and industrial interest. The name change to include manufacture relates to a desire to increase cover of integrated product development aspects, facility layout and scheduling in addition to process / machine control.

A major objective has been to identify fundamental problem areas that can be considered generic across a wide spectrum of multi-disciplinary design activities. Assessment of current state-of-the-art of the integration of adaptive computing strategies with design and manufacturing practice provides an indication of best way forward to ensure short to medium term industrial take-up of the technologies whilst also identifying those areas requiring further medium to long term research effort. Although potential has been realised to a significant extent the technology is still at a formative stage. Extensive applied and fundamental research is required to ensure best utilisation of available computing resource by improving the overall efficiency of the various techniques and to promote the emergence of new paradigms. Complex issues relating to the integration of engineering heuristics with the adaptive processes and the interactive role of adaptive search processes within design / manufacturing team activities must also be considered.

The potential of adaptive computing within the design and manufacture field goes far beyond the optimisation of specific complex systems.   The development of appropriate co-operative strategies involving the various techniques and technologies can provide extensive support in multi-disciplinary design and manufacture domains characterised by uncertainty relating to poor problem definition, multiple objectives and variable constraint.   A breadth first adaptive exploratory approach can enhance innovative problem solving capabilities and provide decision support to establish optimal design direction.   It is also possible that co-operative human / adaptive computing procedures can lead to the emergence of creative design solutions.   This area of widening adaptive computing potential within design is now receiving significant attention from several international research groups and prototype co-operative search tools are gradually emerging.   It is hoped that the papers presented here, in addition to representing state-of-the-art adaptive optimisation procedures, also stimulate the research and application community to explore this greater potential through the development and integration of truly interactive, adaptive computational procedures for design and manufacture.

April 1998

Ian C. Parmee
University of Plymouth

Organisation

Supporting Bodies:

The Institution of Engineering Designers, UK
The Institution of Civil Engineers, UK
The Institution of Mechanical Engineers, UK
The British Computer Society
Society for the study of Artificial Intelligence and Simulation of Behaviour
(AISB), UK;
European Network of Excellence in Evolutionary Computation (EVONET)
CASE Centre for Computer-Aided Engineering and Manufacturing,
Michigan State University
Key Centre for Design Computing, University of Sydney.

ACDM Lecture:

Genetic Programming: Design of Electrical Circuits and Other Complex
Structures by Means of Natural Selection

> Prof J Koza
> Stanford University
> USA

Keynote Presentations:

Adaptive Systems in Design: New Analogies from Genetics and
Developmental Biology.

> Prof J Gero
> Key Centre for Design Computing,
> University of Sydney, Australia

Emergent Computational Tools for Structural Analysis and Design

> Prof P Hajela
> Rensselaer Polytechnic Institute
> USA

Evolutionary Algorithms - On the State of the Art

> Prof H-P Schwefel
> University of Dortmund
> Germany

Contents

Preface vi

Organisation vii

1. Design Issues

1.1 Adaptive Systems in Designing: New Analogies from
Genetics and Developmental Biology.
J. Gero 3

1.2 Emergent Computational Models in Structural Analysis and
Design.
P. Hajela 13

1.3 Exploring the Design Potential of Evolutionary / Adaptive
Search and Other Computational Intelligence Technologies.
I.C. Parmee 27

2. Adaptive Computing in Manufacture

2.1 Tackling Complex Job Shop Problems using Operation
Based Scheduling.
D. Todd, P. Sen 45

2.2 Minimising Job Tardiness: Priority Rules vs. Adaptive
Scheduling.
D.C. Mattfeld, C. Bierwirth 59

2.3 Evolutionary Computation Approaches to Cell
Optimisation.
C. Dimopoulos, A.M.S. Zalzala 69

2.4 A Temperature Predicting Model for Manufacturing
Processes Requiring Coiling.
N. Troyani, L. Montano 85

2.5 Solving Multi-objective Transportation Problems by
Spanning Tree-based Genetic Algorithm.
M. Gen, Y-Z Li 95

3. Generic Issues: Model Representation, Multiobjectives and Constraint Handling

3.1 Optimisation for Multilevel problems: A Comparison of Various Algorithms.
M.A. El-Beltagy, A.J. Keane 111

3.2 Evaluation of Injection Island GA Performance on Flywheel Design Optimization.
D. Elby, R.C. Averill, W.F. Punch III, E.D. Goodman 121

3.3 Evolutionary Mesh Numbering: Preliminary Results.
F. Sourd, M. Schoenauer 137

3.4 Two New Approaches to Multiobjective Optimization Using Genetic Algorithms.
C.A. Coello Coello 151

3.5 Mapping Based Constraint Handling for Evolutionary Search; Thurston's Circle Packing and Grid Generation.
D.G. Kim, P. Husbands 161

4. Structured Representations

4.1 Evolutionary Design of Analog Electrical Circuits using Genetic Programming.
J.R. Koza, F.H. Bennett III, D. Andre, M.A. Keane 177

4.2 Improving Engineering Design Models using an Alternative Genetic Programming Approach.
A.H. Watson, I.C. Parmee 193

4.3 From Mondrian to Frank Lloyd Wright: Transforming Evolving Representations.
T. Schnier, J.S. Gero 207

4.4 A Comparison of Evolutionary-Based Strategies for Mixed Discrete Multilevel Design Problems.
K. Chen, I.C. Parmee 221

5. Aerospace Applications

5.1 Design Optimisation of a Simple 2-D Aerofoil Using Stochastic Search Methods.
W.A. Wright, C.M.E. Holden 233

5.2 Adaptive Strategy Based on the Genetic Algorithm and a
    Sequence of Discrete Models in Aircraft Structural
    Optimization.
    V.A. Zarubin 245

5.3 Multi-objective Optimisation and Preliminary Airframe
    Design.
    D. Cvetkovic, I. Parmee, E. Webb 255

5.4 Evolving Robust Strategies for Autonomous Flight: A
    Challenge to Optimal Control Theory.
    P.W. Blythe 269

6. Other Applications

6.1 Performance of Genetic Algorithms for Optimisation of
    Frame Structures.
    M.R. Ghasemi, E. Hinton 287

6.2 Global Optimisation in Optical Coating Design.
    D.G. Li, A.C. Watson 301

6.3 Evolutionary Algorithms for the Design of Stack Filters
    Specified using Selection Probabilities.
    A.B. G. Doval, C.K. Mohan, M.K. Prasad 315

6.4 Drawing Graphs with Evolutionary Algorithms.
    A.G.B. Tettamanzi 325

6.5 Benchmarking of Different Modifications of the Cascade
    Correlation Algorithm.
    D.I. Chudova, S.A. Dolenko, Yu. V. Orlov, D.Yu. Pavlov,
    I.G. Persiantsev 339

6.6 Determination of Gas Temperature in a CVD Reactor From
    Optical Emission Spectra with the Help of Artificial Neural
    Networks and Group Method of Data Handling (GMDH).
    S.A. Dolenko, A.F. Pal, I.G. Persiantsev, A.O. Serov,
    A.V. Filippov 345

6.7 Multi-Domain Optimisation using Computer Experiments
    for Concurrent Engineering.
    R.A. Bates, R. Fontana, L. Pronzato, H.P. Wynn 355

Chapter 1

Design Issues

Adaptive Systems in Designing: New Analogies from Genetics and Developmental Biology.
J. Gero

Emergent Computational Models in Structural Analysis and Design.
P. Hajela

Exploring the Design Potential of Evolutionary / Adaptive Search and Other Computational Intelligence Technologies.
I.C. Parmee

Adaptive Systems in Designing: New Analogies from Genetics and Developmental Biology

John S Gero
Key Centre of Design Computing
University of Sydney NSW 2006 Australia
email:john@arch.usyd.edu.au

Abstract. This paper introduces the notion that analogies other than Darwinian evolution may be used as the bases for useful designing processes. Two classes of analogies are described: those drawn from genetics and those drawn from developmental biology. Two extensions to the genetic analogy are described. The first utilises concepts from genetic engineering, which is the human intervention in natural genetics, and concepts from reverse engineering. The second utilises concepts from developmental biology which is concerned with how the organism develops in its environment once its genotype has been fixed.

1. Introduction

Designing has long been recognised as a difficult, complex and unusual task. The first mention of design goes back to the code of Hammurabi promulgated around 1950 BC. Science has formed the basis of the technology on which engineering sits. It has provided the necessary theory of material behaviour and the experimental methodology to determine such behaviour. Using theories of material behaviour it has been possible to develop formal methods of analysis of the behaviour of configurations of materials (ie designs) under a variety of environmental conditions. However, science has not had the same success in providing any foundation on which to base the technology of formal design methods. More recently, it has been suggested that designing in its fullest sense maps well onto abductive processes which helps explain why it is so difficult to formalise it. In addition to its abductive nature designing is situated: ie designing cannot be predicted since decisions to be taken depend on where the designer is at any particular time and what the designer perceives the situation to be when he is where he is. We will use the word "designing" to denote the act and the work "design" to denote the results of the act to avoid confusion.

Computational processes which support designing do not necessarily require any theoretical foundation and are usually restricted to some subset of the totality of the activities of human designing. This lack of a need for any theoretical foundation provides enormous flexibility on the source for computationally implementable ideas which may support designing as distinct from analysis.

Computational processes which support designing can be grouped into three categories:

(i)   those founded on empirical evidence of human designing activity;
(ii) those founded on axioms and their derivations; and
(iii) those founded on conjectures of potentially useful processes.

This third category can be broken into two further subcategories:

(a) conjectures based on analogies with perceived human designing processes, and
(b) conjectures based on analogies with other processes (which are clearly not human designing processes).

Adaptive systems fall into this last subcategory. Human designers do not design with large populations or use any of the machinery of formal adaptive systems when they carry out the act of designing themselves.

What we intend to do in this paper is to briefly set the scene for the *genetic analogy* in designing and then to explore two possible research directions in adaptive systems for designing which may prove to be fruitful. The first extends the genetic analogy by introducing two further concepts, whilst the second draws its stimulus from development biology. Rather than provide detailed examples we will describe the ideas at the conceptual level.

2. The Genetic Analogy In Design

The basic genetic analogy in designing utilises a simple model of the Darwinian theory of improvement of the organism's performancethrough the "survival of the fittest". This occurs through the improvement of the genotype which goes to make up the organism. This is the basis of most evolutionary systems. Fundamental to this analogy are a number of important operational aspects of the model:

- the design description (structure) maps on to the phenotype
- separation of the representation at the genotype level from that of the design description level
- the processes of designing map on to the evolutionary processes of crossover and mutation at the genotype level
- performances (behaviours) of designs map on to fitnesses
- operations are carried out with populations of individuals.

In designing terms this maps directly onto the method of *designing as search*. We can describe this notion using the state-space representation of computation:

- state space is fixed at the outset
- state space comprises behaviour (fitness) and structure (phenotype) spaces
- genetic operators move between states in structure space, performance evaluated in behaviour space.

Designing as search is a foundational designing method but one that is restricted in its application to routine or parametric designing. In such designing all the possible variable which could occur in the final design are known beforehand as are all the behaviours which will be used to evaluate designs. Since the goal is to improve the behaviours of the resulting designs, the processes of designing during search map well onto those of optimization. This sits well with our notion of genetic algorithms and genetic programming. They can be readily viewed as robust optimization methodologies. Genetic algorithms and genetic programming have been used successfully as analogies of designing methodologies.

In this paper we will briefly explore other analogies which can be drawn from nature and humans' intervention in nature as possible sources for fruitful ideas on which to base design methodologies.

3. Extending the Genetic Analogy in Designing

There are two extensions of the genetic analogy that we will introduce in outline form here. Firstly, it is a well known hypothesis that certain behavioural characteristics of an organism could be genetic in origin. The field of genetic engineering deals specifically with this issue. *Genetic engineering* in natural systems is the human intervention in natural evolution. We will describe how such notions translate into potentially useful design processes. Secondly, in manufacturing there is a process known as *reverse engineering* where an existing product is explored in such a manner to construct a means by which it could be produced without knowing a priori its production methodology. We will describe how reverse engineering ideas fit into the genetic analogy and how they can be used to enhance nonroutine designing.

3.1. Genetic Engineering and Designing

The practice of genetic engineering in natural organisms involves locating genetic structures which are the likely cause of specified behaviours in the organism [1]. This provides a direct analog with finding significant concepts during the process of designing and giving them a specific primacy. The behaviour of the organism is an observable regularity which maps onto the concept and the structure of the genetic material which causes that behaviour is a representation of that concept, albeit a representation which has to be expressed in the organism for the concept to appear. The practice of genetic engineering is akin to the reverse of synthesis in the sense that one aspect of an already synthesised design is converted into the means by which it could be generated. In fact it is more complex than that since it is the behaviour of the already synthesised design which is the controlling factor but the analogy still holds. Let us examine in a little more detail the concept of genetic engineering.

Consider Figure 1 where the population of designs is divided into two groups (it could be more). One group exhibits a specific regularity whilst the other does not. The goal is to locate an "emergent" common structure in the genotypes of those designs which exhibit this regularity. Here "emergent" means that the structure was not intentionally placed there but could be found and represented for later use. Genetic engineering at this symbolic level uses pattern matching and sequence analysis techniques to locate these genetic structures. The process can be summarised as follows:

- locate emergent properties in the behaviour (fitness) space
- produce new genes which generate those emergent properties -> gene evolution
- introduce evolved genes into gene pool.

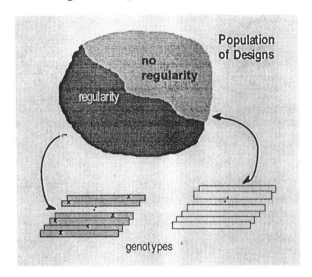

Figure 1. Genetic engineering is concerned with locating groups of genes' regularity, marked as X in the genotypes of those design which exhibit a specific behavioural regularity.

Take as an example the 8 genes shown in Figure 2 represented in the form of state transition rules. These genes are used to form the genotypes of designs within which a regularity is sought.

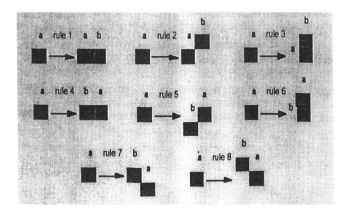

Figure 2. A set of 8 genes in the form of shape transition rules [2].

Figure 3 shows 10 designs produced from those genes. Each design is searched to determine some common regularity.

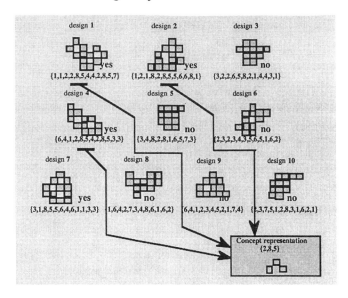

Figure 3. A set of 10 designs produced with the genes in Figure 2 and evaluated according to their regularity ("yes" in this case). Genetic engineering techniques emerge the gene group {2, 8, 5} as being the likely cause of that regularity [2].

The use of genetic engineering concepts allows an evolutionary system to adapt itself in ways different to traditional evolutionary systems when it takes the emergent gene structures and adds them to the alphabet of basic genes from which genotypes can be constructed. The effect of this is to dramatically change the probability landscape of the possible designs which could be produced from the original set of genes. No designs other than those which could have been originally produced are possible, however, the likelihood of designs be selected which exhibit those improved performances will now be increased.

It is possible not only to locate gene structures which map on to good performances, but also to locate gene structures which map on to poor performances. The former evolved genes would be increased in the genotypes whilst the latter would be decreased. The other aspect of genetic engineering concerns the manipulation of these newly found genetic structures or evolved genes. The goal of these manipulations is to improve the resulting performance of the designs. Typical manipulations which have readily modelled computational analogs include:

- gene therapy
- gene surgery
- radiation therapy.

These newly "evolved" genes capture some problem specific characteristics of the genetic representation of the good solutions to that problem. As such they may be able to be re-used in related problems to advantage. Typically each new problem to be solved using optimization techniques is treated anew without taking into account anything which has been learned from previous problems. Genes evolved using genetic engineered provide the basis for learning from previous design episodes and transferring what has been learned to the current design problem.

3.2. Reverse Engineering, the Genetic Analogy and Designing

In the computational model of genetic engineering used in designing the evolved genes are complexes of the original genes. Even when they are mutated they remain complexes of the original genes. As a consequence the boundary of the state space of possible designs is unchanged so that the designs produced are no differentto those which could have been produced using the original genes only. In order to produce novel designs, ie designs which could not have been produced using the original genes only, the evolved genes need to be different to simply being complexes of the original genes. In order to "evolve" such genes differentprocesses are required. We can take ideas from reverse engineering in manufacturing and include them in the genetic analogy.

The concept is analogically similar to that of genetic engineering in that emergent properties are looked for and new genes which generate those properties are produced, although the processes are differentand the result is quite different. The process can be summarised as follows:

- locate emergent design (phenotype rather than fitness) properties
- reverse engineer new genes which can generate those emergent properties -> gene evolution
- introduce evolved genes into gene pool.

The critical differences between this and genetic engineering occur in two places in this process. The first differencesin the locus of emergent properties – these are looked for in the phenotype, ie in the designs themselves rather than in their fitnesses or performances. The second difference is in the means by which "evolved" genes are created.

3.2.1 Locating emergent features

How might emergent features in designs be found? One way which has shown promise utilises the process of re-representation, ie an alternate representation is used to represent the design. In that re-representation it may be possible to locate features which were not placed there by the designer but appeared circumstantially. Figure 4 shows a figure produced by Escher. If the original representation is white images and the re-representation is black images then the angels are placed in the figure whilst the devils are emergent features.

Consider as a further example the following: a design is constructed by producing line segments, where line segments contain no branches. If the design is re-represented as maximal lines (i.e. lines in which line segments may be embedded) it is easy to find lines which are composed of more than one contiguous line segment if they exist by comparing the line segments with the maximal lines. If there is a complete isomorphism between the two then no such features have emerged. If there are maximal lines which are not isomorphic with existing line segments then an emergent feature has been found.

Figure 4. Figure by M. C. Escher "Circle Limit IV"

3.2.2 *Reverse engineering "evolved" genes*

Having located an emergent feature the next step is to reverse engineer a new gene which is capable of producing that emergent feature. This new "evolved" gene is then added to the gene pool. A variety of machine learning-based methods is available for this task. These include inductive substitution of the new representation in the place of the original representation in the design generator, turning constants into variables, and rule-based induction methods.

Evolving genes by reverse engineering is a form of Lamarckism in that characteristics of an organism not directly produced by its genetic makeup are acquired by that organism's genome.

4. The Developmental Biology Analogy in Design

So far the extensions described have all been at the genomic level of an adaptive system. In all of these as in most genetically-based adaptive systems used in designing there is an assumption that the mapping between the genotype and the phenotype is fixed. In natural systems the genotype is expressed through a phenotype through a biological development process which commences with the establishment of a single cell which divides. Further, cell division is a function of not only its genetic programming but also its environment. The normal genetic analogy does not allow for totipotency as occurs in nature at the outset of cell division. One approach is to allow a form of pluripotency to occur as a function of the development environment of the design.

Perhaps more interesting is to specifically model phenotypic plasticity to produce a form of pleiomorphism. This would allow for a form of genotype/phenotype environment interaction during the development of the phenotype. A variety of environmental interactions can be proposed to allow for adaptive mapping between genotype and phenotype. Classes of interactions include the following where "f" is some function:

- phenotype = f(genotype, situation), where situation refers to a state of the environment at some time, or
- $phenotype_t = f(genotype, phenotype_{t-1})$;

both in lieu of :
$$phenotype = f(genotype).$$

Examples of such classes are:

Example 1
Here the phenotype is made up of components but the components themselves are some function of the path taken to reach that component. A simple path function would be that each component is in some way a function of the components it is connected to, ie:

- $phenotype = \{component_1, ... component_i, ... component_n\}$
- $component_i = f(component_{i-1}, path[i-1,i])$.

Example 2
Here the phenotype is developed over some time intermediate periods from a given genotype, during which various intermediate fitnesses control its development in a pleiomorphic sense, ie:

- phenotype = f(genotype, intermediate fitnesses during development).

Example 3
Here the phenotype, as it develops over some time intermediate periods from a given genotype, does so as a function of its expression at the previous time period. This is a crude model of cell division, ie:

- $phenotype_t = f(genotype, phenotype_{t-1})$.

Models such as these provide opportunities to include both problem- and domain-specific knowledge in the evolutionary process.

5. Discussion

The genetic analogy in designing has been based on a model of Darwin's survival of the fittest [3, 4, 5]. This has provided a foundation for a body of important work which has implicitly treated designing as a search method largely akin to optimization. The effect of this in designing terms has been to set up a fixed state-space which is then searched for appropriate solutions. Alternative analogies drawn from both genetics and developmental biology offerthe opportunity to change the state-space of possible designs in some cases, Figure 5.

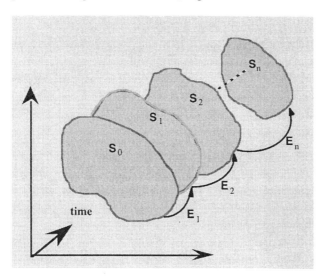

Figure 5. Conceptual model of designing with evolutionary and biological operations (E_i) which modify the state-spaces of possible designs.

The genetic engineering extension to the genetic analogy has been implemented in various environments and has proven to be both interesting and useful in allowing designing which slowly accumulates and represents, albeit implicitly, problem-specific knowledge which aids the solution process. The inclusion of reverse engineering is an obvious extension of genetic engineering to a different locus and has the potential to change the state-space of possible designs. Of interest here is whether the process terminates when applied iteratively.

The introduction of an analogy with developmental biology opens up numerous research paths with possible interest in designing. Concepts from research into natural systems such as switch genes, regulatory genes and gene networks offer a rich ground for fertile ideas.

Acknowledgments

This work is supported by a number of grants from the Australian Research Council.

References

1. Sofer W H, 1991. *Introduction to Genetic Engineering,* Butterworth-Heinemann, Stoneham.

2. Gero J S, Kazakov V, 1996. Evolving building blocks for design using genetic engineering: a formal approach. In: Gero, J S (ed.), 1996. *Advances in Formal Design Methods for CAD,* Chapman and Hall, London, pp 31-50.
3. Holland J, 1992. *Adaptation in Natural and Artificial Systems.* MIT, Cambridge, MA.
4. Goldberg D, 1989. *Genetic Algorithms in Search, Optimization and Machine Learning.* Addison-Wesley, Reading.
5. Koza J, *Genetic Programming.* MIT, Cambridge, MA.

Emergent Computational Models in Structural Analysis and Design

Prabhat Hajela

Rensselaer Polytechnic Institute, Troy, New York 12180
e-mail: hajela@rpi.edu

Abstract. The present paper reviews some recent developments in the applications of soft-computing tools in problems of structural analysis and design. These problems have grown significantly in scope, a development motivated in part by the more easily accessible computational resources. At the same time there is an emerging awareness that existing methods of structural analysis and design are ill-suited to the architecture of new computational machines, and that radically different techniques, more naturally amenable to these machines, must be developed. The paper, to a large extent, deals with the role of soft-computing tools such as genetic algorithms, neural networks, and immune networks, in enhancing the efficiency of solution strategies for increasingly complex structural design problems; to a lesser extent it looks at the role of these methods in the development of a new generation of tools for structural analysis and design.

1. Introduction

There has been considerable recent interest in soft computing principles, with some focus in how they relate to problems of structural analysis and design. This interest has been fostered by a combination of new and projected advances in computational hardware, the need to better adapt computational paradigms to a new computing environment, and the desire to develop tools that can accommodate significant increases in problem complexity. The latter is brought about by an increased emphasis on treating structural synthesis as a multidisciplinary problem, where, the increased level of analysis complexity, and the need to extract the advantages of a synergistic design process, dictate the need for a more comprehensive strategy [1,2]. In general, solution algorithms must account for discreteness in the design space, search for globally optimal solutions, handle a large number of design variables and constraints, and continue to be efficient in an environment where the analysis may be computationally cumbersome and coupled. In an approach where one attempts a solution as a single, large-scale optimization problem, the dimensionality of the design space may overextend the capabilities of traditional mathematical programming methods, and limit the capacity to evaluate the progress of the solution process.

Decomposition based optimization strategies have been proposed for large scale problems wherein the problem is represented by two or more smaller subproblems that are weakly coupled. The form of coupling may be either hierarchic or non-hierarchic, with the latter representative of the more general situation. In such an approach one must have rational methods to identify a topology for decomposition and to account for the interactions between the temporarily decoupled sub-problems. In problems where gradient information is readily available, the sub-problem solutions can be coordinated using such information [3]. However, in those problems where the design variables are a mix of continuous, discrete, and integer type, gradient information is not very useful, and alternative strategies must be investigated. A mixed-variable design space also limits the usefulness of traditional gradient-based optimization algorithms.

The development of design systems like ENGINEOUS [4], which are a combination of algorithmic and heuristic (rule-based) procedures, are typical of efforts that have grown out of a recognition of inadequacies of traditional design techniques. Such methods do suffer from a problem typical of most rule-based systems - the rule base is defined at the outset and does not have the capacity to learn new rules from the problem domain. Preliminary exploration of methods of computational intelligence has been undertaken to account for these inadequacies.

A significant concern in large-scale analysis problems has been the high computational cost of analysis. Early implementations of structural synthesis methods looked towards approximation methods for relief in this area (efficient structural reanalysis and Taylor approximations) [5,6]. While these ideas are still relevant and are broadly described as 'analysis for design', new requirements for developing non-gradient based global approximations have emerged. This latter requirement is in part due to the increasing interest in applying optimization methods to real systems with mixed-variable design spaces. More importantly, it is driven by the need to generate 'mathematical models' for including disciplines such as manufacturability, cost, and maintainability, into the optimization problem. The use of response-surface like approximations based on neural networks and fuzzy logic have been explored in this context.

Genetic algorithms, evolutionary computing, and neural networks have made significant inroads into the analysis and design of structural and multidisciplinary systems [7]. These methods bridge the gap between traditional procedural algorithms and heuristic or rule-based expert systems, and have an established role as search algorithms in generically difficult optimization problems [8-11], function approximation, problem decomposition and structural modeling strategies [12-14], and in computational intelligence and machine learning [15,16]. The scope of applications has grown in both number and complexity, extending to domains of structural control, health monitoring, and manufacture [17-19]. All of these efforts have primarily adopted a combinative approach, using traditional methods of analysis at the basic level, and combining these with soft-computing based representation or search strategies at a higher level, to facilitate synthesis tasks. Subsequent sections of this paper review some key developments that span the application of soft-computing tools in structural analysis and design. Preliminary ideas of how soft-computing can be integrated into the next generation of structural synthesis tools, are also presented.

2.0 Neural Networks in Structural Synthesis

There has been considerable activity in the field of artificial neural networks over the past decade, including a number of applications in the field of structural analysis and design. Among the more widely adapted neural network architectures are the back-propagation (BP) network, counterpropagation (CP) network, the radial basis network, and recurrent networks like the Hopfield model [20]. Neural networks are layers of interconnected neurons, each with an inbuilt activation function that allows the mimicking of the biological neuron, albeit in a very crude manner. There are interconnection weights associated with the neuron connections, and network training simply requires the determination of all interconnection weights and specific characteristics of the activation functions of all neurons in the network. This training is either of a supervised or unsupervised form [20].

Neural networks have been explored as function approximation tools in problems of structural analysis and design, most commonly as a computationally inexpensive replacement for procedural analysis. In such cases, training data is generated from a procedural simulation of the structural system, and a neural network is trained to mimic the input-output relations of this system (generalization). In such use, the neural network may be considered as a response surface approach where the order of the polynomial fitting function does not have to be specified. In fact, the neural network is a special form of response surface where the response function is a nested squashing function; the interconnection weights of the network that have to be learned correspond to the regression parameters in a typical regression model. Such a neural network is ideally well suited for an immersive design model, including one using a virtual reality environment.

The primary differences between the CP and the BP model are in the time and data required for network training, and in the performance of their generalization capabilities. The former requires less computational effort to train, an issue of some importance when one considers modeling of extremely large structural systems. However, its generalization performance is poorer when compared to the BP model. Improvements to the CP network have been implemented that partially circumvent this problem; however, large sets of training data continue to be required for their effective use. An advantage over the BP network is that CP networks provide a pattern completion capability, wherein, upon presentation of an input vector to the network, some components of which may be missing, the network approximates the missing components to produce a relevant output.

In using neural networks for approximate analysis, a set of input-output training patterns must be obtained from the real process that one is attempting to simulate. Determination of the number of such training pairs, and how they should span the intended domain of training, requires experimentation and experience. The same statement is applicable to the selection of the network architecture, i.e., the number of layers and the number of neurons in such layers. While mathematics can be used to assess bounds on such requirements, in a number of engineering problems, such approximations tend to be over conservative. This continues to be an active area for research.

In structural synthesis, the BP networks have been used to approximate procedural structural analyses in a number of problems of varying dimensionality [21,22]. Successful experiments with the BP network have also been performed with the idea of training a neural net to estimate optimal designs directly for a given set of design conditions, and to bypass all the analyses and optimization iterations of the conventional approach [23]. This idea draws on the concept of an "intelligent corporate memory", where several known optimal designs within some domain of variation in design conditions can be used to train a neural network. The estimation of optimal designs for a set of design conditions is the central idea behind the computation of problem parameter sensitivities [24]. Such sensitivity calculations are critical in hierarchical decomposition strategies in optimal design [25]. The BP network has also been used in the modeling of nonlinear stress-strain relations [26], for modeling of nonlinear behavior of reinforced concrete [27] (using experimental data), and to predict hygrothermal and mechanical properties of composites [14], given the environmental conditions, constituent material properties, and volume fractions of constituent materials. Yet another application of such networks that has received attention is in the nondestructive damage assessment of structural systems [17,28]. Before leaving the subject of BP networks, it is important to point out that the interconnection weights of the trained network can be used to assess the influence of each input component on the network outputs, providing meaningful insight into the problem domain. This can be used to better configure the network for training, including amount of training data and network architecture [29], and in a rational approach for problem decomposition in non-hierarchic systems [30].

The CP network has been similarly used as a rapid reanalysis tool in simulated annealing based optimal design with discrete variables [31], and as an inverse mapping capability in system identification problems [28,32]. The former example deals with structural damage assessment where the pattern completion capacity of this network was shown to be useful; incomplete measurements, as would be the case in real on-line damage detection, were used to successfully conduct damage diagnostics. The pattern completion capability has also been studied in the context of decomposition based multidisciplinary design of structural systems. It is also worthwhile to indicate that another network architecture, the radial basis network [33] has also been used for function approximations; this network does provide for the more mathematically rigorous universal approximation property

Networks that have a feedback path between the output and input neurons are described as recurrent networks. Although unconditional stability is not assured in these architectures, they offer a more realistic modeling of the memory process.

Such networks also exhibit a form of learning referred to as self-organization, such as may be required if no known output is available to train the network. The Hopfield network is an example of this form where the network architecture is seen to contain the two principal forms of parallel organization, viz., parallel treatment of inputs and outputs, and extensive connectivity between the neurons. These networks have been used as optimization tools in problems with discrete variables such as the assignment of node numbers in a finite element model so as to minimize the bandwidth of the resulting stiffness matrix [34], and in optimal structural layout problems [9]. While the success of this approach has been demonstrated in a number of such combinatorial optimization problems, the approach is not general enough for widespread use.

3.0 Evolutionary Computing and Genetic Algorithms

The strengths of evolutionary algorithms have been clearly established with reference to optimal search in generically difficult but very realistic structural design problems such as those containing discontinuities or nonconvexities in function behavior, discrete variation in design variables, and where gradient information is generally unavailable. The genetic algorithm is based on an elitist reproduction strategy, where chromosomal representation of designs is evolved using random operations encompassed in operation like crossover and mutation, with bias assigned to those members of the population that are deemed most fit at any stage of the evolution process. In order to represent designs as chromosome-like strings, stringlike representations of the design variables are stacked head-to-tail. Different representation schemes have been adopted, including use of the decimal values of the design variable [35], use of integer numbers [35], or most popularly, a conversion of all design variables into their binary equivalent. In the latter case where the chromosomal string is a number of 0's and 1's, the numerical precision with which the design variable is represented is determined by the string length.

Increased adaptation into the structural design environment has been accompanied by a number of modifications to the basic GA approach. Of these, direct schemes (non-penalty function methods) by which to account for design constraints [36,37], have received some attention. An approach applicable to a case where constraints are linear and the design space convex, has been described in [36]. Other methods, based on strategies that adapt useful features of the feasible designs into the infeasible population, have been proposed [37,38]. In [38], the process of adaptation is through the use of an expression operator, which like the crossover and mutation operations in genetic search, is probabilistic in nature. A similar process of adaptation ("gene-correction" therapy) is also at work in another strategy that is based on immune network simulation [37].

Another area of research that has witnessed activity is in the extension of GA's to large scale design problems. GA's search for an optimal design from among a discrete set of design alternatives, the number of which depend upon the length of the chromosomal string. Large number of design variables, and/or considering a very fine precision in continuous design variable representation contributes to long string lengths which results in increased function evaluations. Methods such as multistage search [39], wherein the granularity of the genetic algorithm based search is varied through a successive increase in the precision, and an approach which assigns significance to the previous generations of evolution in genetic search referred to as directed crossover [39], have been proposed. The latter simply attempts to determine thgrough computations, significant bit positions on the string, and to constrain the crossover operation to these bit locations. A number of applications of both the basic GA and its enhanced forms, in problems of multidisciplinary structural design, structural layout determination, and composites design, are described in [7].

Another approach to enabling better computational efficiency for genetic algorithms has been an attempt to make use of parallel computing hardware. At the very outset, it is important to make the distinction between use of multiprocessor computers to make the analysis involved in GA's more efficient, and the use of such computers to give new slant to the genetic algorithm procedure itself. The latter, termed PGA or parallel genetic algorithm [40,41] is different from the traditional GA in that the former uses few intelligent and active members in the search as opposed to the latter, which use more but passive members to conduct the search. In fact, the power of the PGA stems from a combination of processing speed of parallel hardware and software speed on the inherent parallelism available in the GA.

Evolutionary programming (EP) and evolutionary strategies (ES) are two other procedures that have been motivated by natural evolution [42,43]. These methods, like the genetic algorithm, use transformation operators such as selection, recombination (crossover), and mutation. However, the degree to which each operator is emphasized in the algorithm is where the principal differences can be found. In EP, evolution of populations is done primarily through selection and mutation; recombination is not directly implemented although the contention is that the mutation operation is flexible enough to achieve the characteristics of recombination. In ES, recombination is specifically included, and herein lies the principal difference to the EP. Design representations in both of these approaches are based on real numbers.

4.0 Immune Network Modeling

In biological immune systems, foreign cells and molecules, denoted as antigens, are recognized and eliminated by type-specific antibodies. The task of recognizing antigens is formidable due to the very large number of possible antigens; it is estimated that the immune system has been able to recognize at least 10^{16} antigens. This pattern recognition capability is impressive, given that the genome contains about 10^5 genes, and the immune system must use considerable organization and cooperation among the gene libraries to construct antibodies for all possible antigens that are likely to be encountered. In biological systems, this recognition problem translates into a complex geometry matching process. This process can be simulated using the genetic algorithm approach, and has been the subject of considerable study [44,45].

A matching function that measures the similarities between antibodies and antigens, substitutes for the fitness function of the genetic algorithm. In a typical simulation of the immune system, the fitness of an individual antibody would be determined by its ability to recognize specific antigens, and a genetic algorithm using the matching function as a measure of fitness would evolve the gene structure of the antibody in an appropriate manner. In the context of a binary-coded genetic algorithm, the antibodies and antigens can also be coded as binary strings. The degree of match or complimentarity between an antibody and an antigen string indicates the goodness of that antibody. A simple numerical measure $Z = \sum t_i$ $i=1, N_{string}$ can be defined, where N_{string} is the length of the binary string, and $t_i = 1$ if there is a match at the i-th location of the two strings, and is 0 otherwise. A larger value of Z indicates a higher degree of match between the two strings. Using a traditional GA simulation, a population of antibodies can be evolved to cover a specified pool of antigens, with Z used as the fitness measure for this simulation. The manner in which this pattern recognition scheme is invoked will determine whether the evolved antibodies are 'specialists', i.e. adapted to specific antigens, or generalists that provide the best match to a number of different antigens. From an applications standpoint, generation of both specialist and generalist antibodies is required, and is discussed in what follows.

As stated earlier efficient use of GA's in large scale problems have required adoption of strategies like the directed crossover and multistage search, dynamic parameter encoding (similar in spirit to multistage search) [46], and resorting to the use of design domain decomposition. *The premise of these strategies, in particular the first two, is that the gross schema-patterns that correspond to near-optimal designs begin to assert themselves relatively early in the GA solutions, and identification of these schemas coupled with an opportunistic use of this information, can assist in speeding up the convergence rate of the GA.* The immune network simulation affords another approach that belongs to the category of such schemes. The implementation is relatively simple [47], embedding a cycle of immune network simulation in each GA evolutionary cycle. From the initial population of designs, a small percentage (3%-5%) of the best designs from this population are designated as the antigens, and the entire population is then defined as the starting population of antibodies. A single generation of GA based simulation of the immune network, designed to develop generalist antibodies is then invoked. The best designs from this new population of design are then chosen as antigens, and the process repeated till convergence. It is important to note that this process is somewhat different from the traditional GA implementation. The objective and constraints for the optimization are no longer used in the selection or reproduction stage of the GA evolution. They simply identify the antigens to which the population must adapt.

The immune system simulation process can also be used for constraint handling in GA based optimal design. A typical population of genetic evolution contains a mix of both feasible and infeasible designs. The essential idea behind the use of the immune network approach is to use chromosomal representation of infeasible designs as the gene-library from which antibodies are evolved through the process of immune system simulation. The chromosomal makeup of the feasible designs are selected as the antigens to which the antibodies must be evolved. This simulation co-adapts the infeasible design representations to the structure of feasible designs, and may therefore be considered as a step which reduces constraint violations in the population. As in the previous case, generalist antibodies are required in this simulation and the approach has been shown to be promising [37].

An approach by which large scale structural systems can be accommodated in the formal design process is through decomposition, where a large-scale system is partitioned into an appropriate number of subproblems depending on available computing machines or parallel processors. The GA strategy in each subproblem works with shorter string lengths, and hence smaller population sizes are required in each subproblem. The principal challenge in this approach is to account for subproblem interactions, and the pattern completion. capability of the CP network is one approach to consider this coupling [48]. The modeling of the biological immune system provides an alternative approach to this problem.

In using this approach to account for interactions among temporarily decoupled subproblems, the motivation is to adapt changes in design variables from other subproblems into a particular subproblem with a prescribed frequency in the artificial evolution process. It is important to recognize that the best design information from other subproblems cannot be simply imported into a particular subproblem. Instead, an aggregate of a few best designs of a particular subproblem must be ported over to the other subproblems. A generalist antibody developed using several good designs of a particular subproblem as antigens, works well in this application [30].

An application where the specialist antibody simulation is required is the problem of multicriterion design. Here, using a weighting function approach to multicriteria design, the entire Pareto front can be generated in a single simulation of the GA run. Using the best designs for several different weighting combinations as the antigens, the immune system forces subsets of the population to converge around the optimal designs of each weighting combination. In. analysis, this approach works much like the sharing function approach, maintaining diverse sets of designs during the GA evolution. An application of this approach is presented in [49].

5.0 Computational Intelligence

Evolutionary search algorithms have a significant role in implementing ideas of computational intelligence in structural and multidisciplinary design. Recent research has shown [50] how binary-coded rules in the IF-THEN form can be evolved using the genetic algorithm, based on information derived from a computational domain. In such classifier system type of machine learning approaches the rules may be completely arbitrary in the beginning but evolve into meaningful statements with information extracted from the computational domain. This approach has powerful implications in overcoming problems of a brittle rule-base that were endemic in traditional rule-based systems.

Specific applications of this approach in structural design have been in structural optimization [16,50]. In [16], the approach is used to evolve a set of rules required to solve a minimal weight truss problem with constraints on stresses in the bar elements. Similar applications of classifier systems in turbine design optimization are documented in [51]. The approach has also been used in enhancing the process through which neural networks are used to create function approximations. A rule learning procedure was implemented wherein computational feedback on the performance of a partly trained network was used to determine the amount and location of new training patterns required to improve the network generalization performance [52]. A similar approach to improve the quality of response surface based approximations is presented in [15].

6.0 Research Directions

This review would be incomplete without some reference to future directions for exploration. The role of soft computing tools in structural analysis and design is today on firmer ground than it has ever been before. Computing speeds today are in the MFLOPS-GFLOPS range and predictions indicate TFLOPS performance and better in the next decade. Analysis and design techniques must be revisited from a completely new perspective if such hardware is to be used in the most effective manner. Current algorithms, through all manners of software enhancement and efforts to parallelize, have their origins in serial thinking, and without the required intrinsic parallelism, are victims of the law of diminishing returns when placed on parallel machines. A completely new line of thinking born in the parallel processing environment is required for developing the next generation of structural analysis and design tools, and here is where soft computing tools are likely to have the most visible impact. What must result from new developments are real-time methods to analyze a design artifact - these tools should provide a design engineer with an immersive design capability in which not only search directions and step sizes are computed rapidly, but what-if questions pertinent to the design are available instantaneously. Research targeting these lofty goals has begun, albeit at very rudimentary levels. Specific goals of such research which spans the spectrum of soft computing tools includes the following

Use of neural networks and fuzzy-logic tools to develop response surface-like approximations for discrete and discontinuous functions which are significant in the context of topological design of structural systems, or in an immersive/interactive analysis and design mode, where exploration of discrete concepts is most critical.

The concept of using macro-neural networks (neuroelements) which can be adaptively interconnected during the evolution of design, has been considered. Efforts to expand the use of evolutionary algorithms to large-scale design problems through more efficient implementations in a parallel computing environment are worthy of consideration Similarly, extending these algorithms to better handle fuzzy design constraints and objective criteria through an effective integration with fuzzy-logic techniques must be explored. These are particularly relevant in an environment where manufacturing considerations must be included in the design cycle. Finally, an exploration of entirely different computational paradigms such as cellular automata [53] to both analyze and design structural systems presents distinct possibilities. This approach represents a departure from the traditional procedural analysis. Instead of using fundamental equations of physics pertinent to a problem to analyze a domain, the idea here is to decompose the problem domain into a number of grids, where the property of each cell within this grid evolves or emerges through an interaction with the surrounding cells in the grid. This self-emergent or self-organizing behavior is thought to be significant in the development of the next generation of structural synthesis tools; it is an intrinsically decentralized computational paradigm ideal for multiple parallel processor machines.

7.0 Closing Remarks

The paper has described a subset of attempted applications of soft-computing tools in problems of structural analysis and design. These tools represent significantly improved capabilities for solving generically difficult problems of structural synthesis; more specifically, they overcome difficulties related to problem dimensionality, handling of a mix of discrete, integer, and continuous design variables, accounting for discontinuities or nonconvexities in the design space, and improved capabilities for modeling and design in the absence of gradient information. The last item is particularly relevant in a design for manufacturing environment, where manufacturing and production related constraints have received increased attention.

8.0 References

1. Sobieszczanski-Sobieski, J., 1993. "Multidisciplinary Design Optimization: An Emerging New Engineering Discipline", World Congress on Optimal Design of Structural Systems, Rio de Janeiro, Brazil, August 2-6.

2. Bloebaum, C.L., 1991. Formal and Heuristic Decomposition Methods in Multidisciplinary Design, Ph. D. dissertation , University of Florida.

3. Sobieszczanski-Sobieski, J., 1982. "A Linear Decomposition Method for Large Optimization Problems - Blueprint for Development", NASA TM 83248.

4. Powell, D.J., Skolnick, M.M., and Tong, S.S., 1991. "Interdigitation: A Hybrid Technique for Engineering Design Optimization Employing Genetic Algorithms, Expert Systems, and Numerical Optimization", Handbook of Genetic Algorithsm, ed. L. Davis, Van Nostrand reinhold, New York, pp. 312-331.

5. Noor, A.K., and Lowder, H.E., 1974. "Approximate Techniques of Structural Reanalysis", Computers and Structures, Vol. 4, pp. 801-812.

6. Schmit, L.A., and Farshi, B., 1974. "Some Approximation Concepts for Structural Synthesis", AIAA Journal, 12, 5, 692-699.

7. Hajela, P., 1997. Stochastic Search in Discrete Structural Optimization - Simulated Annealing, Genetic Algorithms and Neural Networks, Discrete Structural Optimization, Springer, New York, pp. 55-134, (ed. W. Gutkowski).

8. Hajela, P., 1990. "Genetic Search - An Approach to the Nonconvex Optimization Problem", AIAA Journal, Vol. 26, No. 7, pp1205-1210.

9. Fu, B., and Hajela, P., 1993. "Minimizing Distortion in Truss Structures: A Hopfield Network Solution", Computing Systems in Engineering, vol. 4, no. 1, pp. 69-74.

10. Hajela, P.and E. Lee, 1997. "Topological Optimization of Rotorcraft Subfloor Structures for Crashworthiness Considerations", Computers and Structures, vol. 64, no 1-4, pp. 65-76.

11. LeRiche, R., and Haftka, R.T.,1993. "Optimization of Laminate Stacking Sequence for Buckling Load MAximization by Genetic Algorithms", AIAA Journal, Vol. 31 (5), pp. 951-956.

12. Hajela, P. and Szewczyk, Z., 1994. "Neurocomputing Strategies in Structural Design - On Analyzing Weights of Feedforward Neural Networks", Structural Optimization, Vol. 8, No. 4, pp. 236-241.

13. Hajela, P. and Berke, L., 1991. "Neurobiological Computational Models in Structural Analysis and Design", Computers and Structures, Vol. 41, No. 4, pp. 657-667.

14. Brown, D.A., Murthy, P.L.N., and Berke, L., 1991. "Computational Simulation of Composite Ply Micromechanics Using Artificial Neural Networks", Microcomputers in Civil Engineering, 6, pp. 87-97.

15. Hajela, P., and J. Lee, 1997. "A Machine Learning Based Approach to Improve Global Function Approximations for Design Optimization", proceedings of the International Symposium on Optimization and Innovative Design, July 28-July 30, Tokyo, Japan.

16. Goel, S., and Hajela, P., 1997. "Classifier Systems in Design Optimization", presented at the AIAA/ASME/ASCE/AHS SDM meeting, Kissimee, Florida, April.

17. Szewczyk, Z., and Hajela, P., 1994. "Neural Network Based Damage Detection in Struc-

tures", ASCE Journal of Computing in Civil Engineering, Vol. 8, No. 2, pp. 163-178.

18. Ku, C.-S., and Hajela, P., 1998. "Neural Network Based Controller for a Nonlinear Aeroelastic System", AIAA Journal, Vol 36, No. 2.

19. Hajela, P. and Teboub, Y., 1996. "Embedded Piezoelectrics in Composites for Damage Sensing and Mitigation - Design Issues", SPIE Far East and Pacific Rim Symposium on Smart Materials, Structures, and MEMS, Bangalore, India, December 11-14.

20. Haykin, S., 1994. Neural Networks - A Comprehensive Foundation, Macmillan Publishing Company, Englewood, New Jersey.

21. Hajela, P. and Berke, L., 1991. "Neurobiological Computational Models in Structural Analysis and Design", Computers and Structures, Vol. 41, No. 4, pp. 657-667.

22. Swift, R. and Batill, S., 1991. "Application of Neural Networks to Preliminary Structural Design", AIAA Paper No. 91-1038, proceedings of the 32nd AIAA/ASME/ASCE/AHS/ASC SDM Meeting, Baltimore, Maryland.

23. Berke, L., and Hajela, P., 1992. "Application of Artificial Neural Networks in Structural Mechanics", Structural Optimization, Vol 3, No. 1.

24. Sobiesczanski-Sobieski, F., Barthelemy. J.-F., and Riley, K.M., 1982. "Sensitivity of Optimum Solutions to Problem Parameters", AIAA Journal, Vol. 20 (9), pp. 1291-1299.

25. Hajela, P. and Berke, L., 1991. "Neural Network Based Decomposition In Optimal Structural Synthesis", Computing Systems in Engineering, Vol. 2, No. 5/6, pp. 473-481.

26. Alam, J., and Berke, L., 1992. "Application of Artificial Neural Networks in Nonlinear Analysis of Trusses", NASA TM.

27. Ghaboussi, J., Garrett, J.H., Jr., and Wu, X., 1991. "Knowledge-Based Modeling of Material Behavior with Neural Networks", Journal of Engineering Mechanics, 117 (1), pp. 132-153.

28. Hajela, P., and Teboub, Y., 1992. "A Neural Network Based Damage Analysis of Smart Composite Beams", AIAA Paper 92-4685, proceedings of the 4th AIAA/NASA/Air Force Symposium on Multidisciplinary Analysis and Optimization, Cleveland, Ohio, September.

29. Hajela, P and Szewczyk, Z., 1994. "Neurocomputing Strategies in Structural Design - On Analyzing Weights of Feedforward Neural Networks", Structural Optimization, Vol. 8, No. 4, pp. 236-241.

30. Lee, J., and Hajela, P., 1996. "Parallel Genetic Algorithm Implementation in Multidisciplinary Rotor Blade Design", AIAA Journal of Aircraft, Vol. 33, No. 5, pp. 962-969.

31. Szewczyk, Z., and Hajela, P., 1992. "Feature Sensitive Neural Networks in Structural Response Estimation", proceedings of the ANNIE'92, Artificial Neural Networks in Engineering Conference.

32. Szewczyk, Z., and Hajela, P., 1993. "Neural Network Based Selection of Dynamic System Parameters", Transactions of the CSME , Vol. 17, No. 4A, pp. 567-584.

33. Fu, B., 1993. An Analog Computing Approach to Structural Analysis and Design Through Artificial Neural Networks, Ph.D. dissertation, Rensselaer Polytechnic Institute, Troy, New York.

34. Hakim, M.M. and Garrett, J.H., Jr., 1991. "A Neural Network Approach for Solving the Minimum Bandwidth Problem", Proceedings of the 28th SES Meeting, Gainesville, Florida.

35. Haftka, R.T., and Gurdal, Z., 1993. Elements of Structural Optimization, Kluwer Academic Publishers, Dordrecht.

36. Michalewicz, Z., and Janikow, C.Z., 1991. "Handling Constraints in Genetic Algorithms", Proceedings of the 4th International Conference on Genetic Algorithms, pp. 151-157.

37. Hajela, P., and Lee, J., 1996. "Constrained Genetic Search Via Schema Adaptation: An Immune Network Solution", proceedings of the First World Congress of Structural and Multidisciplinary Optimization, Goslar, Germany, May-June, 1995, eds. N. Olhoff and GIN Rozvany, pp. 914-920, Pergamon Press.

38. Hajela, P. and Yoo, J., 1996. "Constraint Handling in Genetic Search Using Expression Strategies",AIAA Journal, Vol 34, No. 11, pp. 2414-2420.

39. Lin, C.-Y., and Hajela, P., 1993. "Genetic Search Strategies in Large Scale Optimization", proceedings of the 34th AIAA/ASME/ASCE/AHS/ASC SDM Conference, La Jolla, California, pp. 2437-2447.

40. Manderick, B., and Spiessens, P., 1989. "Fine-Grained Parallel Genetic Algorithm", Proceedings of the Third International Conference on Genetic Algorithms, ed. H. Schaffer, pp. 428-433, Morgan-Kauffmann.

41. Muhlenbeim, H., Schomisch, M., and Born, J., 1991. "The Parallel Genetic Algorithm as a Function Optimizer", Parallel Computing, 17, pp. 619-632.

42. Fogel, L.J., Owens, A.J., and Walsh, M.J., 1966. Artificial Intelligence Through Simulated Evolution, Wiley Publishing, New York.

43. Rechenberg, I., 1973. Evolutionsstrategie: Optimierung Technischer System nach Prinzipien der Bioloischen Evolution, Frommann-Holsboog, Stuttgart.

44. Smith, R. E.; Forrest, S.; Perelson, A. S. 1992: Searching for Diverse Cooperative Populations with Genetic Algorithms. Technical Report CS92-3, University of New Mexico, Department of Computer Science, Albuquerque, NM.

45. Dasgupta, D., and Forrest, S., 1995. "Tool Breakage Detection in Milling Opeartions Using a Negative Selection Algorithm", TR No. CS95-5, Department of Computer Science,

University of New Mexico.

46. Schraudolph, N.N. and Belew, R.K., 1992. "Dynamic Parameter Encoding for Genetic Algorithms," Machine Learning, 9, pp. 9-21.

47. Hajela, P. and Yoo, J., and Lee, J.,1997. "GA Based Simulation of Immune Networks-Applications in Structural Optimization", Journal of Engineering Optimization, Vol. 29, pp. 131-149.

48. Lee, J., and Hajela,P., 1996. "Role of Emergent Computing Techniques in Multidisciplinary Design", proceedings of the NATO ARW on Emergent Computing Methods in Engineering Design, August 1994, Springer Verlag, pp. 162-187, eds. D. Grierson and P. Hajela.

49. Yoo, J., and Hajela, P., 1998. "Immune Network Modeling in Multicriterion Design of Structural Systems," presented at the AIAA/ASME/ASCE/AHS SDM meeting, Long Beach, California, April 20-23.

50. Richards, S.A., 1995. Zeroth-Order Shape Optimization Utilizing a Learning Classifier System, Ph.D. dissertation, Stanford University, California.

51. Goel, S., 1998. Machine Learning Paradigms in Design Optimization: Applications in Turbine Aerodynamic Design, Ph.D. dissertation, Rensselaer Polytechnic Institute, New York, April.

52. Hajela, P. and Kim, B., 1998. "Classifier Systems for Enhancing Neural Network Based Global Function Approximations", submitted to the 7th AIAA/NASA/ISSMO/USAF Multi-disciplinary Analysis and Optimization Meeting, St. Louis Missouei.

53. Wolfram, S., 1994. Cellular Automata and Complexity, Addison Wesley, New York.

Exploring The Design Potential Of Evolutionary / Adaptive Search And Other Computational Intelligence Technologies

I. C. Parmee

Plymouth Engineering Design Centre
University of Plymouth
Devon, UK

Abstract. The paper investigates the integration of evolutionary and adaptive search (ES&AS) strategies with the various stages of the design process (ie conceptual, embodiment and detailed design). The paper primarily attempts to identify the manner in which relevant co-operative ES & AS strategies and related computational intelligence (CI) technologies can provide both a foundation and a framework for design activity that will satisfy the search and information requirements of the engineer throughout the design process whilst also taking into account the many criteria related to manufacturing aspects. Such strategies can support a range of activities from concept exploration and decision support to final product definition, optimisation and realisation and therefore contribute significantly to design concurrency and integrated product development. The objective is to identify overall frameworks to support the various CI technologies in a manner that will ensure their successful integration with design team practice.

1. Introduction

There are many examples of the application of evolutionary and adaptive search algorithms to specific well-defined problems from the engineering design domain . Little research effort, however, has been expended in moving from these well-defined problems to investigate the generic integration of ES/AS with each stage (ie conceptual, embodiment and detailed) of the engineering design process as a whole where a major requirement for success is the development of systems that can capture specific design knowledge through extensive designer interaction. A problem-oriented approach to such ES/AS integration is essential due to the complex iterative processes and human centred aspects evident during the higher levels of design in particular. Interaction with designers and design teams is necessary to identify those areas where ES/AS strategies would be of most benefit and to recognise specific problems relating to their successful integration. Prototype search tools can then be developed as powerful extensions of the design team stimulating innovative reasoning at the higher conceptual levels of the design process; providing diverse, high-performance solutions to support decision-making during embodiment design and acting as powerful global optimisers that can operate efficiently within complex, computationally intensive domains during detailed design. The indication is that co-operative frameworks involving a number of search strategies / optimisation techniques operating concurrently within single or multi-level environments can offer significant utility [1]. The integration of complementary computational intelligence techniques with such strategies can result in overall search and processing capabilities that can support the engineer at each level of engineering design.

The various evolutionary / adaptive search algorithms are well suited to the concurrent manipulation of models of varying resolution and structure due to their ability to search non-linear space with no requirement for gradient information or apriori knowledge relating to model characteristics. These capabilities enable feature / knowledge-based modelling approaches to be utilised alongside more quantitative, mathematically based design descriptions. It is therefore possible to emphasise those design features that have the most significance in terms of manufactureability, economic feasibility and marketing considerations and to promote the emergence of design spaces containing solutions that best satisfy criteria relating to product design, realisation and utilisation. Feature-based modelling here refers to the generation and concurrent evolution of models that simply describe design relationships of particular interest / relevance to these other product development aspects alongside mathematical models that more specifically define the engineering system. The establishment of appropriate information exchange between these models each evolving under differing constraints and criteria may lead to the initial definition of a feasible design domain relating to overall product development and realisation.

Co-operative, adaptive design strategies can also provide an optimal definition process in terms of multi-disciplinary requirements. The potential of concurrent manipulation of both qualitative and quantitative models describing inter-disciplinary requirements to provide an initial feasible design definition requires further investigation. The initial objective is to achieve optimal stage designs which do not involve detailed analysis but contain a considerable amount of information relating to a wide range of design / manufacturing considerations.

2. The Design Process

At the highest level of the design process, conceptual design consists of search across an ill-defined space of possible solutions using fuzzy objective functions and vague concepts of the structure of the final solution. The designer will explore possible ways forward by proposing solutions and partial solutions and assessing their feasibility with regard to those constraints and criteria considered relevant at that time. The entire process is characterised by the use of heuristics, approximation and experimentation with a high degree of flexibility evident in the establishment of domain bounds, objectives and constraint. The design environment itself will evolve with the solutions as the designer / design team gain understanding of the functional requirements and possible structure. Models may be largely qualitative in nature to allow the establishment of initial design direction.

Having established an initial design configuration the design process can proceed to further definition of the subsets describing the whole system. The overall requirements will now be better defined due to the expanding knowledge base of the design team and significant improvement in available data. A degree of uncertainty remains however as engineers experiment again with partial solutions using models that are coarse mathematical representations of the systems under design. Decisions must relate to both qualitative and quantitative criteria many of which cannot be represented mathematically or defy inclusion within a scalar objective function. Insufficient knowledge and a requirement for minimal computational expense during these experimental stages results in unavoidable function approximation. A high degree of engineering judgement is therefore required to best interpret results from such models. Model resolution tends to increase during the many design iterations and eventual convergence upon solutions that are sufficiently low-risk allow continuation into detailed design stages.

Progression to detailed design indicates that the designs achieved during the conceptual and embodiment stages can now be considered to be sufficiently low-risk to allow their detailed design using computationally expensive, complex analysis techniques. A degree of such analysis will already have been evident where it has been necessary to validate results obtained from preliminary design models in order to minimise risk and allow the design process to progress. The uncertainties of the higher levels of the design will now have largely disappeared. Although several criteria may still be relevant they will be quantitative in nature and can be more easily accommodated within the fitness function. As uncertainty has largely been eliminated by a progressive refinement of design direction so the risks associated with the overall design task become less onerous. The major problem that now faces the identification of an optimal design is the considerable computational expense associated with complex analysis.

The above description of design activity assumes little interaction between each stage. Such an assumption is a major simplification which, if realistic, would greatly facilitate the development and introduction of computationally-based support systems for each stage. However, in reality, considerable overlap generally exists with a requirement, even during conceptual design to 'firm up' design concepts by carrying out lower level analysis (albeit perhaps at a reduced resolution). A degree of concurrency is therefore evident with a mixture of design representations being utilised during conceptual and embodiment design (although the lower levels of detailed design are largely deterministic requiring complex analysis alone). We can consider design optimisation, therefore, as a long-term, highly complex process commencing at conceptual / whole system design and continuing through the uncertainties of embodiment / preliminary design to the more deterministic, relatively low-risk stages of detailed design and the eventual realisation of an optimal engineering solution. The task therefore is to identify the optimal design direction at each stage ie that direction that best satisfies those objectives and criteria that appear relevant at that time whilst minimising risk associated with further design development. The following sections explore the potential of advanced adaptive search and computational intelligence techniques within this overall design environment.

3   The Technologies

Population-based evolutionary search techniques such as genetic algorithms (GA) [2] and evolution strategies (ES)[3] have the ability to negotiate high dimensional, multi-modal design spaces to identify high-performance solutions. They require no apriori knowledge of the design model and have been applied to a wide range of design problems providing exploratory capabilities during the earlier stages of design and convergence upon optimal solutions when required. Population-based evolutionary search can therefore provide a basis for a range of design activities relating to search, exploration and optimisation. This is illustrated by the research overviews of sections 4, 5 & 6 describing work at Plymouth Engineering Design Centre (PEDC) that particularly addresses the integration of evolutionary search technologies with generic problems relating to the various stages of design. Other adaptive search algorithms currently under investigation at the PEDC that offer considerable utility to various aspects of the design process include insect colony metaphors [4], simulated annealing [5], tabu search [6] and probabilistic incremental learning (pbil) [7]. It appears from current experience however that such techniques, or components of them, offer a more localised search that can be of considerable use when combined or utilised co-operatively with the exploratory characteristics of population-based strategies.

The utilisation of **genetic programming** [8] techniques for the representation of variable length design grammars is now showing considerable potential particularly within electronic circuit design [9]. Another significant application of GP relates to the generation of improved mathematical representations within preliminary engineering design models. These generally contain approximation in certain areas either due to a lack of knowledge, lack of confidence in available data or the requirement that computational expense must be kept to a minimum to allow rapid design iteration (section 4.3). Either empiric data or the results from more complex, computationally expensive analysis tools can provide a measure of fitness to support the evolution of improved GP-generated representations [10,11]. **Neural networks** [12] can also provide representations of engineering systems. Initial complex analysis generates the necessary training data to allow the generation of neural network representations that will, with appropriate integration, contribute to a significant reduction in computational expense when utilised to provide an approximation to the fitness function. These representations will be non-explicit and therefore must be treated with caution at the later stages of design. Both genetic programming and neural net representations may be of particular benefit where a degree of lower level analysis may be required to improve confidence in results from initial preliminary design work and where such verification must be achieved within a very restricted time-frame. **Fuzzy logic** [13] can also contribute during conceptual / preliminary design via the generation of fuzzy engineering system representations that can subsequently be utilised as fitness functions by appropriate adaptive search algorithms. This provides the means to incorporate uncertainty relating to preliminary design definition within initial design search and exploration. Objectives and constraints can be modelled in fuzzy manner to avoid compromising the design space through the imposition of objectives and constraints that prove inappropriate as the problem knowledge base expands. **Fuzzy technologies** can also provide a qualitative evaluation of the fitness of high performance quantitative solutions identified by adaptive search processes [14]. Such an evaluation, based upon engineer-generated fuzzy rules, can support the designer in the choice of optimal design direction whilst also providing the degree of flexibility required to allow the design space to evolve as the engineer's understanding of the design domain increases (section 5).

A high degree of engineer interaction should be in evidence when all such techniques are utilised during the various stages of design. Experience within the PEDC indicates that a two-way process involving engineer evaluation of results from exploratory design tools incorporating the various CI technologies and the subsequent off-line refinement of design representations, objectives and constraints offers considerable utility. The initial definition of the design space through the setting of variable and constraint bounds will rely, in the main, upon engineer experience and case-based reasoning. However the exploratory nature of the adaptive search algorithms coupled with, say, the fuzzy processing of the results can provide information concerning the overall problem domain that results in re-definition of the design space. Such redefinition through extensive human - computer interaction can result in the identification of innovative solutions whilst also stimulating design creativity. One objective of the regional identification procedures of section 4 and subsequent designer interaction is to support the identification of commonalities between the solutions of significantly differing high-performance design-space regions. Such feedback, through off-line discussion and analysis, may lead to radical reform of the overall problem space. An iterative loop is therefore created which incorporates the machine-based search and processing capabilities as part of the human-based design team activities as suggested in section 7.

As emphasis switches from preliminary design to detailed analysis so the overall search requirement changes from exploration to exploitation and optimisation. Adaptive search processes utilising computationally expensive, high-definition analysis software for evaluation become more self-contained. Designer interaction is required to a far lesser extent. It is possible at this stage to envisage a totally machine-based evolutionary optimisation process. However problems relating to the inherent computational overheads of population-based, adaptive search must be addressed as discussed in section six. The lack of robustness of advanced analysis tools can also cause problems when such tools are autonomously manipulated by population based search.. This may re-introduce a necessity for engineer involvement to ensure the validity of, say, GA generated solutions.

The inclusion of **intelligent agent-based technologies** within this adaptive AS/ CI / design team framework is discussed in section 7. It is first intended to introduce the manner in which research at the PEDC is addressing generic problems relating to the various stages of the design process.

4. Design Decomposition

4.1. Region Identification

A strategy for design decomposition / problem reduction adopted at the PEDC involves the identification of high performance (HP) regions of conceptual / preliminary design spaces and the subsequent extraction of relevant information regarding the characteristics of the designs within those regions. HP regions are rapidly identified during a coarse-based evolutionary search with maximum region cover in terms of number of solutions being a major objective. Subsequent qualitative and quantitative evaluation of the solutions of each region can provide information for off-line evaluation by the designer. Such information should enhance domain knowledge and support the definition of "good" designs in terms of specific features of each region and "feature commonalities" between regions. This leads to the improvement of system representation / fitness function through the refinement of inherent heuristics and approximations; the introduction /removal of constraints and variables and a modification of objective weightings. Such actions could radically alter the design environment and lead to innovative or perhaps even creative design solutions.

Simple variable mutation regimes that encourage diversity during the early stages of a GA search promoting the formation of solution clusters in the better areas of the design space were initially introduced [15,16,17]. Populations from selected generations are extracted and stored in a final clustering set thereby removing the need to maintain such clusters through successive generations. The objectives of these initial COGAs (cluster-oriented genetic algorithms) have been to identify a maximum number of regions; improve set cover ie solution density of each region; improve the robustness of the techniques and minimise the number of test function evaluations. An adaptive filter has been introduced to prevent low performance solutions passing into the clustering set (ie the final set comprising of the extracted populations of pre-selected generations). Extracted solutions are first scaled and a

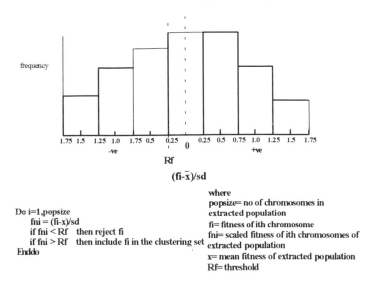

Single Threshold:

Distribution of results from the extracted population of generation 10:

$(fi-\bar{x})/sd$

where
popsize= no of chromosomes in
extracted population
fi= fitness of ith chromosome
fni= scaled fitness of ith chromosomes of
extracted population
x= mean fitness of extracted population
Rf= threshold

Do i=1,popsize
    fni = (fi-x)/sd
    if fni < Rf   then reject fi
    if fni > Rf   then include fi in the clustering set
Enddo

Figure 1: The Adaptive Filter

threshold value [Rf] is introduced as shown in figure 1. Solutions are either rejected or accepted into the clustering set depending upon their scaled fitness in relation to Rf which can be preset or varied at each of the selected generations. Varying Rf introduces an investigatory aspect in initial experimental runs which can provide information concerning the relative nature of differing regions of the design space. Figure 2 shows application to an eleven dimensional model of the cooling hole geometries of a gas turbine blade. A two dimensional hyperplane relating to two selected parameters is shown. For clarity, in this case, only one cluster relating to a particular internal geometry is shown in order to illustrate the degree of problem reduction that can be achieved by varying the Rf value. Such variation allows the engineer to explore the extent of the high-performance regions. Relative solution fitness is represented by colour variation of the points. High values of Rf result in the identification of individual solutions of very high performance as shown in graph 4. A prototype design tool utilising these techniques has now been introduced at Rolls Royce plc for evaluation by turbine engineers.

Alternative strategies involve the adoption of a geographically structured GA after Davidor [18]. A basic model defined by an NxN network is established with a local neighbourhood consisting of a location plus its eight neighbours ie a 3x3 grid. A mechanism based upon phenotypic similarity is introduced to identify when a sub-population is converging upon a particular area of the search space and to mark that area as 'tabu'. The sub-population can then be re-initialised [19].

An increase in problem dimensionality greatly increases the probability of premature convergence upon a few high-performance regions of a design space and the resulting

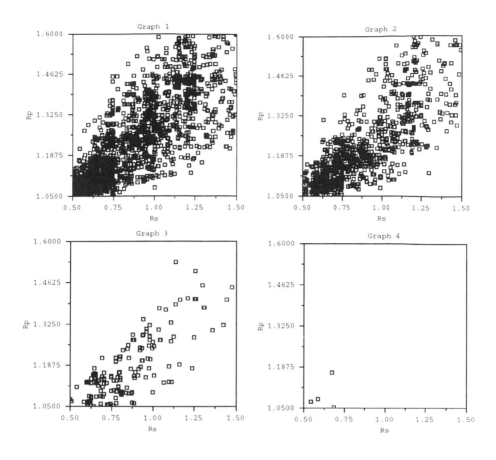

Figure 2: Two dimensional hyperplanes of the 11 dimensional blade cooling design space showing inlet / outlet pressure ratio (Rp) against cooling hole perimeter ratio (Rs). The sequence (graph 1 to graph 4) shows the degree of problem decomposition related to gradual increase of Rf value

non-identification of other regions of high potential. Strategies to overcome this problem have been investigated [19] where the bounds of potential regions are initially identified from the clustering of a few points and the fitness of individuals within these roughly bounded regions is suppressed. This allows search to continue and other regions to be identified. Individual search agents are left behind in the roughly defined regions in order to identify their true extent based upon the minimum fitness bound defined by the adaptive filter. These agents implement simple line searches from the centroid of the clusters and from the boundaries to establish the true extent of the regions. Although showing considerable potential this strategy has yet to be tested upon the turbine blade cooling hole geometry problem.

34

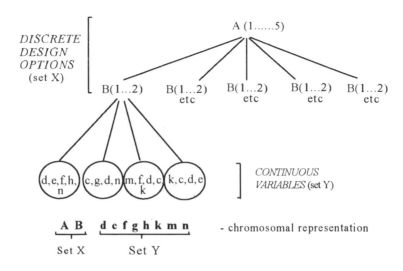

Figure 3: Simple design hierarchy

4.2. Whole System Design

Other research within the Centre concerns search across design hierarchies of discrete design configurations which are further described by dependant continuous variable sets. Achieving an efficient search across such hierarchies is not a trivial task. Many discrete paths have differing sets of dependant variables as illustrated in figure 3. The search space therefore comprises of differing continuous design spaces each related to a discrete design option. A search strategy is required that efficiently samples each continuous design space in order to rapidly determine the discrete design paths of high potential. The objective is to provide a more extensive search of available design alternatives within budget and time constraints. This results in the identification of regions containing competitive solutions that may have been overlooked during the problem decomposition processes of traditional heuristic design. The strategies should enable the engineer to rapidly survey the potential of diverse regions of the hierarchy thus avoiding premature concentration of search effort within previously known regions.

A dual agent strategy (GAANT) has been developed that maintains the combination of discrete and continuous variables within a single chromosome. In order to maintain the information exchange evident during traditional crossover it is necessary to speciate the populations in terms of like discrete configurations and restrict crossover to members of the same species. Elements of the Ant Colony metaphor [4] have been integrated for the manipulation and continuous improvement of the discrete set whereas a genetic algorithm operates within the continuous variable sets. Speciation prevents disruption of high performance parameter sets in earlier generations as a result of crossover of the continuous variables between differing discrete configurations. Such disruption leaves little opportunity for the evolution of better continuous variable sets for any particular configuration.

Appropriate communication between the two search agents results in overall improvement of the whole system solutions. Results from this co-operative strategy show satisfactory robustness and significant improvement in search diversity and solution fitness when compared to other fixed-length structured representations. Of equal importance is the simplicity of the chromosomal representation which is essential if complex, real-world design hierarchies are to be considered.

GAANT and variations of the algorithm have been utilised in two design domains. The first concerns the initial design stages of large scale hydropower systems [20] where several potential sites exist with a number of discrete options relating to mode of operation and major components. Continuous variables include pressure tunnel lengths, dam height, generation period, powerhouse depth etc. This relatively simple hierarchy contains twenty possible discrete paths with related continuous design spaces described by five to seven variables. Results shows that GAANT can provide an efficient search across the hierarchy and can identify a number of high performance design solutions at an acceptable level of computational expense.

The second domain is that of thermal power systems [21]. In this case discrete parameters relate to feed heater number and layout and steam line tapping points and configuration. Continuous variables include friction coefficients, heat conductances and parallel feed heater fractional flow rate. In addition there are eight control setting variables related to various feed water and gas flow rates, turbine control, pressures etc. Very significant improvements in predicted power outputs have been achieved in addition to significant reductions in design lead time. The GAANT strategy has been successfully integrated with design team practice within the UK power industry and research is underway to further improve overall performance.

4.3. Variable Length Hierarchies

The concepts behind the GAANT representation have also been applied to the manipulation of variable-length multi-level mathematical function representations. The objective has been to improve the calibration of preliminary design models to empiric data or to results from a more in-depth analysis (FEA or CFD). This is achieved by identifying those areas of coding where insufficient knowledge or the requirement of keeping computational expense to a minimum has resulted in unavoidable function approximation. A contributing factor may be the inclusion of empirically derived coefficients (ie discharge, drag etc). The objective is to evolve improved coding within these areas in order to achieve a better calibration with existing empiric data or results generated from a more in-depth, computationally expensive analysis. If this is possible then the element of associated risk would be correspondingly lessened whilst rapid design iteration can still be achieved utilising these simple, but more representative models. Initial research indicated that genetic programming (GP) [8] when utilised for system identification can achieve these objectives to a certain extent. GP manipulation of engineering relationships has provided some reasonable results related to the generation of formula for pressure drop in turbulent pipe flow and also energy loss associated with sudden contraction or sudden expansion in incompressible pipeflow [10]. However, it soon became apparent that the problems associated with the crossover of continuous coefficients between differing discrete functional structures was causing similar problems to those identified in the previous section relating to the successful crossover of useful information within the fixed-length design hierarchies. As previously noted the exchange of information from continuous design spaces to unrelated discrete design configurations does not promote the formation of high performance variable parameter combinations.

The success of the GAANT strategy stimulated experimentation to assess the benefits of a similar approach to the manipulation of the variable length 'design' hierarchies describing the mathematical functions. A classification system relating to the functional forms within the variable length hierarchies that would subsequently allow useful exchange of information between similar representations is required The variable lengths of the discrete decision trees and the related number of possible discrete system / function configurations eliminates possible classification in terms of identical structure. A classification in terms of complexity by arbitrarily ranking the mathematical functions and real numbered terminals has been introduced ie terminals =1.0; addition or subtraction = 1.1; multiply or divide = 1.2. Individual node complexity (NC) is then calculated from the weighted values of the nodes / terminals below it. The complexity of the tree decreases with depth and the root node NC(0) gives an overall rating of the whole tree. Trees can now be classified in terms of similar complexity and appropriate crossover regimes can be introduced between similar classes to reduce semantic disruption. In a similar manner to thatof the GAANT model elements from two search algorithms are introduced. A genetic programming operator manipulates the discrete functional structure whilst a genetic algorithm searches the continuous coefficient space.

A return to the engineering problems previously addressed utilising the canonical GP form indicates a significant improvement in performance relating to a reduction of overall population size and required calls to the fitness function resulting in a much reduced CPU time; a reduction in computer memory requirement; better accuracy in terms of correlation with empiric data or data generated from more complex analysis techniques and an ability to fit higher dimensional surfaces [11]. Although encouraging, further improvement is required in order to generate meaningful multi-variable models. It is expected that further improvement will be achieved as research continues relating to the GAANT concept and the successful manipulation of design grammars.

5.0. Multiple Solutions and Qualitative Assessment

Although many criteria may be represented quantitatively many will be qualitative in nature eg those relating to in-house design and manufacturing preference etc. Many multi-objective optimisation techniques rely upon a clear quantitative representation of the criteria and assume a high confidence in their validity whereas other approaches fuzzify quantitative criteria in order to soften the objectives and lead the search to satisfactory solutions. In general however, these approaches only consider quantitative criteria. There remains a requirement to include the designer in the assessment loop to both provide qualitative judgement and to assess the relevance of the included criteria. One attempt to achieve this utilises an evolutionary niching approach to first identify good solutions before introducing a fuzzy logic interface which assesses the solutions in terms of designer-generated qualitative rules concerning relationships between the defining parameters [14]. The result is a quantitative evaluation from the GA generated solutions which is enhanced by a crisp qualitative rating from the fuzzy interface. Additional information concerning solution and variable sensitivity and degree of constraint violation is also extracted and presented to the engineer.

We must consider however the validity of single solutions identified from preliminary design models that are coarse representations of the system under design. This could be a too specific approach at this stage as subsequent, more in-depth analysis may prove such solutions erroneous. An alternative is to return to the regional identification procedures introduced in section 4.1. In this case the characteristics of high-performance bounded regions can be assessed by the designer with the assistance of complementary soft computing techniques such as the fuzzy interface already mentioned. This approach has already been utilised for the identification of regions characterised by low gradient and thus providing robust design solutions [15,16]. Research is continuing in this area to further assess the extent and nature of relevant engineering information that can be extracted from these regions.

6.0 Co-operative Strategies for Detailed Design

The engineer's requirements from adaptive search strategies change considerably as we progress to detailed design or wish to introduce low-level analysis to verify results from preliminary design tools. The emphasis is upon the identification of a single high-performance solution as opposed to the achievement of a number of alternative designs. The major problem is the considerable computational expense associated with complex analysis and population-based search.. It is essential that the number of calls to the system model is minimised if extensive integration is to be considered feasible.

Appropriate representation of the component under design and communication between co- evolving representations of differing resolution can introduce significant speed-up. The 'injection island' strategy [22] is a prime example. The co-operative co-evolution of differing representations of a design problem distributed across a number of separate 'evolution islands' intoduces a significant reduction in the number of calls to the evaluation function. Good lower resolution solutions are 'injected' into higher resolution evolution processes. This co-evolutionary, multi-level approach offers great potential as a framework for the support of single component design through both preliminary and detailed stages. Research [23] at Plymouth is investigating the combination of computationally inexpensive preliminary design models with more highly definitive finite element techniques within a single injection island framework. The intention is to evolve the design from initial configuration through to product realisation [24].

Work in this area has concentrated upon the structural analysis of concrete flat plates where a finite element analysis (FEA) is ultimately required to provide a low-risk design solution. The problem domain is high-dimensional with circa 400 elements of the plate being able to vary in depth and has conflicting objectives ie the minimisation of stress violation and the minimisation of weight. Initial research concentrated on the sequential introduction of plate representations of increasing resolution during an evolutionary search process. Various adaptive search techniques were introduced but the CHC GA [25] has been found to provide best performance across the range of plate resolutions. However, satisfactory convergence was initially only possible with plate representations of up to eighty varying elements. This was sufficient to allow the evolution of specific areas of the plate and the achievement of improved plate designs. The research relates to a real-world design and is supported by an international manufacturer of building materials. Integration of these initial techniques has led to the evolutionary design and subsequent mass manufacture of these building components.

Introduction of the injection island techniques whilst still utilising the CHC GA has subsequently led to successful concurrent search of several plate resolutions culminating in the required 400 element variable depth representation. A very significant reduction in the overall number of calls to the plate model is evident in addition to further significant reductions in

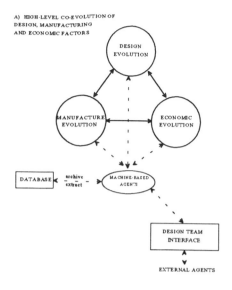

A) HIGH-LEVEL CO-EVOLUTION OF
DESIGN, MANUFACTURING
AND ECONOMIC FACTORS

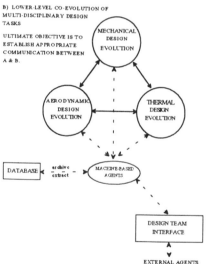

B) LOWER-LEVEL CO-EVOLUTION OF
MULTI-DISCIPLINARY DESIGN
TASKS

ULTIMATE OBJECTIVE IS TO
ESTABLISH APPROPRIATE
COMMUNICATION BETWEEN
A & B.

Figure 4: High and low level
co-evolutionary processes

plate weight. Early comparison of the various AS algorithms upon this plate problem indicated better performance from certain AS techniques during the earlier generations than from others and this has led to the establishment of a multi-agent approach with appropriate search agent introduction / removal strategies within the framework of the injection island architecture. Two differing AS techniques (CHC GA and probabilistic incremental learning (Pbil) [7]) are added to or removed from the co-evolving search processes dependant upon their performance. Indications are that such a strategy can offer significant utility although further research is required to investigate the interaction between the differing search agents.

The injection island / CHC approach is now being integrated with industrial design practice and it is expected that this will result in a machine-based, stand-alone process that can take single high-resolution component representations from preliminary through detailed design to manufacture within an acceptable period of time and with significant improvement in component performance. Research related to muti-agent integration is continuing.

7.0 Co-operative Computational Intelligence Systems

A common element of the PEDC research described in the preceding sections is the utilisation of co-operative strategies that utilise two or more search techniques to overcome the complexities related both to the various design spaces and to the structure of the problem representation. The research described indicates that such co-operative strategies can be of significant benefit within the adaptive search domain. PEDC research has shown that such co-operation can overcome problems related to constraint satisfaction [26], mixed discrete / continuous variable design hierarchies [8], design space decomposition [9] and computational expense related to complex analysis during detailed design [24]. In addition, although each of the related computational intelligence technologies of section 3 offer potential in their stand-alone application to specific design problems far greater utility in the form of their satisfactory integration with current design practice can be achieved by an investigation of overall co-operative structures that utilise their particular characteristics in a more generic manner. This integration of co-operating CI techniques offers major potential especially in achieving meaningful designer / machine interaction.

It is suggested that a major role of machine based **agent technologies** [27, 28,29] within the design domain will relate to this communication aspect. As the utilisation of ES / AS techniques moves to meaningful integration with the design process and current design team practice so the need for co-operative strategies will increase. This will involve co-evolving processes each of which may describe:
* individual elements of the engineering system
* aspects related to individual disciplines (ie multi-disciplinary co-evolution)
* high-level aspects relating to design, manufacture and marketing etc (ie integrated product development).

Irrespective of the level at which processes operate there will be a requirement for communication both between each process and with the design team. Such communication may relate to, for instance, the degree of constraint / criteria satisfaction in each domain or degree of convergence. A capability for state recognition will be required to identify the need to :
* redistribute search resource
* terminate search or instigate new search processes
* relax, harden, eliminate or introduce constraints and criteria
* communicate current status or seemingly relevant data to the design team or conversely collect and distribute relevant data via the design team interface
* in the same manner, archive relevant information to an on-line database or extract data required by the search processes in order to continue

This suggests an intelligent agent approach integrated both with machine-based and design team activities and a structure similar to that shown in figure 4. The structure of figure 4a is aimed primarily at high level concept formation to establish an initial design brief that takes into account the many factors relating to overall product development. Evaluation models utilised at this stage may be simple rule bases concentrating on features relevant to each area. Appropriate machine-based communication between the co-evolving sets in addition to machine / external communication should resolve conflicts resulting in the evolution of 'best' design scenarios. Figure 4b describes the manipulation of relatively simplistic preliminary design representations in order to establish an overall initial feasible domain. Current co-evolution / agent related research at the PEDC is firmly based within these high-level design domains. The initial concurrent manipulation of relatively simplistic design descriptions should provide an indication of the overall feasibility of the approach. Scaleability issues can subsequently be investigated. Although ambitious, the development of such overall co-evolutionary strategies will support meaningful integration with design team practice The designers themselves can be regarded as external agents extracting information from the system and processing such information off-line. Such off-line processing allows the dynamic re-definition of the problem domain via the re-introduction of reformed variable bounds, constraint, criteria and objective functions. These systems will utilise the related CI components of section 3.

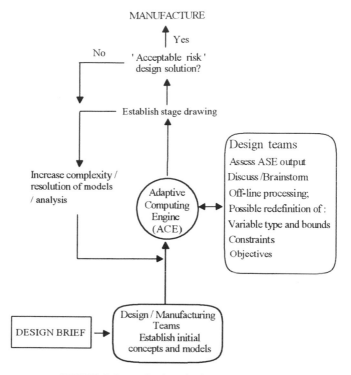

FIGURE 5: Integrating the Adaptive Computing Engine
with Design Team Practice

In many ways this suggested combination of evolutionary / CI techniques and strategies relates more to current design practice rather than natural evolution by attempting to model design team interaction, information exchange and processing. With appropriate internal / external communication this will result in concurrent, iterative improvements in system / product design. The objective is to provide an interactive facility which results in a far broader search of design , manufacturing ,economic and marketing alternatives within an acceptable timeframe during all stages of design. A greater range of variable design parameters can be combined with manufacturing objectives and constraint to significantly increase the probability of achieving an optimal, acceptable design definition within a shorter timespan and with the added possibility of competitive innovation. The design team can rapidly assess design concepts by utilising the search and decision support capabilities of the suggested adaptive computing engine as illustrated in figure 5. Off-line discussion based upon solutions generated by the search processes redefines the problem space whilst increasing the engineer's knowledge-base with regard to the overall design problem. It is intended that initial establishment of such frameworks will be restricted to the simpler representations utilised during conceptual design. Having established basic feasibility then future work will address lower level co-evolutionary design tasks.

8.0 Summary

The paper attempts to define the search, exploration and optimisation requirements of the engineering designer and draws upon previous research experience within the PEDC to identify CI techniques and strategies that can best support design and manufacturing activity. The current contribution of specific technologies are recognised and overall structures are presented which, with relevant research and development, may provide the degree of integration required to ensure that the potential of such advanced computational techniques is fully realised. Benefits should include a reduction in design lead time and in any necessary design change in addition to the identification of innovative and competitive design solutions. There is no suggestion here that design can be represented as a computable function rather that exploratory tools can be developed that enhance the creative capabilities of the engineering designer during the earlier stages of design whilst also providing powerful optimisation capabilities when required.

It is very apparent that there is significant utility in the application of evolutionary search and optimisatiom techniques to complex, routine design problems. However, meaningful integration with generic design practice must involve the development of overall strategies that address wider aspects such as problem structure, model representation, variable objectives and constraints, machine-based communication and design team interaction. Although some of these aspects can be addressed by introducing co-operating adaptive search algorithms further involvement of other established and emerging computational intelligence technologies will be required to ensure full potential is realised. The importance of design team interaction with adaptive systems cannot be overstressed. The emergence of innovative and creative design solutions will be greatly enhanced by the development of truly interactive systems that allow concurrent and efficient machine and human manipulation of design information.

9. Acknowledgements

The paper relates to research supported by the UK Engineering and Physical Science Research Council and by several UK industrial organisations including Rolls Royce plc, British Aerospace and Nuclear Electric. The author thanks these organisations for their continuing support.

10. References

1. Parmee I. C. ,1997, Strategies for the Integration of Evolutionary / Adaptive Search with the Engineering Design Process. In: Dasgupta D. & Michelewicz Z. (eds), *Evolutionary Algorithms in Engineering Applications;* Springer-Verlag.

2. Goldberg D. E., 1989, *Genetic Algorithms in Search, Optimisation & Machine Learning.* Addison - Wesley Publishing Co., Reading, Massachusetts.

3. Rechenburg I.,1984, The Evolution Strategy: A Mathematical Model of Darwinian Evolution. *Synergetics: from Microscopic to Macroscopic Order;* Springer Series in Synergetics Vol 22; pp 122 - 132.

4. Colorni A., Dorigo M., Maniezzo V., 1981, Distributed Optimisation by Ant Colonies. In: Varela F, Bourgine P. (eds); *Proceedings of First European Conference on Artificial Life,* Paris.

5. Kirkpatrick S., Gelatt C. D, Vechi M. P., 1983, Optimisation by Simulated Annealing. *Science,* Volume 220, No. 4598.

6. Glover F., 1989, Tabu Search - Part I, *ORSA Journal on Computing,* Vol. 1, No. 3.

7.   Baluja S., 1994, Population Based Incremental Learning: A Method for Integrating Genetic Search Based Function Optimization and Competitive Learning. Technical Report, School of Computer Science, Carnegie Mellon University, Pittsburgh, CMU-CS-94-194.

8.   Koza, J. R., 1992, *Genetic Programming - on the Programming of Computers by Means of Natural Selection*. The MIT Press, Massachusetts,.

9.   Koza J.R., 1998, Evolutionary Design of Analog Electrical Circuits Using Genetic Programming . In:Parmee I. C. (ed); *Adaptive Computing in Design and Manufacture*, Springer Verlag.

10.   Watson A. H., Parmee I. C., 1996, Identification of Fluid Systems using Genetic Programming. *Proceedings of Fourth European Congress on Intelligent Techniques and Soft Computing*, Aachen, Germany.

11.   Watson A. H., Parmee I. C., 1998, Improving Engineering Design Models using an Alternative Genetic Programming Approach. In: Parmee I. C. (ed); *Adaptive Computing in Design and Manufacture*, Springer Verlag.

12.   Hajela P., Lee J., 1994, Role of Emergent Computing Techniques in Multidisciplinary Rotor Blade Design.. In: Grierson D. E., Hajela P. (eds); *Emergent Computing Methods in Engineering Design*; NATO ASI series F: Computer and Systems Sciences, Vol 149; Springer Verlag.

13.   Zadeh L. A., 1965, Fuzzy Sets. *Journal of Information and Control*, vol. 8, pp 29-44.

14.   Roy R, Parmee I. C, Purchase G. Integrating the Genetic Algorithm with the Preliminary Design of Gas Turbine Cooling Systems. In: Parmee I. C. (ed); *Proceedings of 2nd International Conference on Adaptive Computing in Engineering Design and Control*, PEDC, University of Plymouth, 1996.

15.   Parmee I. C., Denham M. J., 1994, The Integration of Adaptive Search Techniques with Current Engineering Design Practice. In: Parmee I. C. (ed); *Proceedings of Adaptive Computing in Engineering Design and Control*; University of Plymouth, UK; pages 1-13.

16.   Parmee I. C., 1996, The Maintenance of Search Diversity for effective Design Space Decomposition using Cluster-oriented Genetic Algorithms (COGAs) and Multi-agent Systems (GAANT). In: Parmee I. C. (ed); *Proceedings of 2nd International Conference on Adaptive Computing in Engineering Design and Control*, PEDC, University of Plymouth.

17.   Parmee I. C., Cluster-Oriented Genetic Algorithms (COGAs) for the Identification of High-Performance Regions of Design Spaces. Proceedings of EvCA96 Conference, Moscow, June 24-27 1996.

18.   Davidor Y., Yamada Y. N., 1993, The Ecological Framework: Improving GA Performance at Virtually Zero Cost. In: Forest S. (ed); *Proceedings of the Fifth International Conference on Genetic Algorithms*, Morgan Kaufman.

19.   Parmee I. C., Beck M. A., 1997, An Evolutionary, Agent-Assisted Strategy for Conceptual Design Space Decomposition, In: D. Corne and J.L. Shapiro (eds.), Evolutionary Computing: Selected Papers from the 1997 AISB International Workshop, Springer Lecture Notes in Computer Science, No. 1305, Springer, ISBN 3-540-63476-2, pp. 275–286.

20.   Parmee I. C., 1996, The Development of a Dual-Agent Strategy for Efficient Search   Across Whole System Engineering Design Hierarchies. *Parallel Problem Solving from Nature IV, Lecture Notes in Computing 1141*, Springer Verlag.

21.   Chen K., Parmee I. C., 1998, A Comparison of Evolutionary-based Strategies for Mixed-discrete Multi-level Design Problems. In: Parmee I. C. (ed); *Adaptive Computing in Design and Manufacture*, Springer Verlag.

22.   Punch W. F., Averill R. C., Goodman E., Ding Y., Lin S., 1995, Using Genetic Algorithms to Design Laminated Composite Structures. *IEEE Expert*.

23.   Vekeria H. D., Parmee I. C., 1996, Reducing Computational Expense Associated with Evolutionary Detailed Design. In: *Proceedings of International Conference on Evolutionary Computing '97*; Indianapolis.

24.   Parmee I. C., Vekeria H., 1997, Co-operative, Evolutionary Strategies for Single Component Design. In: Back T. (ed); *Proceedings of Seventh International Conference on Genetic Algorithms,* pp 529-536

25.   Eshelman L. J. The CHC Adaptive Search Algorithm: How to Have Safe Search When Engaging in Nontraditional Genetic Recombination. In G.J.E Rawlins (editor), Foundations of Genetic Algorithms and Classifier Systems. Morgan Kaufmann, San Mateo, CA, 1991.

26   Bilchev G., Parmee I. C., 1995, Constrained Optimisation with an Ant Colony Search Model. In: Parmee I. C. (ed); *Proceedings of 2nd International Conference on Adaptive Computing in Engineering Design and Control*, PEDC, University of Plymouth, 1996.

27.   Talukdar S., deSouza P., Murthy S. , 1993, Organisations for Computer-based Agents. Engineering Design Research Centre, Carnegie Mellon University, Pttsburgh, USA.

28.   Clearwater S., Hogg T., Hubermann B. Cooperative Problem Solving. Computation: The Micro and Macro View. B. A. Hubermann, ed.; *World Scientific*, pp. 33-70, 1992.

29.   Wooldridge M., Jennings N. R., 1995, Intelligent Agents: Theory and Practice. *Knowledge Engineering Review*, 10(2).

Chapter 2

Adaptive Computing In Manufacture

Tackling Complex Job Shop Problems using Operation Based Scheduling.
D. Todd, P. Sen

Minimising Job Tardiness: Priority Rules vs. Adaptive Scheduling.
D.C. Mattfeld, C. Bierwirth

Evolutionary Computation Approaches to Cell Optimisation.
C. Dimopoulos, A.M.S. Zalzala

A Temperature Predicting Model for Manufacturing Processes Requiring Coiling.
N. Troyani, L. Montano

Solving Multi-objective Transportation Problems by Spanning Tree-based Genetic Algorithm.
M. Gen, Y-Z Li

Tackling Complex Job Shop Problems Using Operation Based Scheduling

D S Todd and P Sen

Engineering Design Centre and Dept. of Marine Technology
University of Newcastle Upon Tyne, NE1 7RU, UK.
email: d.s.todd@ncl.ac.uk, pratyush.sen@ncl.ac.uk

Abstract. Scheduling is a combinatorial problem with important impact on both industry and commerce. If it is performed well it yields time and efficiency benefits and hence reduces costs. Genetic Algorithms have been applied to solve several types of scheduling problems; Flow Shop, Resource, Staff and Line Balancing have all been tackled. However Job-shop Scheduling is the most common problem of interest. The Job-shop Scheduling Problem (JSP) involves placing jobs onto a set of machines with the aim of minimising makespan, the total time to complete all jobs. The standard JSP model is relatively simple and cannot cope with many real world situations. This paper outlines a new method, using an operation based construction, which expands the JSP to increase its functionality and applicability. This allows for much more complex job shop problems involving flexible machines, setup and maintenance times, deadlines and multiple optimisation criteria. The method is fully explained and demonstrated using a 20 job-4 machine example.

1. Introduction

The Job-shop Scheduling Problem (JSP), first attempted with GAs by Davis [1], is one of the hardest [2] and most commonly encountered scheduling problems. Since Davis' work there have been many attempts at the simple JSP with a single criterion [3-9]. More complex approaches by Liang [10] and Sridhar [11] used a weighted sum approach to combine disparate metrics to give a single fitness measure. Murata [12] used a random weight system to generate a range of trade-off solutions in a flowshop problem, while Hamada [13] addressed optimization with constraints. However, none of these approaches combine multiple criteria considerations and the increased job shop complexity which is considered in this paper.

A classical JSP is defined as follows:

      i) There are k machines and n different *jobs*.
      ii) Each job is composed of a set of m or less *stages*.
      iii) Each stage requires a specific machine.
      iv) Each stage has a fixed processing time.
      v) The stages must be carried out in a specific order.

vi) There are no precedences between different jobs.

vii) Operations cannot be interrupted.

viii) Each machine can process only one job at a time.

ix) Neither release times nor due dates are specified.

The aim is to determine the operation sequences on the machines in order to minimise the makespan. To demonstrate this problem a 4 job 3 machine example is used (Table 1.1). M1 denotes that machine 1 is needed to perform the operation and the following number indicates the time required to perform it. The ordering of the jobs on each machine is very significant. This is demonstrated in Figure 1.1. The first case is randomly generated and has a makespan of 38. The effect of optimisation can be clearly seen in the second example. The makespan has been reduced to 31. This effect becomes more significant as problem size increases. The final example shows what can be achieved if the flexibility of the machines is increased, in that the same stage of a job can be performed on more than one machine, an extension which is discussed later. In this case machine 2 has been modified to allow it to perform machine 1's jobs as well as its own. This reduces the makespan to 24.

	Stage 0	Stage 1	Stage 2	Stage 3	Stage 4
Job 0	M1 ,5	M0, 3	M1, 4		
Job 1	M2, 3	M1, 6	M0, 5	M1, 3	M2, 4
Job 2	M0, 4	M2, 3	M1, 6		
Job 3	M1, 4	M2, 5	M1, 3	M0, 5	

Table 1.1 - Job Shop Problem

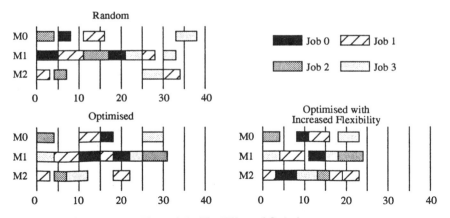

Figure 1.1 - The Effect of Ordering

2. Complex Scheduling

Scheduling plays an important role in most manufacturing environments. It can lead to reduced lead times and inventory levels as well as improved integration between different functions within an organisation such as manufacturing and assembly . This increasing level of interaction facilitates the use of Concurrent Engineering techniques to increase process efficiency. However integration requires a systematic consideration of manufacturing to deadlines. Deadlines will occur internally but often a company will have to pay compensation for every day or week a product is delivered late to the customer.

It is clear from the above description that the standard JSP does not realistically represent these problems and hence is very limited in its application. At this point a specification for the Complex JSP was developed to allow for these additional requirements.

The Complex JSP implements the following additions to the Simple JSP:

i) Each stage requires a specific *operation*.
ii) Each stage has a *standard processing time*.
iii) A job can visit the same machine one or more times.
iv) There are interdependencies between different jobs.
v) Release times and due dates are specified.
vi) Machines can perform several operations.
vii) Machines perform operations with different efficiencies.
viii) Machine maintenance times can be specified.
ix) Machine setup times for operations can be specified.
x) Due date penalties can be specified.
xi) Multiple scheduling criteria can be optimised.

The Complex JSP model has a much greater level of complexity and range of possible applicability. There are still n-jobs to be tackled. Each of these jobs has m stages but instead of specifying a specific machine an operation is specified for each stage. The motivation for operation based scheduling is outlined more fully in section 3. Each stage is also given a standard processing time which is the time it takes to perform that operation on a 'standard' machine. The stages must be processed in a specific order as before but a stage can involve going on a specific machine more than once, consecutively if necessary. Jobs are not given precedences but interdependencies can be specified. For example if a job requires the completion of other jobs before it can start this can be specified within the problem. Operations cannot be interrupted unless it is for maintenance. It is possible to specify whether a job should wait until after maintenance. The jobs also have release times and due dates. The release times allow scheduling to be integrated with previous events such as the completion of a design task or the delivery of materials which prevent a job starting. Due dates can also be specified and the severity of penalties on these due dates can also be set. Hard due dates can be given large penalties but ones where slippage is allowed can be less heavily penalised. Machines can be flexible and are able to perform more than one type of operation. The time it takes the machine to process a given stage is dependent on its efficiency at performing the operation required at that stage. With the machines performing different operations machine setup times must also be considered within the schedule plans. Finally, scheduling is not just about minimising makespan, in complex scheduling it also involves the consideration of several conflicting criteria. This is discussed further in section 4.

3. Operation Based Scheduling

Schedulers have commonly used the classical job shop approach which assumes that any given stage of a job can be undertaken at only one machine and the scheduling task is to order the handling of the stages on the given machines. This is often not the case in practice as a workshop may have several machines of the same type or various machines which can carry out the same operations. Therefore the operations needed for any one stage of a job could be carried out on a number of different machines. There are also other complications involved in this type of model. If a machine is set up for one type of operation and then switched to another type of operation there is some setting up involved. Additionally different types of machines performing the same operation may also take different amounts of time (e.g newer machines will operate faster than older ones). These factors indicate that the scheduler required must be capable of handling operations and not jobs.

4. Multiple Criteria Scheduling

Scheduling has commonly been tackled using Linear Programming, Monte-Carlo and Heuristic methods [14]. The aim of such procedures has been to minimise makespan, the total time from the start of the first task to the end of the last. However, other important factors exist within a scheduling task. Regulated allocation of resources, limits on the allowed number of tasks in progress, the completion of tasks by specific deadlines, interrelationships between tasks and conflicts for shared resources are just a few examples. Solutions are sought which strike a balance between these potentially competing requirements.

As a simple example a workshop with limited space may be considered. One objective is to limit the amount of work in progress, the other is to minimise the time taken to complete a set of tasks. The trade-off can be clearly shown on a trade-off curve (Figure 4.1). We can either minimise makespan (1) or minimise the maximum level of work in progress (3). However a compromise solution (2) may be more preferable. The solutions to the problem will be confined within a feasible region which is defined by system limitations and imposed constraints. Hence, the aim of scheduling in a multi-criteria context is to generate the trade-off boundary. To do this a Multiple Criteria GA (MCGA) may be used. This is explained elsewhere [15]. The generated boundary can then be used to choose the solution that has the most desirable balance of criteria for the given set of circumstances. The most desirable balance is obviously dictated by the priorities in question.

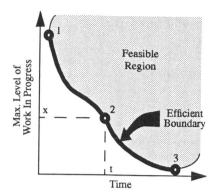

Figure 4.1 - Trade-off in Scheduling

5. Multiple Criteria Scheduler

5.1. Complex Job Shop Scheduling

The complexity of the scheduling problem can no longer be represented by a single table showing jobs, stages, machines and times as in Table 1.1. Several tables relating to different aspects of system are needed to fully specify the problem and these are explained below. The system uses the user defined clock tick as a time unit. The clock tick can be of any real-time duration (1 minute, 10 minutes, 1 hour, etc.). However, the higher the definition (smaller the clock time) the longer the scheduler will take to run for a given length schedule as there will necessarily be more clock cycles.

The problem in question is defined in terms of several variables: Number of Machines, Number of Operations, Number of Jobs, Maximum Number of Stages, Maximum Number of Dependencies and Maximum Number of Maintenance Periods. It also has a variable which indicates whether job splitting is allowed. Job Splitting occurs when a job is on a machine and scheduled maintenance has to take place before the job is complete. This results in the job being stopped and restarted after maintenance. If job splitting is not allowed the scheduled job will not be started until after maintenance has ceased.

The first table used is the machine operating efficiency table (Table 5.1). This is used to show the efficiency of machines when performing the different operations. Those with an efficiency of 1.0 perform operations in the standard specified time. Those with an efficiency greater than 1 perform it faster whilst those with less perform it slower. Machines with efficiency 0 cannot perform the operation. For example machine 0 can perform Op0 with 100% efficiency and Op3 with 50% efficiency but cannot perform any other operations.

	Op0	Op1	Op2	Op3	Op4	Op5
M0	1.0	0	0	0.5	0	0
M1	0	0.5	0	0	0.8	0
M2	0	1.0	0.8	0	0	0.8
M3	0.5	0.8	0	0.3	0	1.0

	No.	N/A 0	N/A 1	N/A 2	N/A 3
M0	2	50-60	100-110		
M1	1	20-25			
M2	4	10-15	25-30	34-37	50-52
M3	2	20-35	60-64		

Table 5.1 - Machine Operating Efficiency Table

Table 5.2 - Machine Maintenance Times

The machine maintenance times are shown in Table 5.2. If the machine requires new parts to be fitted or to be stripped down and cleaned then routine maintenance times can be allocated. For each machine the number of periods is noted and for each of these periods a time range is specified. For example, machine 3 has 2 periods where it is not available, 20-35 clock ticks and 60-64 clock ticks

If a machine is able to perform several different types of operation the transition between operation types may require some setup time. These times are specified in Table 5.3. The current operation is on the left and the next operation reads down from the top. If a transition from Op3 to Op4 takes place 2 clock units will be required to complete this transition. Inclusion of this facility allows the MCGA to experiment with performing all of one type of operation on one machine, hence no setup time would be needed, or changing machine operation types. The effects of this can only be found through simulation, since it is case dependent.

In Table 5.4 each job has a number of stages each of which is specified in terms of two numbers. The first is the operation type which has to be performed at that stage. The second is the standard time required at that stage.

	Op0	Op1	Op2	Op3	Op4	Op5
Op0	0	4	2	2	3	1
Op1	3	0	1	5	4	2
Op2	3	2	0	2	3	1
Op3	4	3	2	0	2	1
Op4	3	5	2	1	0	2
Op5	5	4	4	1	2	0

	No.	S0	S1	S2	S3	S4
J0	2	2,4	4,10			
J1	4	1,6	2,3	5,8	0,3	
J2	3	4,5	1,3	2,8		
J3	4	0,7	2,7	4,3	5,1	
J4	5	1,4	0,5	1,7	3,9	4,6
....	....	....	....	....	....	....

Table 5.3 - Operation Setup Times

Table 5.4 - Job-Stage Table

If Table 5.4 and Table 5.1 are brought together the effect of efficiency can be demonstrated. For an example let job 3-stage 0 be chosen from Table 5.4. This stage requires operation 0 to be performed. Using Table 5.1 it can be seen that machine 0 and machine 3 can perform this operation. If machine 3 is chosen this machine performs this operation with efficiency 0.5. The total time taken for this operation is given by:

$$\text{Time} = \frac{\text{Standard Time}}{\text{Efficiency}}$$

Hence, on machine 3, job 3-stage 0 will take 7/0.5 = 14 clock ticks.

Similarly, job 3-stage 0 will only take 7 clock ticks on machine 1, because the efficiency is 1. If the result of this calculation is a decimal it is rounded up to the nearest clock tick in this implementation. This means that over estimates will be provided. If a shorter clock tick unit is used this error becomes smaller but as mentioned earlier the simulation will take longer.

The final table (Table 5.5) used specifies the earliest start times, dependencies and deadlines for each job. The column labelled 'start' shows the earliest time that job can commence. This may be due to the delivery date of raw material or the date for completion of a part design. The next column specifies the number of dependencies that this job has. For example job 3 has 2 dependencies, and these relate to jobs 7 and 8 (from the following columns). This means that before job 3 can start both job 7 and job 8 must be completed. Finally, the last two columns specify the deadlines and penalties for the jobs. The deadline is the clock time at which the job is due. After this point penalty costs are incurred. Job 3 has a stricter deadline than the other jobs because it has a 5 unit cost penalty. Job 4 incurs no penalties whatever time it finishes.

	Start	No.	D0	D1	D2	Deadline	Penalty
J0	10	3	6	8	9	100	1
J1	0	0				50	1
J2	5	0				90	1
J3	0	2	7	8		100	5
J4	0	1	16			0	0
....	....	....	...	...	...	....	....

Table 5.5 - Job-Stage Table

5.2. String Configuration in MCGA

Due to the nature of this scheduling problem it is not possible to use a string which is partitioned into machine sections each containing a prioritised, fixed length, list of stages. This is because the possible interchange of stages between machines would mean that although strings would remain the same length overall, the boundaries between machine partitions would change as genes move around in the string. This could also cause the GA to introduce duplicates during crossover. The system used here still partitions the string into machines but it holds multiple copies of stages which occur on several machines. By using a group of activation genes inserted at the start of the string it activates/disables the duplicated stages within the string. This technique was developed from the ideas of the Structured GA [16].

5.3. Selective Gene Activation

The structured GA techniques are based on a binary string. However if the concept is extended to use an integer string a different type of activation can be initiated. The string is the same as before but instead of being able to turn on and off sets of genes the upper level genes can now choose which genes they activate (Figure 5.1).

Structure

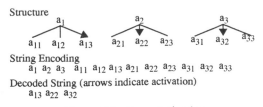

String Encoding
a_1 a_2 a_3 a_{11} a_{12} a_{13} a_{21} a_{22} a_{23} a_{31} a_{32} a_{33}

Decoded String (arrows indicate activation)
a_{13} a_{22} a_{32}

Figure 5.1 - Gene Activation

5.4. String Construction

In the context of the scheduler, selective gene activation can be used to choose which jobs are performed on which machines. This technique is demonstrated here using a simple example. Table 5.6 is a table of the machine operating efficiencies and Table 5.7 shows the Job-Stage data. Each stage of a job is now defined in terms of the operation that needs to be performed at that stage. Operations can be thought of as specific tasks (e.g drilling or milling in a machine shop) and it is usually the case that more than one machine can perform this operation. Operations can still be restricted to a single machine (Op0, Op1, Op2 in this example) so the simple JSP can still be tackled. The data can be interpreted as follows: Job 0-Stage 1 involves Op4 which can be performed on either Machine 0 or Machine 2.

	Op0	Op1	Op2	Op3	Op4	Op5
M0	1.0	0	0	1.0	1.0	0
M1	0	1.0	0	1.0	0	1.0
M2	0	0	1.0	0	1.0	1.0

Table 5.6 - Machine Operating Efficiency Table

	No.	S0	S1	S2	S3	S4
J0	3	1,5	4,3	1,4		
J1	5	2,3	1,6	3,5	1,3	4,4
J2	3	0,4	2,3	1,6		
J3	4	5,4	2,5	1,3	0,5	

Table 5.7 - Job-Stage Table

The string is made up of n+1 distinct segments, where n is the number of machines and the extra segment is made up of a set of activation genes. Each of these activation genes will correspond to one of the machine choices (i.e. stages which can be performed by more than one machine) so in this case there will be 4 activation genes. This is because only operation types 3,4 and 5 can be performed on more than one machine, the other operations being limited to a single machine, as Table 5.6 shows. In each of the machine segments there is a gene for each of the stages which are to be performed on that machine, including those which could appear on two or more machines. Hence the string is a concatenation of these segments:

Machine Choices - Machine 0 - Machine 1 - ..... - Machine i - ..... - Machine n

In the case of the example we have:

Machine Choices = j0s1, j1s2, j1s4, j3s0

Machine 0 = j0s1, j1s2, j1s4,j2s0, j3s3

Machine 1 = j0s0, j0s2, j1s1, j1s2, j1s3, j2s2, j3s0, j3s2

Machine 2 = j0s1, j1s0, j1s4, j2s1, j3s0, j3s1

To translate this into a string which can be manipulated by the MCGA several encoding steps take place. First the machine choice section is constructed by concatenating the machine choices in ascending job-stage order. In this example there are four machine choices: j0s1, j1s2, j1s4, j3s0. This means that there are four integer activation genes in the machine choice section. Taking the first gene, j0s1, this is of type Op4. Looking in Table 5.6 this can go either on machine 0 or on machine 2. Hence the value of the gene will be restricted to either 0 or 2. For j1s2, this can go on machines 0 or 1, hence this gene will be restricted to either 0 or 1 and so on for the other machine choices. Following this each machine section is constructed. These sections are similar to the Travelling Salesman Problem in construction. Taking Machine 0 as an example there are 5 job stages so this string section will be 5 genes long. Each job is simply represented by its job number and its stage number is not used. For this example a sample string would be 0,1,1,2,3. Any permutation of these five genes is legal; for example 1,2,0,1,3 or 3,1,2,1,0. The order of the jobs form a preference list for that machine with the first job being attempted given highest priority. The stage numbers do not need to be included due to the implied ordering of jobs. In the case of machine 1 there are two 1's, but in order for j1s4 to be started j1s2 must already have been completed, hence stage 2 must come first. This encoding is repeated for all machines until the string is complete. The initial process of constructing the string is done automatically from the input datafile. The user does not need to specify machine choice restrictions, segment lengths and job contents or machine sections.

As the system uses an indirect notation a schedule builder is used. This was done to allow the extra functionality required. The schedule builder requires a preference list for each machine in order to carry out processing. Within these lists each job stage should only appear once.

To demonstrate the decoding process a sample string from the above example is decoded. The string is first separated into machine choice and machine sections.

Machine Choice	Machine 0	Machine 1	Machine 2
0, 0, 0, 2	3, 1, 2, 1, 0	2, 1, 1, 0, 3, 3, 0, 1	3, 1, 0, 1, 3, 2

The machine choice string is now used to activate the corresponding jobs in each of the machine lists. In all other lists these jobs will be removed. For example, the first gene in the machine choice list is for job 0 stage 1. This job can either go on machine 0 or machine 2. In this case machine 0 has been selected hence job 0 stage 1 must be removed from machine 2. Looking at the Machine 2 there is only one job 0 so this is removed. If there is more than one stage of the same job on the machine the implicit job ordering is used. For example, the second machine choice gene belongs to job 1 stage 2. Machine 0 has been chosen so the other machine location for this job, machine 1, has to be removed. It can be seen that machine 1 has three job 1 genes. By referring to Table 5.6 and 5.7, job 1 stage 2 is the second possible job 1 stage to appear on machine 1, the first being job1 stage 1. Hence the second 1 is removed from the machine 1 list. If this is continued for all machine choice sections the following machine preference lists result:

Machine 0 Preference List:	3, 1, 2, 1, 0
Machine 1 Preference List:	2, 1, 0, 3, 0, 1
Machine 2 Preference List:	3, 1, 3, 2

All the extra jobs have been removed from the machine sections and no duplicates exist. Using these lists the schedule builder can now be invoked to build a schedule and collect statistics on the data.

5.5 Crossover and Mutation Operators

A complete string operator cannot be used on the string because it is separated into several distinct sections. Instead crossover is applied to each of the individual sections in the undecoded string. The machine choice section is crossed using a simple two point crossover. This means that two random points are selected within the machine choice string section and the genes between these points are swapped between partners. This is valid as this section is not order based.

Within each of the machine job list sections a new operator is employed based on a modification of the enhanced edge crossover operator [17]. In these sections of the string the two-point crossover operator cannot be used because it would introduce repeated jobs. The modified operator is slightly more complex, but it ensures that repeated jobs are not created and jobs are not lost during crossover. The procedure is shown in Figure 5.2.

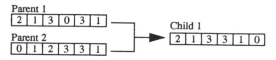

Figure 5.2 - Crossover in Machine Sections

Using parent 1 the process is started from job 2. Then the following jobs in both parents, 1 and 3, are found and a choice is made between them at random. If 1 is chosen the next jobs are 3 and 2. Job 2 only appears once and has already been used so job 3 must be used. Following job 3 in the parent strings are jobs 0 and 3, and 3 is chosen at random. Following the second job 3 is job 1 in both cases so this is chosen. Job 0 is the first job in parent 1 unused so is inserted at the end. This process is repeated from the first job in parent 2 to create another child. This operator tries to consider the immediately following job and also the overall ordering by trying to select the earliest jobs in cases of conflict. If two identical strings are present the children will be the same as the parents.

Mutation is a much simpler operator. Again mutation is performed on separated sections of the string. In the machine choice section, if a gene is chosen it is randomly reset to another machine choice. For example if there is a choice of three machines 0, 2 and 3 for a specific job and the current machine selected is 2 it would change to either 0 or 3. In the machine job list sections a gene selected for crossover will simply be swapped with another randomly selected gene in that section.

Due to the combinatorial complexity of the problem a small local search is initiated which behaves like an intelligent mutation operator. This search chooses a first job and then a second is picked randomly and the two are swapped. If this swap provides improvement in at least one of the scheduling criteria without loss in any of the others then the swap is accepted. If this is not the case the jobs are swapped back and another job is picked randomly to swap with the first. This process is repeated up to 5 times if a good swap is not found then the string is left untouched. These swaps occur within each machine section.

5.6. Schedule Builder

The schedule builder takes the strings provided by the MCGA and decodes them into machine preference lists as outlined before. These lists are then used to create the schedule. The builder will always create the same schedule for a given input string. The builder starts by resetting the clock to zero. Then the system enters a loop which is terminated when all jobs have finished. This loop starts by looking at each machine to determine its current state, either setup, maintenance, working or idle. If the machine has finished a job, the job is marked complete, its finishing time is noted and the machine set to idle. If the machine is either working, in setup or in maintenance the time to completion of this task is decremented by 1 clock tick. If the machine is working, then the time of utilisation of that machine is incremented.

Following this the preference list of each machine which is idle is examined to find the highest priority job which is available and has not yet been performed. Availability is defined in terms of several factors. A job is available if:

i) All previous stages of that job are complete.

ii) All dependencies of that job are satisfied.

iii) The machine is set up for that job.

iv) If job splitting is not permitted the job must be completed
    before any maintenance begins.

In the cases of i), ii) and iv) the next job in the preference list of the machine is examined to see if it is available. If iii) is the case the machine begins its setup routine for the operation of that job. The time required for setup is determined by the current machine operation and the operation required by the new job and is found in the operation setup time table (Table 5.3). If no job on the list is available the machine is left in its idle state. If a job is found then the time required to complete that job is calculated and it is placed on the machine. The clock is then incremented and the process returns to the start of the loop.

When all jobs are complete the code exits the loop and various statistics can be calculated such as makespan, penalty costs, average job times, machine utilisations etc. These are passed back to the MCGA for analysis. The code will also generate plans showing each machine, what jobs and stages are being performed, any setups, maintenance times, idle times and also statistics from the run.

6. Scheduler Testing

The scheduler was tested to check for correct operation by examining the results of the MCGA. In this example there are 4 machines, 20 jobs with a maximum of 5 stages and 6 types of operations. It includes all of the complex features discussed earlier. The search space contains at most $3^{12}x2^{23}x16!x20!x35!x35!$ possible combinations. This is calculated by finding the number of machine choices, there are 12 stages with 3 choices (3^{12}) and 23 stages with 2 choices (2^{23}). This is then multiplied by the number of possible combinations in each of the machine sections (16!, 20!, 35!, 35!). The search space will actually be smaller than this due to the fact that some of the jobs are repeated in the machine sections. This is the same as having the same city occurring twice or more times in the Travelling Salesman Problem. This test was run using two criteria, makespan and average job time, the latter being defined as the sum of time taken by each job divided by the number of jobs. The aim of the average job time criterion is to reduce the number of jobs which are in progress at one time. This criterion may be used in small workshops where storage space is at a premium.

For the tests a population of 500 was run over 100 generations. A typical run on this example takes about 25 minutes. The graphs showing the development of the trade-off solutions over the 100 generations are shown in Figure 6.1. The effect of job splitting was also tested. Figure 6.2 shows the final population of two runs, with and without job splitting.

Firstly the schedule plans generated by the system were examined for validity. The job list was checked to ensure all jobs were completed. The timings of the jobs were checked as were the setups when required. The MCGA Scheduler generates an average of 72 solutions per test over a 20 test run. It was found that multiple solutions would generate the same trade-off values even though they were genotypically different. This is why the graphs do not appear to have all of the 72 solutions present. The schedule plans reveal that, in general, in order to reduce makespan the average job time would increase. This is because the scheduler has to be more efficient with resources, so job stages need to be moved around so that they fit into gaps within the schedule. To minimise average job times, the schedules which group several job stages together were favoured. This leads to large gaps in the schedule where jobs will not commence until necessary resources are free. In the extreme case minimising average job time is equivalent to a 'no-wait' job shop where no gaps are allowed between job stages.

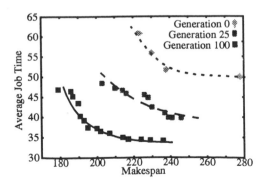

Figure 6.1- Generation of Trade-off Solutions

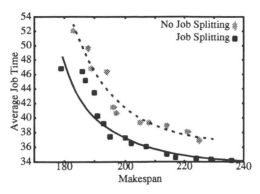

Figure 6.2 - Job Splitting vs No Job Splitting

As mentioned earlier Figure 6.2 shows the effect of job splitting. Clearly using job splitting has a positive effect on both makespan and Average Job Time. If job splitting is not permitted then jobs are held until such time as they can be completed without any interruption. However if there are short jobs which can be completed these will be selected to fill in the gaps. The problem arises when there are long jobs and short gaps between maintenance periods. The jobs are continually put back until large enough windows exist. This sometimes leads to long delays especially on machines with lots of scheduled maintenance. In this case the best schedules are those where as many jobs as possible are placed onto other machines effectively bypassing the troublesome machine. The current algorithm does not allow the jobs to be removed from machines by other jobs of higher priority. This is a possible extension for future work.

7. Conclusions

This paper has outlined the use of operation based scheduling and evolutionary techniques to solve complex scheduling problems. The simple job shop model was thought not to be detailed enough for application to many scheduling problems. For this reason a complex scheduling definition was drawn up incorporating many new features, such as the use of multi-operation machines, job dependencies and setup periods.

Following this a description of operation based scheduling along with the data required to perform it is given. The construction of the MCGA Scheduler is outlined and the string representation, crossover and mutation operators and the schedule builder are also presented.

The MCGA Scheduler is found to be a useful and flexible optimisation tool providing insights into the processes involved in complex scheduling. Further investigations are being carried out on more real world examples with the aim of improving efficiency and the effectiveness of genetic operators. Additionally, system functionality continues to be enhanced in order to increase the range of problems which the scheduler can tackle.

8. References

1. Davis L, 1985. Job Shop Scheduling with Genetic Algorithms. *Proceedings of The First International Conference on Genetic Algorithms*. Lawrence Erlbaum Associates, Hillsdale, NJ, pp136-140.
2. Tsujimura Y, Cheng R, Gen M, 1997. Improved Genetic Algorithms for Job-Shop Scheduling Problems. *Engineering Design and Automation* 3(2):133-144.
3. Nakano R, Yamada T, 1991. Conventional Genetic Algorithm for Job Shop Problems. *Proceedings of the Fourth International Conference on Genetic Algorithms*. Morgan Kaufmann Publishers, pp474-479.
4. Syswerda G, Palmucci J, 1991. The Application of Genetic Algorithms to Resource Scheduling. *Proceedings of the Forth International Conference on Genetic Algorithms*. Morgan Kaufmann Publishers, pp502-508.
5. Bruns R, 1993. Direct Chromosome Representation and Advanced Genetic Operators for Production Scheduling. *Proceedings of the Fifth International Conference on Genetic Algorithms*. Morgan Kaufmann Publishers, pp352-359.
6. Fang H-L, Ross P, Corne D, 1993. A Promising Genetic Algorithm Approach To Job-Shop Scheduling, Re-Scheduling and Open-Shop Scheduling Problems. *Proceedings of the Fifth International Conference on Genetic Algorithms*. Morgan Kaufmann Publishers, pp375-382.
7. Gen M, Tsujimura Y, Kubota E, 1994. Solving Job Shop Scheduling Problems by Genetic Algorithm. *IEEE International Conference on Systems, Man and Cybernetics*, Vol.2. pp1577-1582.
8. Reeves C R, 1995. A Genetic Algorithm for Flowshop Scheduling. *Computers in Operations Research*, 22(1):5-13.
9. Cheng R, Gen M, Tsujimura Y, 1996. A Tutorial Survey of Job-Shop Scheduling using Genetic Algorithms - I.Representation. *Computers and Industrial Engineering* 30(4):983-997.
10. Liang S J, Lewis J M, 1994. Job Shop Scheduling Using Multiple Criteria. *Proceedings of the Joint Hungarian-British Mechatronic Conference*. pp77-82.
11. Sridhar J, Rajebdran C, 1996. Scheduling in Flowshop and Cellular Manufacturing Systems with Multiple Objectives - A Genetic Algorithmic Approach. *Planning Production and Control* 7(4):374-382.
12. Murata T, Ishibuchi H, Tanaka H, 1996. Multi-Objective Genetic Algorithm and it's Application to Flowshop Scheduling. *Computers in Industrial Engineering* 30(4):957-968.
13. Hamada K, Baba T, Sato K, Yufu M, 1995. Hybridizing a Genetic Algorithm with Rule-Based Reasoning For Production Planning. *IEEE Expert* October:60-67.
14. Muth J F, Thompson G L, 1963. *Industrial Scheduling*. Prentice Hall Inc., Englewood Cliffs, New Jersey.
15. Todd D S, Sen P, 1997. A Multiple Criteria Genetic Algorithm for Containership Loading. *Proceedings of the Seventh International Conference on Genetic Algorithms*. Morgan Kaufmann, ISBN 1-55860-487-1, pp674-681.
16. Dasgupta D, McGregor D, 1992. Structured Genetic Algorithms: The Model and First Results. Department of Computer Science, University of Strathclyde, UK, 1992.
17. Starkweather T, McDaniel S, Mathias K, Whitley D, Whitley C, 1991. A Comparison of Genetic Sequencing Operators. *Proceedings of The Third International Conference on Genetic Algorithms*. Morgan Kaufmann, ISBN 1-55860-308-9, pp69-76.

Minimizing Job Tardiness: Priority Rules vs. Adaptive Scheduling

Dirk C. Mattfeld and Christian Bierwirth

Department of Economics, University of Bremen
email: dirk@uni-bremen.de

Abstract. This paper addresses job tardiness for non-deterministic job shop scheduling. A comparative study shows that a GA consistently outperforms different priority rules regardless of the workload and the objective pursued.

1 Introduction

The bulk of research in production scheduling is concerned with priority rule based on-line control of manufacturing systems. Although successful, research on Genetic Algorithms (GA) in this field is predominately confined to static scheduling problems. The transfer of GA experiences to production control was firstly approached in [1] and has been recently taken up by [2].

With the spread of automated manufacturing systems the control problem of dynamically assigning jobs to machines receives increasing attention. Due to a lack of adequate optimization methods, control approaches based on priority rules are emphasized [3–5]. Typically a priority rule is designed to pursue a certain objective. In this paper we concentrate on the objective of minimizing job tardiness which is of particular importance because tardy deliveries of products discourages costumer satisfaction and may also result in penalty costs.

Whenever jobs are in danger to become tardy, control can seek to complete most jobs in time at the expense of a few jobs delayed badly. On the other hand, it also can seek to complete all jobs with little delays only. Some priority rules are particularly devoted to minimize job tardiness in the former or the latter direction. We investigate whether adaptive scheduling can outperform highly specialized rules in both directions by just switching the fitness function.

The paper is organized in 5 sections. In Sect. 2 we present a prototypical manufacturing system for which we choose suitable parameters in order to model different production scenarios. In Sect. 3 we discuss the different logic of four priority rules which are compared in a simulation experiment. Then, we introduce a Genetic Algorithm in Sect. 4 which is applied to the same scenarios. The results obtained are comparatively discussed in Sect. 5.

2 The Manufacturing System

2.1 Modeling the System

Throughout the paper we consider an idealized manufacturing system consisting of m machines. The machines are fully connected by an automatic material handling system and therefore the passing times of jobs between machines are neglected. The machines are dedicated to perform particular tasks, thus no machine can substitute any other machine. Jobs arrive at the system at unforeseen points in time which are called their *release dates*. Since a job release is considered as a non-deterministic event, the specific processing requirements of a job are unknown in advance. We concentrate on the following three types of processing requirements.

1. Each job J_i defines a technological order μ_i in which it is routed through all machines or a subset thereof. The k-th operation of job J_i, dedicated to be processed by machine $M_{\mu_i(k)}$, is denoted as o_{ik}.
2. The time span necessary to process operation o_{ik} is given by p_{ik}.
3. For each job J_i a *due date* d_i is announced, i.e. a job defines a time window between its release date and its due date, called its allowance.

Breaking a due date, although highly undesirable, often cannot be circumvented. In this situation the *tardiness* T_i of job J_i denotes the deviation of its actual completion time C_i from the intended due date, i.e. $T_i = C_i - d_i$. A job which is completed within its allowance has a tardiness $T_i = 0$. Minimizing the mean tardiness $\overline{T}$ of all n jobs involved in the manufacturing system is often of high economic significance because tardy jobs typically cause penalty costs.

$$\overline{T} = \frac{1}{n} \sum_{i=1}^{n} \max(0, C_i - d_i),$$

$$RMST = \left(\frac{1}{n} \sum_{i=1}^{n} \max(0, (C_i - d_i)^2 \right)^{\frac{1}{2}}.$$

If the penalty costs increase disproportionate to the tardiness of jobs it is even useful to minimize the root-mean-square of tardiness ($RMST$), see [6]. Here, each job tardiness figure is squared, then summed over the number of jobs completed before taking the square root. This value tends to penalize systems with a few jobs that are very late more than those with many jobs that are little late.

2.2 Describing System States

At each point of time the manufacturing system modeled above is determined by the state of its m machines which are either idle or busy processing some job. Jobs which await processing are queued in front of their dedicated machine. The upper Gantt chart in Fig. 1 shows a snapshot of the manufacturing system.

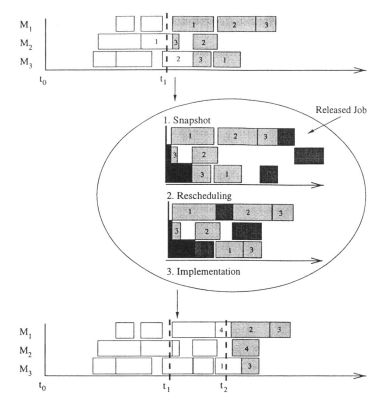

Fig. 1. Job scheduling on a rolling time basis.

At time t_1 machine M_1 is idle whereas machines M_2 and M_3 are occupied by operations belonging to jobs J_1 and J_2 respectively. Job J_3 is in the queue of M_2. The remaining operations of these jobs are already scheduled for processing but have not been started yet. They appear in light grey shade in the chart.

The state description of the system at t_1 can be used to derive an associated deterministic optimization model for scheduling the manufacturing system. The focus of attention in Fig. 1 is on the snapshot procedure which filters the already scheduled but so far unprocessed operations into a new disposition space. Two adjustments must be made. Since some machines are busy at t_1 they are not immediately available. Therefore machine *lead times* are introduced which express the earliest point in time a machine becomes available again (depicted by black shadings in the two central Gantt charts). Accordingly, the subsequent operation of jobs which are currently under process (e.g. job J_2 on M_3) cannot be started before their predecessor operation is completed. Consequently all jobs get a modified release date larger than or equal to t_1, for details see [1]. The resulting problem is solved and its solution is implemented in the manufacturing system as far as possible with respect to a forthcoming snapshot (shown at t_2

in Fig. 1). By repeating this procedure a production program is rescheduled on a rolling time basis [7].

With respect to the introduction of release dates, due dates, and machine lead times the optimization model derived in each period is an extension of the well known static job shop scheduling problem. Of course, any algorithm capable to solve this extended model may be used to optimize the system performance. For a survey see [8] and for a recent GA approach [9].

2.3 Generating System Load

The *workload* of a manufacturing system is regarded as a crucial aspect of its overall performance. The workload is defined by the number of operations in the system which await processing. Obviously the workload observed in a system is influenced by two independent forces. First, of course, the *arrival process* of new jobs released effect the workload of the manufacturing system. The shorter the *inter-arrival times* of jobs are, the larger the workload gets. However, a second force counteracts to the increase of workload. Whenever a job is completed, it leaves the manufacturing system immediately, i.e. the workload decreases. Actually the workload of a system results from the ratio of the job arrival rate to the job completion rate. Advantageous scheduling accelerates the job completion process and therefore it influences the workload positively.

We model the arrival process of jobs in order to generate different workload situations for the manufacturing system. The mean inter-arrival time of jobs is determined by the mean processing time of jobs $\bar{p}$, the number of machines m in the system and a desired *machine utilization rate U*. If $\bar{p}$ is given, we can calculate the mean inter-arrival time by $\lambda = \bar{p}/(mU)$. For the purpose of comparing alternative scheduling methods under different workload situations we simulate a manufacturing system with parameters typically used in literature [10].

1. The manufacturing system consists of $m = 6$ machines.
2. The technological orders μ_i of jobs are generated at random. The jobs are routed through 2 to 6 machines (on average 4 machines).
3. The processing times of operations are uniformly distributed in the range of $[1, 19]$. The mean processing time of jobs is given by $\bar{p} = 4 \cdot 10 = 40$.
4. The inter-arrival times of jobs are exponentially distributed to the mean of λ by using utilization rates of $U = 0.65, 0.75, 0.85$, and 0.95.
5. The allowance of job J_i is set to two times the processing time it requires, i.e. $d_i = r_i + 2\sum p_{ij}$.

Recall that the utilization rate approximates a certain workload situation by neglecting the influence of scheduling. A utilization of $U \geq 1.00$ leads to an excessive workload which cannot be handled anymore. With $U = 1.00$ the jobs are released so fast that already the processing times fully occupy the system. A continuous increase of workload can theoretically be avoided if scheduling exhausts the machine capacities totally, i.e. the machines never become idle. The utilization rates used in this paper are slightly lower. Rates of 0.65 and 0.75 represent moderate workload scenarios whereas challenging scenarios can be expected from utilization rates of 0.85 and 0.95.

Table 1. Performance of priority rule based control.

	$\overline{T}$				$RMST$			
	0.65	0.75	0.85	0.95	0.65	0.75	0.85	0.95
SLACK	13.44	37.08	94.53	226.96	24.35	52.16	*112.52	*249.22
COVERT	*8.62	*21.92	60.43	175.14	*22.09	*50.09	135.25	394.17
SPT	9.82	23.24	*59.83	*171.66	29.56	61.36	146.97	399.55
FCFS	16.54	42.74	108.76	271.32	31.06	63.31	138.44	315.35

3 Priority Rules

Priority rules, which are important components of shop floor control systems, have been intensively studied over the last 30 years. Roughly speaking, a strategy that selects a job among those awaiting processing each time a machine gets idle, is called a priority rule. We consider three rules which are known to reduce job tardiness effectively and another rule (FCFS) for reasons of comparability.

SPT: The *shortest processing time* rule selects the operation with the shortest imminent processing time. Jobs waiting in a queue may cause that their dedicated successor machines run idle. SPT alleviates this risk by reducing the length of the queue in the fastest possible way.

SLACK: The *slack* of a job is defined as the time span left within its allowance assumed that its remaining operations are processed without any delay. The rule gives priority to the job with minimal slack, i.e. the smallest potential waiting time available.

COVERT: The *cost over time* rule combines the ideas of SPT and SLACK. It prioritizes jobs according to the largest ratio of expected job tardiness to operation processing time. In this way COVERT retains SPT performance but seeks to respect the due dates if jobs are late.

FCFS: The *first come, first serve* rule implements a fair scheduling authority. Since this rule does not favor a decisive property of a job, the outcome of this rule is close to what can be expected from random decisions.

The performance of these rules applied to our manufacturing system model is shown in Tab. 1. The system is simulated for a total of 1.000 jobs, of which the first and the last 200 jobs are discarded in order to analyze the system in a stable constitution. Each simulation is replicated ten times. The average values are reported whereof the best are labeled (*).

The results for the $\overline{T}$ objective confirm the dominance of SPT reported in literature for high utilization rates (U=0.85 and 0.95). SPT excessively delays jobs with above average operation processing times, and therefore the rule fails for minimizing $RMST$. Here, SLACK outperforms its rivals for heavily loaded systems, whereas the rule produces poor results concerning $\overline{T}$. The differences in result obviously stem from the "anti-SPT" logic of the SLACK rule, for an example see [5]. COVERT combines features of SPT and SLACK and succeeds for moderately loaded manufacturing systems (U=0.65 and 0.75). Here, COVERT dominates SPT as well as SLACK for both objectives pursued. For high utilization rates COVERT approximates the SPT behavior leading to very similar results compared to those obtained from SPT.

Since FCFS acts similar to a random strategy its outcome allows us to access the efficacy of the three rules already discussed. For the $\overline{T}$ objective FCFS is clearly outperformed by all other rules. This proves that even the SLACK rule effects the reduction of $\overline{T}$ positively to some extend. Turning to $RMST$, for low utilization rates of $U = 0.65$ and 0.75 we observe the same rank. Amazingly, for $U = 0.85$, FCFS performs better than SPT. Finally for $U = 0.95$, FCFS outperforms both, SPT and COVERT. Obviously, for this scenario the SPT logic, which tends to delay a few jobs badly completely fails. $RMST$ penalizes these jobs heavily which leads to poor results even compared to those of random-like scheduling by FCFS.

Summarizing, non of the rules presented cuts out its rivals with respect to both objectives and a varying workload. Therefore in the following we investigate whether a GA can show a more consistent performance by outperforming the priority rules at the same time.

4 Genetic Algorithm

In order to solve deterministic scheduling problems resulting from a snapshot of a manufacturing system (see 2.2) we employ a GA for static scheduling which is briefly described in the following.

4.1 Static GA Scheduling

We use a permutation representation consisting of the operations of all jobs to be scheduled. A schedule is build by assigning starting times t_{ik} to the operations o_{ik} occuring in the permutation. The operations are scheduled non-delay as already described in the context of priority rules. Whenever a machine runs idle, the operation is selected from its queue which occurs at the leftmost position of the permutation. The decoding procedure is given below.

1. Build the set of all beginning operations, $A = \{o_{i1} \mid 1 \leq i \leq n\}$.
2. Determine the earliest possible starting time t' of all operations in A, $t' \leq t_{ik}$ for all $o_{ik} \in A$.
3. Build set B from all operations in A which have their earliest possible starting time at t', $B = \{o_{ik} \in A \mid t_{ik} = t'\}$. .
4. Select operation o_{ik}^{*} from B which occurs leftmost in the permutation.
5. Delete operation o_{ik}^{*} from A, $A = A\backslash\{o_{ik}^{*}\}$.
6. If $o_{i,k+1}^{*}$ exists, insert it into A, $A = A \cup \{o_{i,k+1}^{*}\}$.
7. If $A \neq \emptyset$ goto Step 2, else terminate.

Table 2. Performance of adaptive scheduling vs. priority rules.

	$\overline{T}$				$RMST$			
	0.65	0.75	0.85	0.95	0.65	0.75	0.85	0.95
FCFS	16.54	42.74	108.76	271.32	31.06	63.31	138.44	315.35
best rule	8.62	21.92	59.83	171.66	22.09	50.09	112.52	249.22
GA	8.04	20.61	54.52	144.63	20.89	42.98	95.34	202.37

Obviously, for this decoding procedure the relative order of operations given in a permutation is a crucial aspect of inheritance. Therefore we use the well known order based crossover (OX) which preserves this characteristic appropriately. In this way the GA evolves favorable precedence relations among the operations involved.

The GA parameterization is like the widespread standard: The crossover rate is set to 0.6. Since the crossover operator produces implicit mutations, we refrain from using an additional mutation operator. Proportional selection is used. The GA terminates if no further improvement has been gained within the last ten generations. The population size is set to 50 individuals.

4.2 Dynamic GA Rescheduling

The GA is applied to a series of deterministic problems resulting from snapshots taken at the arrival time t of jobs, see Fig. 1. For each GA run the initial population is biased by falling back on information previously processed. The final population of the most previous problem establishes the initial population of the current problem: First, all operations started before t, including those which are still processed, are deleted in the permutations of the population. Then the operations of new jobs are inserted at random positions in all permutations of the population. The idea behind this procedure is to retain information obtained in the previous run which has a good chance to be useful in the new context. We have already discussed this technique in [1]. Recently [2] confirm that this feature increases the efficiency of the GA significantly.

The performance of the GA approach has been tested on the same parameter setting of the manufacturing system as used for the priority rule based simulations. Again, each simulation is replicated ten times. Since the GA is a probabilistic method, ten runs are carried out for each simulation resulting in a total of 100 runs for each utilization factor considered. The averaged results are shown in Tab. 2.

The first line in the table shows the FCFS results taken from Tab. 1, because these values indicate the quality level we can expect from a random population. The GA roughly halves the FCFS results for $\overline{T}$ in all workload scenarios. For the $RMST$ objective the GA approximately reduces the values to 2/3 of the FCFS results.

The GA outperforms the best priority rule in all cases. The improvement gained by the GA increases from a few percent ($U = 0.65$) to about 20% ($U =$

0.95) for both objectives considered. Thus, for a small workload priority rules can control the manufacturing system surprisingly well, whereas for heavily loaded systems they show significant weaknesses.

5 Discussion

Priority rule based as well as GA based scheduling seeks an overall sequence of machine allocations which minimizes a tardiness objective. For $\overline{T}$ this sequence should ensure that no machine runs idle. For $RMST$ the completion of late jobs is of predominant importance.

As we have seen, for both objectives pursued an appropriate rule exists. The COVERT rule combines features of these rules which works well under relaxed workload conditions only. Our GA approach finds improving sequences compared to those built by priority rules independently of the workload and the objective pursued.

Different to priority rules the GA can anticipate consequences of potential decisions to a certain extent. Thus, a major finding drawn from the computational results is that anticipation is useful, especially under heavy workload conditions. High utilization rates lead to long queues of jobs waiting in front of the machines. The longer a queue grows, the more potential decisions are anticipated by the GA based scheduling.

Unfortunately, the advantage taken from anticipation is paid by a high computational effort. Long queues result in large problem sizes which come along with considerable computer run-times. Recall that 1000 job arrivals (as simulated) result in 1000 subsequent GA runs. For a utilization rate of $U = 0.65$ run-times of at most a few minutes are needed, whereas $U = 0.95$ already requires some hours of runtime on a Sparc Ultra/2 workstation. The more favorable anticipation gets, the more expensive this feature is to achieve. Nevertheless, in the practise of manufacturing systems, rescheduling intervals are provided such that the GA runtime will not become a bottleneck even under heavy load.

In this paper we have dealt with a prototypical manufacturing system by pursuing two simplified objectives. Due to a changing environment the objective pursued in practise is typically much more complex and therefore a mix of priority rules is used. As we have shown, the GA-based rescheduling method is capable of adapting to different needs. This feature may be flexibly used to address the demands of read-world scheduling.

References

1. Bierwirth C, Kopfer H, Mattfeld D C, Rixen I, 1995. Genetic Algorithm based Scheduling in a Dynamic Manufacturing Environment. In: Palaniswami M et al. (eds), 1995. *Proceedings of the 2nd IEEE Conference on Evolutionary Computation.* IEEE Press, New York, pp 439-443.

2. Lin S-C, Goodman E D, Punch W E, 1997. A Genetic Algorithm Approach to Dynamic Job Shop Scheduling Problems. In: Bäck T (ed), 1997. *Proceedings of the 7th International Conference on Genetic Algorithms*. Morgan Kaufmann, San Mateo, pp 481-489

3. Panwalkar S S, Iskander W, 1977. A Survey of Scheduling Rules. *OR* 25:45-61,

4. Blackstone J H, Phillips D T, Hogg G L, 1982. A State-of-the-Art Survey of Dispatching Rules for Manufacturing Job Shop Operations. *Int J Prod Res* 20:27-45.

5. Haupt R, 1989. A Survey of Priority Rule-Based Scheduling. *OR Spek* 11:3-16.

6. Russell R S, Dar-El E M, Taylor B W, 1987. A Comparative Analysis of the COVERT Job Sequencing Rule using various Shop Performance Measures. *Int J Prod Res* 25:1523-1540.

7. Raman N, Talbot F B 1993. The job shop tardiness problem: A decomposition approach. *EJOR* 69:187-199

8. Błażewicz J, Domschke W, Pesch E, 1996. The job shop scheduling problem: Conventional and new solution techniques. *EJOR* 93:1-30.

9. Fang H L, Corne D W, Ross P M, 1996. A Genetic Algorithm for Job-Shop Problems with Various Schedule Quality Criteria. In: Fogarty T (ed), 1996. *Proceedings of the AISB Workshop*. Springer, London, pp 39-49.

10. Holthaus O, Rajendran C, 1997. Efficient dispatching Rules for Scheduling in a Job Shop. *Int J Prod Ec* 48:87-105.

Evolutionary Computation Approaches to Cell Optimisation

C Dimopoulos and AMS Zalzala

Department of Automatic Control and Systems Engineering,
University of Sheffield, Mappin Street, Sheffield S1 3JD, UK.
(Email:rrg@sheffield.ac.uk)

Abstract. This paper examines a cellular manufacturing optimisation problem in a new facility of a pharmaceutical company. The new facility, together with the old one, should be adequate to handle current and future production requirements. The aim of this paper is to investigate the potential use of evolutionary computation in order to find the optimum configuration of the cells in the facility. The objective is to maximise the total number of batches processed per year in the facility. In addition, a two-objective optimisation search was implemented, using several evolutionary computation methods. One additional objective is to minimise the overall cost, which is proportional to the number of cells in the facility. The multi-objective optimisation programs were based on three approaches: The weighted-sum approach, the Pareto-optimality approach, and the Multiobjective Genetic Algorithm (MOGA) approach.

Keywords: cellular manufacturing, multi-objective optimisation, evolutionary computation.

1. Introduction

Cellular manufacturing is an application of group technology to manufacturing optimisation problems (Wemmerlov and Hyer, 1989), aiming to divide the plant in a certain number of cells. Each cell contains machines that process similar type of products. The application of cellular manufacturing in a plant minimises makespan, reduces the set-up time of the machines, and improves the quality of the products (Singh, 1989). There are two major optimisation problems associated with cellular manufacturing, namely the cell-formation problem and the cell-layout problem

In the case study of this paper, the aim is to find the best configuration of cells in order to maximise the total number of batches processed in the plant per year. One distinct characteristic of this case is that there is only one type of machinery in the plant, the reactor. However, reactors are grouped in cells due to the cross-contamination of products. Only one batch can be processed at a time in a group of reactors that stand close together. We define this optimisation problem as a 'numerical' cell-formation problem, due to its distinctive nature.

Evolutionary programming was used as a guide in our search for the optimum solution. The algorithm tests a population of potential solutions in parallel, in order to find the best configuration of cells. Domain knowledge was incorporated both to the genetic representation of the solutions and the design of the genetic operators.

The trend in manufacturing optimisation is to consider the reduction of cost as one of the most significant objectives. Using the traditional optimisation methods, it is very difficult to incorporate more than one objective in the optimisation process. Therefore, cost is either considered separately, or not considered at all. Evolutionary computation provides the means of implementing multi-objective optimisation in an easy and efficient way. We have performed multi-objective optimisation for this case study, using three different approaches: The weighted-sum approach, the Pareto-optimality approach, and the Multiobjective Genetic Algorithm (MOGA) approach. The minimisation of cost was used as a second objective in the multi-objective optimisation search. By combining partial preference information in the form of a goal vector, with the Pareto-optimality approach, a local search was performed in certain areas of the solution space.

2. Cell-Formation Problem

The objective in this problem is to identify machine families that process similar parts and to group these families into cells. The traditional approach to this problem is the selection of cells by direct observation, which is generally possible for the simplest cases. Although this is a very old method and has obvious limitations, it is still used by companies throughout the world. Another way to determine the configuration of cells, is to code components according to their features. Families of components are then formed, according to the similarity of their code. Each family determines a group of machines that will form a cell. There are various coding and classification methods that have been proposed (Bennett, 1986). Each of them uses different features or combination of these features. The main drawback of these methods is that they do not divide the plant directly into cells. The components are grouped in easily identified families, and this data is used as a guide for the cell-formation. However, the actual machines that will form each cell are derived directly from the data.

The most popular method for solving the cell-formation problem is the Production Flow Analysis (PFA) method (Burbidge,1975). PFA is a technique that assumes the physical shape of the components is less important than the route that they have to follow in order to be manufactured. PFA is mainly concerned with manufacturing methods, and the aim is to identify family parts that follow similar routes during their processing. Any group of machines that process a part-family will form an independent cell. The data required for PFA is contained in the process route cards and in the list of machines.

The classic PFA method is inefficient for large problems, due to the fact that the grouping of parts into families is implemented manually. A number of alternative methods based on PFA have been proposed. Array Based Clustering uses the part-incidence matrix to identify the potential cells of the plant. This matrix contains all the information about the route that each part has to follow in order to be processed. The machines and the parts are grouped into families, by performing a series of row and column manipulations. Rank-Order Clustering (King, 1980) uses a slightly different way for the implementation of array clustering, by assigning binary weights in each row and column. Single Linkage Cluster Analysis (Mcauley,1972) is a method which seeks to find a measure of similarity between machines, tools, and every other feature of production. The part-families are then formed, based on this similarity. Mathematical programming methods, like the one proposed by Boctor (1991), address the formation of cells as an optimisation problem, where the objective is to maximise the total sum of similarities between each pair of components.

Due to the nature of the cell-formation problem, many artificial intelligence methods have been proposed for the search of the optimum configuration. Kao and Moon (1991) used the Artificial Neural Network (ANN) to form part families based on design features. Fuzzy logic has also been applied by Xu and Wang (1989).

In recent years there is a growing interest in the use of Genetic Algorithms for this optimisation problem. Modified Genetic Algorithms, like the one proposed by Falkenauer (1993), use problem-specific chromosome representation and purpose-based genetic operators to determine the formation of cells. Unlike classic GAs, these algorithms proved to be very effective in finding the best configuration of cells, because they incorporate domain knowledge in their search for the optimal solution.

3. EC Approach to Manufacturing Optimisation

The proposed algorithm searches for the best configuration of cells in the plant, so that the total number of batches produced per year will be maximised. Cost is also introduced in the problem as a second objective. The algorithm is modified according to different multi-objective optimisation methods, and several alternative solutions are reported.

3.1 Problem Description

Work load projections on the utilisation of an existing pilot plant facility, at the factory of a pharmaceutical company, indicate that it will not be adequate to handle current and future production requirements. Therefore, the company decided to build a new facility, in order to accommodate their future needs. The company produces a number of different products. The products are classified as in the following example; PROD 5-4 : product batch that requires 5 days of processing, and occupies 4 reactors while being processed.

Based on prior and projected work knowledge, and statistical data generated in-house by the company, a basis for modelling including 15 products was developed. This basis is presented in a relevant table in the Appendix (Table 1). The constraints of the problem are the following:

$$2 \le a_n \le 6, \ 2 \le n \le 6, \text{ and } \sum_{k=1}^{n} a_k \in [12,13]$$

where a_n : integer number representing the total number of reactors in the nth cell, and n : integer number representing the total number of cells. The company proposes the following scheduling cases:

CASE1. A list of products is to be produced over a period of a year. Batches are processed in strict sequence, but there's no restriction in the type of batch that a cell can process. Only one batch can be processed at a time in a cell, due to cross-contamination of the products.

CASE2. The same list of products is to be produced in a year's time. Batches are again processed in strict sequence, but in this case, each cell processes a certain type of batch. There are small, medium and large-sized batches. A cell can process either small and medium-sized batches or large and medium-sized batches. Only one batch can be processed at a time in a cell, due to cross-contamination of the products.

3.2 Chromosome Representation

Many researchers have argued that the binary representation of the solutions is inefficient for a series of optimisation problems (Davis,1987). De Jong (1985) believes that when a search space is best represented by complex structures like arrays, trees, integers, etc., the programmer should not try and linearise them in binary strings. Instead, it would be better to work directly on them.

The modified GA described here was designed to fit the case study problem. It incorporates domain knowledge not only to the representation of the solutions, but to the design of the genetic operators as well. A potential solution of the problem has the following representation:

$$\left\{ a_1, a_2, a_3, \dots a_n \right\}$$

This representation has no fixed length, a fact that makes the design and the function of the genetic operators a difficult task. On the other hand, this representation has the advantage of carrying both the variables of the problem in one chromosome, one variable being explicit and the other implicit.

3.3 Genetic operators

The regular crossover and mutation operators, produce most of the times illegal offspring when applied to the previous type of chromosomes. Instead of using penalty functions, which is time consuming and inefficient, two purpose-based operators were produced to fit this particular problem.

3.3.1 Same-Total Operator

This is a crossover-type operator. Suppose that the two following chromosomes have been selected for genetic alteration: { 4   6   2 } and { 2   2   4   4 }. The algorithm performs a serial search in the first chromosome and finds a number or a sum that belongs to the space [4,9]. The algorithm then searches in the second chromosome to find a number or a sum of adjacent genes which is equal to the previous number. The genes that represent the same total in the two chromosomes exchange positions. For our example, the operator will function as follows:

Search ⇒ { 4   6   2 }   GENE: 1; TOTAL: 4,
Search ⇒ { 2   2   4   4 } GENES: 1,2; TOTAL: 4,
Yielding: { 2   2   6   2 } and { 4   4   4 }.

None of the chromosomes violates the constraints, while the offspring are totally different from the parents. In some special cases the application of this operator does not alter the shape of the parent chromosomes. The algorithm compensates for these cases by having an increased probability for this particular operator.

3.3.2 Decomposition Operator

This is a mutation-type operator. Preliminary research showed that small-sized cells perform better than the large-sized ones because they are less affected by the cross-contamination factor. Large-sized cells suffer from low utilisation of their reactors, because they cannot process more than one batch at a time. The idea of the decomposition operator is to guide the search for the optimal solution towards a small-sized cell area. Therefore, when a large-sized cell is selected for genetic alteration, the following offspring is produced:

Selected Gene	Offspring after Decomposition
4	2 2
5	3 2
6	2 2 2

3.4 Fitness Evaluation

The measure of fitness for the solutions of the algorithm, is the total number of batches processed per year in the plant. In order to find this number, we simulate the annual manufacturing production for each of these solutions. For the case of multi-objective optimisation, a cost model was constructed, related to the total number of cells in the plant. The cost increases linearly with the number of cells.

3.5 Multiobjective Optimisation methods

3.5.1 Weighted-Sum Approach

In this approach a weight is given for every objective to be optimised, which determines the importance of the objective. If the objective function for the objective i is $f_i(x)$, then the overall objective function will be:

$$F(x) = \sum_{i=1}^{k} w_i \cdot f_i(x)$$

where the weights $w_i \in [0,1]$ and $\sum_{i=1}^{k} w_i = 1$. By assigning different weight vectors for the objectives, the algorithm will converge to different solutions.

3.5.2 Pareto-Optimality Approach

The concept of Pareto-Optimality is defined as follows (Morad,1997):
Consider the vector optimisation problem:

$$\min_{x \in X} f(x) = \min_{x \in X} \{ f_1(x), f_2(x), \ldots, f_m(x) \}$$

where: $x = n$ dimensional vector of decision variables and $X =$ the set of all feasible solutions subject to constraints. A decision vector $x_\mu \in X$ is said to be Pareto-optimal, if and only if there is no other $x_\nu \in X$ such that $f(x_\nu) = (v_1, v_2, \ldots, v_m)$ dominates $f(x_\mu) = (u_1, u_2, \ldots, u_m)$. In other words, there is no x_ν such that: $\forall i \in \{1, \ldots, m\}, v_i \leq u_i$. This solution is called a *non-inferior* or *non-dominated* solution. If we use a simple GA for a multi-objective optimisation problem, the first generation will normally evolve a population of solutions. Some of them will be non-dominated, and they will form the so-called *dominant front* of the solutions. All these solutions are considered as rank '1' individuals, and they will be assigned with the same fitness value for the next generation. We then remove from the population the dominant front, and the new dominant front will be the solutions of rank '2'. They will all be assigned with the same fitness value, but of course lower than the rank '1' solutions. This procedure will go on until all the solutions are assigned a fitness value, and then the evaluation of the new population will start.

3.4.3 Multi-Objective Genetic Algorithm (MOGA) Approach

In this scheme the ranking of each individual corresponds to the number of individuals in the current population by which is dominated (Fonseca and Fleming, 1993). In that way, the dominant front of the solutions is assigned the same rank, while the rest of the solutions are assigned a lower ranking according to the population density of the region of solutions that dominate them. The basic advantage of this method is that it can perform local search, by combining Pareto dominance with partial preference information in the form of a goal vector. In this way, the ranking mechanism can exclude objectives that already satisfy their goals. If fully unattainable goals are specified, then we have the basic Pareto ranking, because no objective will ever be excluded from comparison.

4. Results

In the weighted-sum approach, several different weights were given to the objectives of the optimisation for both scheduling methods. Some examples of the results are given in Figures 1-4.

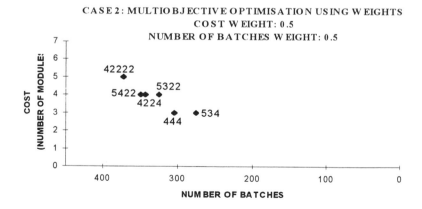

Figure 1 : Weighted-sum approach - Results 1

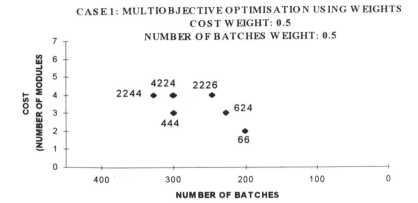

Figure 2 : Weighted-sum approach - Results 2

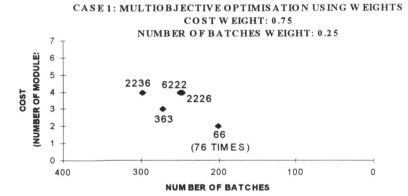

Figure 3 : Weighted-sum approach - Results 3

The technique of fitness sharing was used in the Pareto-Optimality approach, to prevent the premature convergence of the algorithm to a single solution (Goldberg,1987). This technique leads to the formation of stable sub-populations (species) of solutions. The results are presented in Figures 5 and 6.

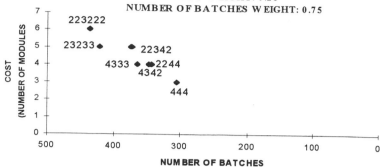

Figure 4 : Weighted-sum approach - Results 4

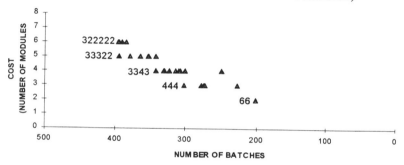

Figure 5 : Pareto -Optimality approach - Results 1

78

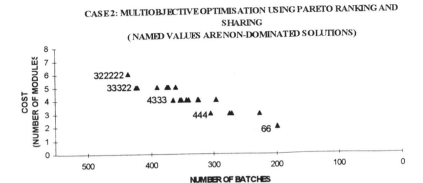

Figure 6 : Pareto -Optimality approach - Results 2

The same technique was used in the case of the MOGA approach. The algorithm converged to a number of alternative solutions that are presented in Figures 7 and 8. The same algorithm was modified in order to be able to perform local search in the solutions' search space. Results were obtained using different goal vectors. A graphical presentation of results is given in Figures 9 and 10, while the values for various parameters of the program can be found in Table 2.

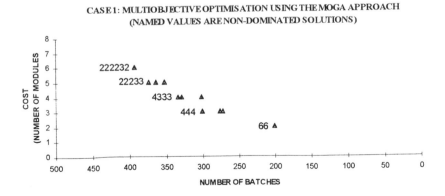

Figure 7 : MOGA approach - Results 1

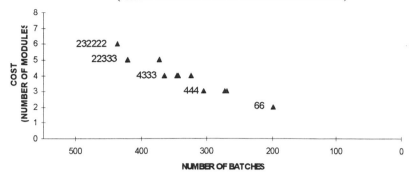

Figure 8 : MOGA approach - Results 2

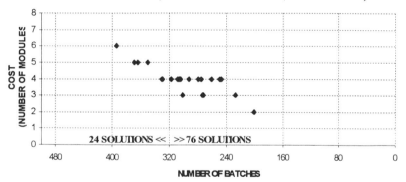

Figure 9 : Local Search using the MOGA approach

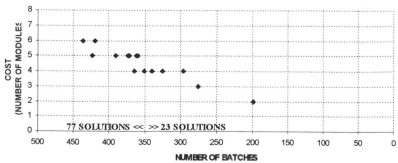

Figure 10 : Local Search using the MOGA approach

4.3 Comparisons

Considering the previous results, CASE2 scheduling method outperform CASE1, because of the grouping of products and cells in certain families. A comparison of the two scheduling methods in terms of their fitness is given in Figure 11. There are no significant differences in the results of the three multi-objective optimisation approaches.

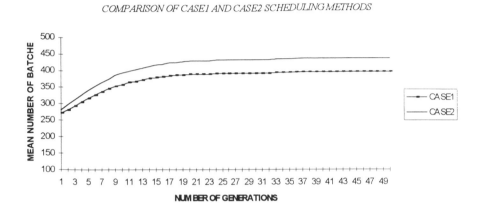

Figure 11 : Comparison of scheduling methods

5. Discussions

Results showed that the evolutionary approach to this numerical cell formation problem outperformed the traditional approach in terms of efficiency and number of alternative solutions produced, especially if purpose-based operators and chromosome representation is used. In addition, the evolutionary approach to multi-objective optimisation provides the means of incorporating more than one objective to the optimisation process. Different multi-objective optimisation methods were used, and all performed equally well.

6. Conclusions

Most of the manufacturing optimisation problems cannot be solved easily and efficiently using traditional optimisation methods. This paper describes an evolutionary algorithm designed to solve a numerical cell formation problem. Results showed that when the plant is divided to a large number of small-sized cells, the total number of batches processed in the plant per year is increased. In addition, if these cells are grouped according to the type of the products that they process, the performance of the plant is improved furthermore. Multiobjective optimisation is a major consideration in a manufacturing plant, since the cost factor is a critical issue in every aspect of the production. Evolutionary algorithms provide the means of implementing multi-objective optimisation, using a number of different approaches.

References

1. Bennett, D., *Production Systems Design*, Butterworths, 1986.
2. Boctor, F. 'A linear formulation of the machine-part cell formation problem', *International Journal of Production Research*, Vol 29, part 2, pp 343-356, 1991.
3. Burbidge, J.L., *The Introduction of Group Technology*, Heineman, London, 1975.
4. Davis, L., and Steenstrup, M. *Genetic Algorithms and Simulated Annealing: An Overview*, Genetic Algorithms and Simulated Annealing, Morgan Kaufmann Publishers, Los Altos, CA, 1987.
5. De Jong, K.A., Genetic Algorithms, *A 10 Year Perspective*, Proceedings of the first International Conference on Genetic Algorithms, Lawrence Erlbaum Associates, Hillsdale, NJ, 1985.
6. Falkenauer, E., 'New representation and operators for GAs applied to Grouping problems', *Research Report No CP 106-P4*, Research Centre for Belgian Metalworking Industries, 1993.
7. Fonseca, C.M., and Fleming P.J., *Genetic Algorithms for Multiobjective Optimisation: Formulation, Discussion and Generalisation*, Proceedings of the fifth International Conference on Genetic Algorithms, Morgan Kaufmann Publishers, San Mateo, CA, 1993.
8. Goldberg, D.A. and Richardson, J., *Genetic Algorithms with Sharing for Multimodal Function Optimisation*, Proceedings of the second International Conference on Genetic Algorithms, Lawrence Erlbaum Associates, Hillsdale, NJ, 1987.
9. Goldberg, D.E., *Genetic Algorithms in Search, Optimisation, and Machine Learning*, Addison-Wesley, Reading, MA, 1989.
10. Holland, J.H., *Adaptation in Natural and Artificial Systems*, University of Michigan Press, Ann Arbor, 1975.
11. Kao, Y. and Moon, Y.B., 'A unified group technology implementation using the back propagation learning rule of neural networks', *Computers and Industrial Engineering*, Vol 20, No 4, pp 425-437, 1991.
12. King, J.R., 'Machine-component grouping in production flow analysis: an approach using a rank order clustering algorithm', *International Journal of Production Research*, Vol 18, No 2, pp 213-232, 1980.
13. McAuley, J.,'Machine Grouping for Efficient Production Production', *Production Engineer*, pp 53-57, 1972.
14. Michalewicz, Z., *Genetic Algorithms + Data Structures = Evolution Programs*, (second edition), Springer-Verlag, 1994.

15. Morad, N., *Integrated production planning and scheduling in cellular manufacturing using Genetic Algorithms*, (Doctoral Dissertation), University of Sheffield, Dept. of Automatic Control and Systems Engineering, Sheffield, 1997.
16. Singh, N., *Computer Integrated Design and Manufacturing*, Wiley, 1996.
17. Wemmerlov, U. and Hyer, N.L., *Cellular Manufacturing in the U.S. Industry: A Survey of Users*, International Journal of Production Research, 27(9), 1989.
18. Xu, H., and Wang, H.P., 'Part family formation for GT applications based on fuzzy mathematics', *International Journal of Production Research*, Vol 27, No. 9, pp 1637-1651, 1989.

Appendix

Product	Batch Time (Days)	No. of Reactors	% Product	Preparation Time
PROD 5-4	5	4	4%	0.33 Days
PROD 4-4	4	4	4%	»
PROD 3-4	3	4	4%	»
PROD 2-4	2	4	4%	»
PROD 5-3	5	3	11%	»
PROD 4-3	4	3	4%	»
PROD 3-3	3	3	4%	»
PROD 2-3	2	3	21%	»
PROD 5-2	5	2	11%	»
PROD 4-2	4	2	6%	»
PROD 3-2	3	2	11%	»
PROD 2-2	2	2	6%	»
PROD 3-1	3	1	6%	»
PROD 2-1	2	1	2%	»
PROD 1-1	1	1	2%	»

Table 1 : Modelling data

Variable	Value
Number of generations	50
Maximum number of cells in the plant	6
Prob. of a chromosome to be selected for	0.4
Prob. of a gene to be selected for	0.05
Number of chromosomes in each	100
Maximum number of reactors in a cell	6
Minimum number of reactors in a cell	2

Table 2 : Values for various parameters

A Temperature Predicting Model for Manufacturing Processes Requiring Coiling

Nando Troyani* and Luis Montano**

* Departamento de Mecanica, Universidad de Oriente, Apdo. 4327,

Puerto La Cruz, Venezuela e-mail ltroyani@reacciun.ve

** CEGELEC Automation Inc., Pittsburgh, Pennsylvania, USA

Abstract. A model for predicting temperature evolution for automatic controling systems in manufacturing processes requiring the coiling of bars is presented. Although the method is of a general nature, the presentation in this work refers to the manufacturing of steel plates in hot rolling mills. The predicting strategy is based on a mathematical model of the evolution of temperature in a coiling and uncoiling bar and is presented in the form of a parabolic partial differential equation for a shape changing domain.. The mathematical model is solved numerically by a space discretization via geometrically adaptive finite elements which accomodate the change in shape of the domain, using a computationally novel treatment of the resulting thermal contact problem due to coiling. Time is discretized according to a Crank-Nicolson scheme. Since the actual physical process may take less time than the time required by the process controlling computer to solve the full mathematical model a special predictive device was developed, in the form of a set of least squares plynomials, based on the numerical solution of the mathematical model.

Key words and phrases. temperature prediction, thermal contact, finite elements, metal coiling, Coilbox.
Work partially supported by CEGELEC Automation Inc., Pittsburgh, PA, USA

1. Introduction

In order to minimize heat loss to the environment in the transfer table, the stage between rougher rolling stands and finishing rolling stands in a hot mill, and for reasons of energy consumption, metallurgical uniformity and rollability, the Coilbox [1], a Stelco Inc. patent, has been succesfully used in some industrial instalations worldwide. By means of this device the hot bar coming out of the rougher is coiled and uncoiled, Figure 1.1, prior to processing it in the finisher, resulting in a substantial heat loss rate reduction.. To preset the parameters of the finishing rolling stands in order to make the finishing process more accurate and efficient it is

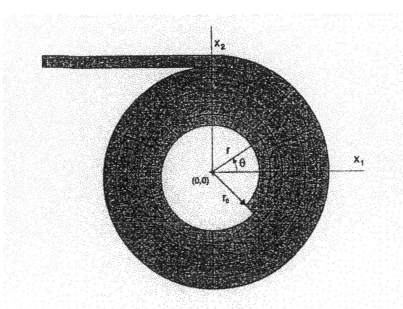

Figure 1.1 Finite element discretization and coordinate sytems in a partially coiled bar

necessary to have an estimate of the temperature distribution along the bar. Here, a brief description of the mathematical model of the process is given, the reader is referred to Troyani [2] for a full account. Based on the numerical solution of the mathematical model of the heat transfer process, an on-line fast temperature prediction system (OFTPS) was developed and is presented in this work. The OFTPS is a necessary device since the actual physical process of coiling and uncoiling takes less time than the numerical solution of the mathematical model for the particular computers used to run the software for automatic control of hot mills.

The model accounts for the different speeds of the Coilbox operation during both coiling and uncoiling, Figure 4.1. This is considered a critical issue since this aspect of the operation directly affects the time of exposure to the environment.

The mathematical model for the problem is stated in section 2. It consists of the non-linear parabolic partial differential equation of heat conduction together with the non-linear convective and non-linear radiative boundary conditions, initial conditions in the form of an initial distribution of temperature, as well as, the equations which describe the motion and change of shape of the bar as it coils and uncoils. The Lagrangian approach is used to keep track of the evolving temperature field in a shape changing domain.

In order to accomodate the shape changing domain of the bar the model is solved using geometrically adaptive finite elements for the space discretization. Time discretization was achieved via a Crank-Nicolson scheme. The problem was treated as a first approach as two dimensional.

Special attention was given to the issue of continuosly changing boundary conditions due to thermal contact. The novel numerical treatment given to the thermal contact problem which arises as a result of adjacent coiling wraps coming into contact is described in section 3.
The numerical results for a 0.0254 m thick, 85.34 m long steel bar as well as a full decription of the OFTPS are given in sections 4 and 5 respectivelly.

2. The Mathematical Model

The equation governing the temperature evolution in the bar is the well known non-linear parabolic partial differential equation of heat transfer by conduction [3] given by:

$$\nabla \cdot [k(T) \nabla T] = c(T) \rho (T) \frac{\partial T}{\partial t} \quad \forall\, x \in \Omega(t) \tag{1}$$

where, k, T, c, ρ, t and $\Omega(t)$ represent thermal conductivity, temperature, heat capacity, density, time and time dependent domain respectively, and $x \equiv (x_1, x_2)$.
The boundary conditions are the combined convective and radiative boundary conditions given by:

$$-k \frac{\partial T}{\partial \eta} = h_{cr}(T - T_\infty) \quad \forall\, x \in \partial \Omega(t) \tag{2}$$

where $h_{cr} = h_c + h_r$. The indicated derivative represents the outward normal derivative. h_c, h_r, T_∞ and $\partial \Omega(t)$ represent convective heat transfer film coefficient, equivalent radiative heat transfer film coefficient, temperature at a sufficient distance from the boundary of the bar and time dependent boundary of the bar respectively.

With T^* an initial distribution of temperature , an initial condition is assumed in the form:

$$T(x,0) = T^*(x) \quad \forall\, x \in \Omega(0) \tag{3}$$

The motion of the bar is described with the following equations based on the coordinate systems of Figure 1.1.
1. Prior to any coiling the motion is given by:

$$x_1(t) = x_1^0 + \int u(t) dt \tag{4}$$
$$x_2(t) = x_2^0 \tag{5}$$

x_1^0, x_2^0 and $u(t)$ represent initial coordinate value of bar points and speed function respectively.

2. Once coiling starts the motion is given by:

2.1 The x_1 coordinate of points which have yet to be coiled change according to expression in eq (4) above
and the x_2 coordinate changes according to:

$$x_2(t) = x_2^0 + \frac{\delta\theta(t)}{2\pi} \tag{6}$$

$\theta(t)$ represents the total angle of arc of coiled bar up to time t, and δ the thickness of the bar.

2.2 points which have coiled (crossed line $x_1 = 0$ at least once), and here for simplicity the switch is made to cylindrical coordinates, move according to:

$$x_1(t) = r(t)Cos\,\theta(t) \tag{7}$$

$$x_2(t) = r(t)Sin\theta(t) \tag{8}$$

where $r(t)$ is given by the following expressions:
for points on the inner surface of the wraps

$$r(t) = r_0 + \frac{\delta\theta(t)}{2\pi} \tag{9}$$

for points on the middle plane of the wraps

$$r(t) = r_0 + \frac{\delta}{2}\left(\frac{\theta(t)}{\pi} + 1\right) \tag{10}$$

for points on the outer surface of the wraps

$$r(t) = r_0 + \frac{\delta}{2}\left(\frac{\theta(t)}{\pi} + 2\right) \tag{11}$$

r_0 represents the radius of the bar point closest to the center of the coil.

3. Position of the coiled bar for uncoiling is achieved by the following 180°rotation:

$$x_1 \rightarrow -x_1$$

$$x_2 \rightarrow -x_2$$

4. Uncoiling is controlled by expressions similar to those in numerals 1 and 2 above.

The numerical solution of the mathematical model was achieved using a space discretization through finite elements and time discretization was resolved using a Crank-Nicolson implicit scheme. In order to comply with the curving of the bar during coiling the 8-node isoparametric element was used [4].

3. The Contact Problem

Of major concern in the model is the numerical handling of the thermal contact which arises as a result of adjacent wraps coming into contact during coiling and losing contact during uncoiling. This issue which appears in many applications has been treated in a number of different ways by various authors, Nakajima [5], Sridhar et al. [6] and Tseng et al. [7], just to name a few. In all these works some form of actual contact is used in the modelling of the interfacial heat transfer process. A totally different approach was used to deal with the thermal contact aspect of the problem based on the following considerations.
First, it is assumed that the outer surface of any wrap is at an infinitesimal distance from the inner surface of the next outer wrap (zero distance in the computations).
Second, the surfaces facing each other are assumed to exchange heat according to a film coefficient of heat transfer which is made consistent with the thermal contact conductance. In effect, the contacting surfaces are assumed to exchange heat according to a boundary condition of the form:

$$-k\frac{\partial T}{\partial \eta} = h_{tccf}(T - T_\infty) \qquad \forall\, x \in \partial\Omega(t) \tag{12}$$

where h_{tccf} is defined as the thermal contact conductance consistent film coefficient, T is the heat
emmiting surface temperature, and T_∞ is the heat receiving surface temperature. The form of Eq. (12) is completely analogous to the form in Eq. (2), and so the complex thermal contact problem has been reduced to a simple boundary condition.

4. Numerical Results and Discussion

A double precision FORTRAN code FETAHBRF.FOR (finite element thermal analysis hot bar rougher to finisher) was developed, tested and run in a VAX 4000/200. Reference [8] reports, from experimental measurements, that a 0.027 m thick carbon steel bar exposed to the environment loses temperature at an average rate of 2.256 °C/s in the range of 1120.0 °C-980.0 °C to be compared with an everage rate of 2.289 °C/s for the same range using the present approach. For the coiled portion of a 0.0254 m thick bar Stelco reports an average loss of temperature of 0.0556 °C/s to be compared with 0.0544 °C/s using the present approach.

The speed schedule of the bar is shown in Figure 4.1. Figure 4.3 shows the calculated evolution of temperature for a 0.0254 m thick 85.34 m long bar, for the Rimming 0.06% carbon steel., in three groups of three curves each, corresponding to nine nodal points, based on the initial temperature distribution shown in Fig. 4.2.

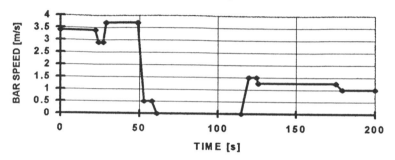

Figure 4.1. Bar speed schedule

The upper set corresponds to three points located in the upper surface, middle plane and lower surface respectively at a distance of 0.305 m of the, initially, head end of the bar. The middle set corresponds to three points similarly located through the thickness as the ones just mentioned except that they are located in the center, lengthwise, of the bar. The lower set corresponds to three points also similarly located as the ones already mentioned except that they belong to the , initially, tail end of the bar, and at 0.305 m of the end.

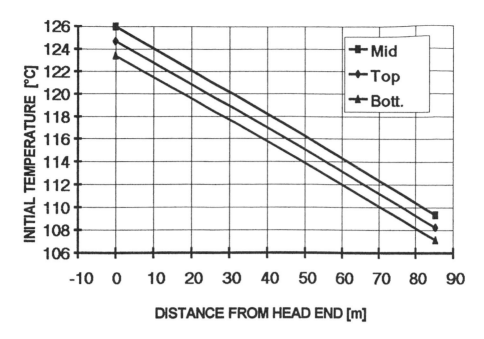

Figure 4.2. Initial distribution of temperature

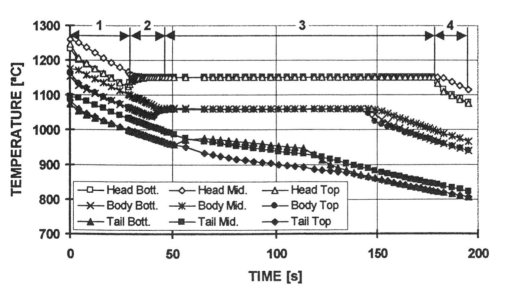

Figure 4.3. Temperature evolution at nine selected points.

Note that as the bar is coiled and uncoiled, what was originally the head end of the bar becomes the tail end, and viceversa. The value of $3.4 \ W / m^{2 \circ} C$ for the contact conductance was used from reference [9]. This value corresponds to a surface roughness of 2.54 μm at an interface pressure of 5.2 atm. and 204.0 °C.

Each set of three curves exhibits four stages:

1. The typical initial exponential temperature loss for the uncoiled parts.
2. A short transition stage corresponding to the process of "temperature soaking" through the thickness of the bar. This stage coincides with actual coiling of a given sector of the bar.
3. A third stage where the temperature within the thickness of the wrapped bar is practically uniform, and temperature loss is reduced to an exponential curve with large relaxation time.
4. The last stage is similar to the first one in that there is full exposure to heat loss to the environment and corresponds to uncoiled parts of the bar.

5. On Line Fast Temperature Prediction System

The OFTPS is based on the solution of the described model for 5 different thickness (0.0125 m, 0.01875 m, 0.025 m, 0.03125 m and 0.037 m) covering the range of bar thicknesses, 0.0125m to 0.037 m, and it predicts the temperature at the nine points described above. For this purpose the curves in Fig. 3 are divided, as already explained in the previous section, in four identifyable sectors, for each thickness. For each sector, corresponding to each curve, and for each thickness a least squares approximating polynomial giving the temperature as a function of time was developed in the form

$$T = a_0 + a_1 t + a_2 t^2 + a_3 t^3 + a_4 t^4 \tag{13}$$

A total of 180 polynomials chosen so that the least squares norm is minimized.

The estimated temperature $T(x_j, t)$ for point x_j and time t from a given initial temperature T_{in} (determined through pyrometers in the mill) is obtained by adding to the initial temperature a sum of temperature changes ΔT_i calculated from the appropiate polynomials, in the following form:

$$T(x_j, t) = T_{in} - \sum_{i=1}^{4} \Delta T_i \tag{14}$$

Each of the four temperature changes correspond to each of the four stages in figure 4.

The OFTPS system is based on using expression (14) in connection with the following steps.

1. Given an initial temperature for a given point, an initial fictitious time is calculated through a standard Newton-Raphson procedure from the appropiate approximating polynomial of the form in expression (13).

2. A final time for the first stage is calculated through the speed schedule of the coilbox.
3. Corresponding to this final time a final temperature for the stage T_{sf} is computed using (13) again.
4. The change of temperature corresponding to the first stage is then simply

$$\Delta T_1 = T_{in} - T_{sf} \tag{15}$$

5. For thicknesses other than the five specified above a linear interpolation is used.

The other three temperature changes calculations in expresion (14) above, are performed in a totally similar fashion, keeping in mind that the final time and final temperature of a given stage calculation are the initial time and the initial temperature of the next stage calculation.
Clearly, the proposed temperature prediction system is independent of the initial fictitious time.

6. Conclusions

A novel and efficient temperature prediction mathematical model and the corresponding finite element solution is presented for manufacturing processes requiring coiling. The definition of a consistent film coefficient of heat transfer with the thermal contact conductance to treat the complex contact problem results in a particularly efficient computational scheme since the need for remeshing and node renumbering at each computational coiling step is eliminated. This novel definition also provides the added advantage of producing the minimum possible bandwith of the resulting equations with the clear advantage of minimizing CPU time. Herein lies the computational advantage of the present approach over previous ones which, typically, treat the contact problem in more traditional fashion. An on-line learning fast temperature prediction system based on the model and the solution was also presented.

7. References

1. W. Smith, 1981. The Coilbox: A New Approach to Hot Strip Rolling, *AISE Year Book* , pp 432-436.
2. N. Troyani, 1996. Nonlinear Geometrically Adaptive Finite Element Model of the Coilbox, *Numerical Heat Transfer*, Part A, pp 849-858.
3. M.N. Ozisik, 1993. *Heat Conduction*, Chap. 1, 2nd Ed., Wiley.
4. E. B. Becker, G. F. Carey and J.T.Oden, 1891. *Finite Elements: An Introduction*, Chap. 5, Prentice Hall.
5. K. Nakajima, 1995.Thermal Contact Resistance Between Balls and Rings of Bearings under Axial, Radial

and Combined Loads, *Journal of Thermophysiscs and Heat Transfer*, vol. 9, n. 9, pp 88-95.

6. M. R. Sridhar, M. M. Yovanovich, 1996. Elastoplastic Contact Conductance Model for Isotropic Conforming Rough Surfaces and Comparison with Experiments, *Journal of Heat Transfer-Transactions of the ASME*, vol. 118, pp 3-9.

7. A. A. Tseng, S. R. Wang, 1996. Effects of Interface Resistance on Heat Transfer in Steel Cold-Rolling, *Steel Research*, vol. 67, pp 44-51.

8. F. Hollander, 1967. A Model to Calculate the Complete Temperature Distribution in Steel During Hot Rolling, *AISE Year Book*, pp 46-78.

9. M. E. Barzelay, K. N. Tong and G. F. Holloway, 1955. *NACA Tech. Note* 3295.

Solving Multi-objective Transportation Problems by Spanning Tree-based Genetic Algorithm

Mitsuo Gen and Yin-Zhen Li

Dept. of Industrial & Systems Engineering
Graduate School of Engineering
Ashikaga Institute of Technology, Ashikaga, 326 Japan
Email:{gen yzli}@genlab.ashitech.ac.jp

Abstract. In this paper, we present a new approach which is spanning tree-based genetic algorithm for solving a multi-objective transportation problem. The transportation problem as a special type of the network optimization problems has the special data structure in solution characterized as a transportation graph. In encoding transportation problem, we introduce one of node encodings based on a spanning tree which is adopted as it is capable of equally and uniquely representing all possible basic solutions. The crossover and mutation were designed based on this encoding. Also we designed the criterion that chromosome has always feasibility converted to a transportation tree. In the evolutionary process, the mixed strategy with $(\mu + \lambda)$-selection and roulette wheel selection is used. Numerical experiments show the effectiveness and efficiency of the proposed algorithm.

Keywords: Multi-objective Optimization, Transportation Problem, Spanning Tree, Genetic Algorithm

1 Introduction

In more real-world cases transportation problem can be formulated as a multi-objective transportation problem. Because, the complexity of the social and economic environment requires the explicit consideration of criteria other than cost. Examples of additional concerns include: average delivery time of the commodities, reliability of transportation, accessibility to the users, product deterioration, among others.

Recently, there are many investigation of evolutionary approachs to solve the variety transportation problems. Michalewicz *et al.* [7] [9] firstly discussed the use of genetic algorithm (GA) for solving linear and nonlinear transportation problems. They used these problems as an example of constrained optimization problems, and investigated how to handle such constraints with GA. The matrix representation was used to construct a chromosome and designed the matrix-based crossover and mutation in their investigation. Gen *et. al.* [2] further extended Michalewicz's works to bicriteria linear transportation problem and bicriteria solid transportation problem. They embedded the basic idea of criteria space approach in evaluation phase so as to force genetic search towards exploiting the nondominated points in the criteria space.

However, in more real-world transportation problems to be solved it will have a large-scale problem. For the problem with m origins and n destinations the matrix-based representation for a solution requires $m \times n$ memories in the evolutionary process. It requires so much memories in the development of implementation, and will be spent more computational time.

In this paper, we present a new approach of the genetic algorithm for solving multi-objective transportation problem. Transportation problem as a special type of the network optimization problems has a special data structure in solutions characterized as a transportation graph. We focus on this special graph and utilize the Prüfer number [1] encoding based on a spanning tree which is adopted as it is capable of equally and uniquely representing all possible trees. Using the Prüfer number representation the memory only requires $m + n - 2$ on the problem. Transportation problems have separable set of nodes from origins and destinations. From this point, we design the criterion for feasibility of the chromosome. We also design crossover and mutation based on new encoding and to generate feasible solutions, *i.e.*, transportation spanning trees, by added criterion of feasibility. And the mixed strategy with $(\mu + \lambda)$-selection and roulette wheel selection is used. The proposed algorithm can find Pareto optimal solutions for multi-objecitve transportation problem in criterion space. Finally, the numerical experiments will be shown the effectiveness and efficiency of the proposed spanning tree-based genetic algorithm for solving multi-objective transportation problem.

2 Multiobjective Transportation Problem

Assuming there are m origins, n destinations, and Q objectives, the multi-objective transportation problem can be formulated as follows:

$$\min \quad z_q(x) = \sum_{i=1}^{m} \sum_{j=1}^{n} c_{ij}^q x_{ij}, \quad q = 1, 2, \cdots, Q$$

$$\text{s. t.} \quad \sum_{j=1}^{m} x_{ij} \leq a_i, \quad i = 1, 2, \cdots, m$$

$$\sum_{i=1}^{n} x_{ij} \geq b_j, \quad j = 1, 2, \cdots, n$$

$$x_{ij} \geq 0, \quad \forall \, i, j$$

where a_i is the amount of homogeneous products for the i-th origin which are transported to n destinations. b_j is the demand of homogeneous products for the j-th destination. c_{ij}^q is the coefficients of the q-th objective functions which are associated with transportation of an unit of the product from source i to destination j. x_{ij} is the unknown quantity to be transported from origin i to destination j.

The first set of constraints stipulates that the sum of the shipments from a source can not exceed its supply, the second set requires that the sum of the shipments to a destination must satisfy its demand. The above problem implies that the total supply $\sum_{i=1}^{m} a_i$ must be equal to total demand $\sum_{j=1}^{n} b_j$. When total supply is equals to total demand (i.e., total flow), the resulting formulation is called a balanced transportation problem. In this paper, we assume a balanced transportation problem. Because the unbalanced transportation problem can be converted to a balanced transportation problem after including a dummy origin or a dummy destination.

Transportation problem have some characteristics in solution:

(1) it has $m + n - 1$ basic variables and they correspond to some cell in the transportation tableau,

(2) the basis must be transportation tree, that is, there must be at least one basic cell in each row and in each column of the transportation tableau, and

(3) the basis should not contain a cycle. It is illustrated with one of basis solution in Figure 1.

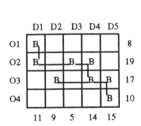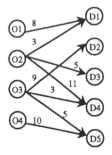

Figure 1: Illustration of a basis on the transportation tableau and the transportation graph.

There are separable set of nodes from origins and destinations on the transportation problem. We denote that origins $1, 2, ..., m$ as the component of the set $O = \{1, 2, ..., m\}$ and denote that destinations $m+1, ...m+n$ as the component of the set $D = \{m+1, ..., m+n\}$. Obviously, the transportation problem has $m + n$ nodes and $m \times n$ edges.

In the sense of multi-criteria mathematical programming, criteria are usually conflicting another in nature and the concept of optimal solution gives place to the concept of Pareto optimal(efficient, nondominated, or noninferior) solutions, for which nonimprovement sacrificing on another objective functions.

3 Spanning Tree-based Genetic Algorithm

3.1 Representation

The genetic representation is a kind of data structure which represents the candidate solution of the problem in coding space. Usually different problems have different data structures or genetic representations. Transportation graph in Figure 1 can be represented as a spanning tree such as in Figure 2.

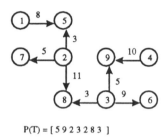

P(T) = [5 9 2 3 2 8 3]

Figure 2: A spanning tree and its Prüfer number

For a complete graph with p nodes, there are $p^{(p-2)}$ distinct labeled trees. This means that we can use only permutation of $p-2$ digits in order to uniquely represent a tree with p nodes where each digit is an integer between 1 and p inclusive. For any tree there are always at least two leaf nodes.

Prüfer number is one of the node encoding for the tree. It can be used to encode a transportation tree. The constructive procedure for a Prüfer number according to a tree is as follows:

procedure: convert tree to Prüfer number

Step 1. Let i be the lowest numbered leaf node in tree T. Let j be the node which is the predecessor of i. Then j becomes the rightmost digit of Prüfer number $P(T)$, $P(T)$ is built up by appending digits to the right; thus, $P(T)$ is built and read from left to right.

Step 2. Remove i and the edge (i,j) from further consideration. Thus, i is no longer considered at all and if i is the only successor of j, then j becomes a leaf node.

Step 3. If only two nodes remain to be considered, then $P(T)$ has been formed, stop; else return to Step 1.

Each node in the transportation problem has its quantity of supply or demand which are characterized as constraints. Therefore to construct a transportation tree, the constraint of nodes must be considered. From a Prüfer number, an unique transportation tree also is possible to be generated by the following procedure:

procedure: convert Prüfer number to transportation tree

Step 1. Let $P(T)$ be the original Prüfer number and let $\bar{P}(T)$ be the set of all nodes that are not part of $P(T)$ and designed as eligible for consideration.

Step 2. Repeat the process (2.1)–(2.5) until no digits left in $P(T)$.

> **(2.1)** Let i be the lowest numbered eligible node in $\bar{P}(T)$. Let j be the leftmost digit of $P(T)$.
>
> **(2.2)** If i and j are not in the same set O or D, then add the edge (i,j) to tree T. Otherwise, select the next digit k from $P(T)$ that not included in the same set with i, exchange j with k, and add the edge (i,k) to the tree T.
>
> **(2.3)** Remove j(or k) from $P(T)$ and i from $\bar{P}(T)$. If j(or k) does not occur anywhere in the remaining part of $P(T)$, then put it into $\bar{P}(T)$. Designate i as no longer eligible.
>
> **(2.4)** Assign the available amount of units to $x_{ij} = \min\{a_i, b_j\}$ (or $x_{ik} = \min\{a_i, b_k\}$), where $i \in O$ and $j, k \in D$ to the edge (i,j) (or (i,k)).
>
> **(2.5)** Update availability $a_i = a_i - x_{ij}$ and $b_j = b_j - x_{ij}$ (or $b_k = b_k - x_{ik}$).

Step 3. If no digit remain in $P(T)$, then there are exactly two nodes , i and j still eligible in $\bar{P}(T)$ for consideration. Add edge (i,j) to tree T and form a tree with $m + n - 1$ edges.

Step 4: If no available amount of units to assign, then stop. Otherwise, there are remaining supply r and demand s, add edge (r,s) to tree and assign the available amount of units $x_{rs} = a_r = b_s$ to edge. If there exists a cycle, then remove the edge that assigned zero flow. A new spannning tree is formed with $m + n - 1$ edges.

3.2 Initialization

The initialization of a chromosome (*i.e.*, a Prüfer number) is performed from that randomly generated $m + n - 2$ digits in range $[1, m + n]$. However, it is possible to generate an infeasible chromosome which is not adapted to generate a transportation tree.

Prüfer number encoding is not only capable of equally and uniquely representing all possible spanning tree, but also explicitly contains the information of node degree that any node with degree d will appear exactly $d - 1$ times in the encoding. This means that when a node appears d times in Prüfer number, the node exactly have $d+1$ connections with other node.

We design the handling for feasibility of the chromosome with the following criterion: denote that s_O and s_D are the sum of connections of nodes which are included in set O and D respectively from $P(T)$. Also we denote that $\bar{s}_O$ and $\bar{s}_D$ are the appearing times of those nodes in $\bar{P}(T)$ and included in set O and D, respectively. If $s_O + \bar{s}_O = s_D + \bar{s}_D$, then $P(T)$ is feasibility. Otherwise $P(T)$ is infeasible.

3.3 Genetic Operators — Crossover and Mutation

Crossover: the one-cut-point crossover operation is used as illustrated in Figure 3. For avoiding unnecessary decoding from which an infeasible chromosome(*i.e.* a Prüfer number) may be generated after crossover operator, we add the criterion for feasibility of the chromosome. A Prüfer number via this criterion is always feasible and can be decoded into a corresponding transportation tree.

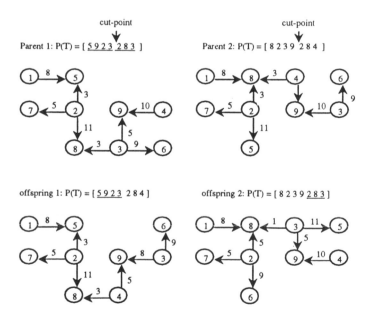

Figure 3: Illustration of crossover and offspring

Mutation: the inversion mutation and displacement mutation are used. The inversion mutation selects two positions within a chromosome at random and then inverts the substring between these two positions as illustrated in Figure 4. The displacement mutation selects a substring at random and inserts it in a random position as illustrated in Figure 5.

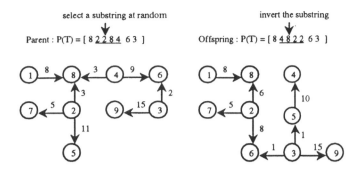

Figure 4: Illustration of inversion mutation

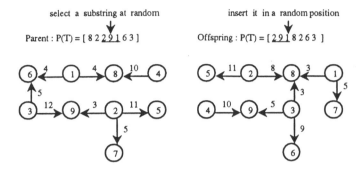

Figure 5: Illustration of displacement mutation

These two mutation operators always generate feasible chromosomes if the parents are feasible, because the criterion $s_O + \bar{s}_O = s_D + \bar{s}_D$ is unchanged after these operations.

3.4 Evaluation and Selection

In this approach, the evaluation procedure also consists of the following two steps:

(1) convert a chromosome into a tree, and

(2) calculate each objective function.

It is given as follows:
procedure: evaluation
begin
    $T \leftarrow \{\emptyset\}$;
    $p \leftarrow 0$;
    define $\bar{P}(T)$ according to the $P(T)$;
    repeat

select the leftmost digit from $P(T)$, say i;
select the eligible node with the lowest numbered from $\bar{P}(T)$, say j;
if $i, j \in O$ or $i, j \in D$ then
    select next digit from $P(T)$ not in the set with j, say k;
    exchange k with i and take $i \leftarrow k$;
end
assign the flow: $x_{ij} \leftarrow \min\{a_i, b_j\}$, $i \in O, j \in D$;
$T \leftarrow T \cup \{x_{ij}\}$;
$z_q(T) \leftarrow z_q(T) + c_{ij}^q x_{ij}$, $q = 1, 2, \cdots, Q$;
update available amount:
    $a_i \leftarrow a_i - x_{ij}$;
    $b_j \leftarrow b_j - x_{ij}$;
remove i from $P(T)$;
remove j from $\bar{P}(T)$;
if i does not occur anywhere in remaining $P(T)$ then
    put i into $\bar{P}(T)$;
end
$p \leftarrow p + 1$;
until($p \le m + n - 2$)
assign the flow for $r, s \in \bar{P}(T)$: $x_{rs} \leftarrow \min\{a_r, b_s\}$, $r \in O, s \in D$;
$T \leftarrow T \cup \{x_{rs}\}$;
$z_q(T) \leftarrow z_q(T) + c_{rs}^q x_{rs}$, $q = 1, 2, \cdots, Q$;
repeat
    if $a_i > 0, \forall\ i$ and $b_j > 0, \forall\ j$ then
        assign the flow: $x_{ij} \leftarrow \min\{a_i, b_j\}$, $i \in O, j \in D$;
        $T \leftarrow T \cup \{x_{ij}\}$;
        update available amount:
            $a_i \leftarrow a_i - x_{ij}$;
            $b_j \leftarrow b_j - x_{ij}$;
    end
    $z_q(T) \leftarrow z_q(T) + c_{ij}^q x_{ij}$, $q = 1, 2, \cdots, Q$;
until(no available amount)
if there exists a cycle then
    find edge with zero flow and remove it;
end
end

In multi-criteria optimization context, usually the Pareto optimal solutions are characterized as the solutions of the multi-objective programming problem. In this stage, the procedure for Pareto solutions consists of the following two steps:

(1) evaluate chromosomes by the objective function,

(2) select Pareto solutions based on objective function values.

Let $E(t)$ be the Pareto solution set generated up to current iteration t, then the procedure for Pareto solutions is given as follows:

procedure: Pareto solutions
begin
    $t \leftarrow 0$;
    $E(t) \leftarrow \{\emptyset\}$;
    while(**not** termination condition) **do**
    begin
        count the number of current chromosomes i_size;
        $k \leftarrow 1$;
        repeat
            evaluate a chromosome T_k;
            obtain the solution vector $\mathbf{z}_k = [z_1(T_k) \ z_2(T_k) \ \cdots \ z_Q(T_k)]$;
            register Pareto solutions into $E(t)$ and
            delete non-Pareto solutions in $E(t)$;
            $k \leftarrow k + 1$;
        until ($k \leq i\_size$);
        $t \leftarrow t + 1$;
    end
end

As the fitness function for survival, the weighted sum method is used to construct the fitness function. The multi-objective functions z_q, $q = 1, 2, \cdots, Q$ are combined into one overall objective function at hand.

Handling for fitness function:

(1) Choose the solution points which contain the minimum z_q^{min} (or the maximum z_q^{max}) that corresponding to each objective function value, and then compare with the stored solution points at the previous generation and select the best points to save again.

$$z_q^{min(t)} = \min_k \{ z_q^{min(t-1)}, z_q^{(t)}(T_k) \mid k = 1, 2, \cdots, i\_size \}, \quad q = 1, 2, \cdots, Q$$

$$z_q^{max(t)} = \max_k \{ z_q^{max(t-1)}, z_q^{(t)}(T_k) \mid k = 1, 2, \cdots, i\_size \}, \quad q = 1, 2, \cdots, Q$$

where $z_q^{max(t)}$ $(z_q^{min(t)})$ is the maximum (minimum) value of objective function q in current generation t, and i_size is the number of the chromosomes from the population and generated offspring in current generation t.

(2) Solve the following equations to get weights for the fitness function:

$$\delta_q = z_q^{max(t)} - z_q^{min(t)}, \quad q = 1, 2, \cdots, Q$$

$$\beta_q = \frac{\delta_q}{\sum_{q=1}^{Q} \delta_q}, \quad q = 1, 2, \cdots, Q$$

(3) Calculate the fitness function value for each chromosome as follows:

$$eval(T_k) = \sum_{q=1}^{Q} \beta_q z_q(T_k), \quad k = 1, 2, \cdots, i\_size$$

In the selection procedure, we will use the mixed strategy with $(\mu + \lambda)$-selection and roulette wheel selection can enforce the best chromosomes into the next generation. This strategy selects μ best chromosomes from μ parents and λ offspring. If there are no μ different chromosomes available, then the vacant pool of population is filled up with roulette wheel selection.

3.5   Overall Procedure

Let $P(t)$ be a population of chromosomes, $C(t)$ be the generated chromosomes in current generation t, and $E(t)$ be the Pareto solution set generated up to current generation t. The overall procedure is summarized as follows:
overall procedure:
begin
    $t \leftarrow 0$;
    initialize $P(t)$;
    evaluate $P(t)$;
    determine $E(t)$;
    while (**not** termination condition) **do**
    begin
        recombine $P(t)$ to generate $C(t)$;
        evaluate $C(t)$;
        update $E(t)$;
        select $P(t+1)$ from $P(t)$ and $C(t)$;
        $t \leftarrow t+1$;
    end
end

4   Numerical Experiments

The proposed spanning tree-based genetic algorithm is implemented in C language and run on an NEC EWS 4800 under the EWS-UX/V release 4.0 UNIX OS to do our simulations.

Firstly, to confirm the effectiveness of the spanning tree-based genetic algorithm on the transportation problem, the five numerical examples were used in the computational studies. The examples were randomly generated where the supplies, demands are generated randomly and distributed uniformly over [10, 100] and the transportation costs are over [5, 25].

The simulations were carried out on each example with their best mechanism (parameter setting). Table 1 shows average simulation results from 20 times running by the spanning tree-based genetic algorithm (called t-GA) and matrix-based genetic algorithm (called m-GA). From Table 1 we known that the optimal solution was found accurately by the t-GA with more short time than m-GA in each running time. This means that t-GA is more appropriate than m-GA for solving transportation problem.

Table 1: Average results by the m-GA and t-GA approach for TP

problem size:	parameter		matrix-based GA		tree-based GA	
$m \times n$	pop_size	max_gen	percent	ACT(s)	percent	ACT(s)
3×4	10	200	100(%)	1.02	100(%)	1.07
4×5	20	500	100(%)	12.75	100(%)	8.77
5×6	20	500	100(%)	18.83	100(%)	9.99
6×7	100	1000	100(%)	1038.25	100(%)	413.63
8×9	200	2000	100(%)	4472.19	100(%)	848.11

percent: the percentage of running times that obtained optimal solution
ACT: the average computing time

The performance of this approach for solving multi-objective transportation problem was tested on four numerical examples. First and second examples are bicriteria transportation problem, in which one is having 3 origins and 4 destinations, and another one is having 7 origins and 8 destinations where the supplies, demands and coefficients of objective functions shown in Table 2.

For first example, the parameters for the algorithm are set as follows: population size $pop\_size = 30$; crossover probability $p_c = 0.2$; mutation probability $p_m = 0.4$; maximum generation $max\_gen = 500$; and run by 10 times. The real Pareto optimal solutions can be found at all times. The obtained solutions (143, 265), (156, 200), (176, 175), (186, 171), (208,167) were known as Pareto optimal solutions which are formed in Pareto frontier with the exact extreme points (143, 265) and (208, 167). The average computing time, on the t-GA and m-GA are 5.45 seconds and 10.42 seconds respectively.

The other two examples are randomly generated having three objectives. The supplies and demands also are generated randomly and distributed uniformly over [10, 100] and the coefficients of each objective function are over [5, 25]. The problem size and the simulation results with the best mechanism ($p_c = 0.2$ and $p_m = 0.4$ for t-GA) by the proposed algorithm are shown in Table 3. The average computing time was computed on 10 running times.

Figure 6 shows that the comparison between the m-GA and t-GA with parameters $p_c=0.4$, $p_m=0.2$ for the m-GA and $p_c=0.2$, $p_m=0.4$ for the t-GA on example 2. From Figure 6, we can see that the obtained results by the t-GA are better than those by the m-GA in the sense of Pareto optimality because most of them are not dominated by those obtained in the m-GA. Therefore, we can get concluded that they are closer to the real Pareto frontier than those by the m-GA.

Table 2: Supplies, demands and coefficients of objectives for bicriteria transportation problem

From O_i To D_j	c^1_{ij}								c^2_{ij}								a_i
	8	9	10	11	12	13	14	15	8	9	10	11	12	13	14	15	
1	1	2	7	7	8	10	9	2	4	4	3	4	5	8	9	10	10
2	1	9	3	4	3	5	7	1	6	2	5	1	7	4	12	4	8
3	8	9	4	6	4	1	6	9	2	9	1	8	9	1	4	0	12
4	2	4	5	5	3	2	3	2	3	5	5	3	2	8	3	3	16
5	5	4	5	1	9	9	1	6	1	4	12	2	1	5	4	9	21
6	8	3	3	2	2	3	6	7	2	23	4	4	6	2	4	6	15
7	1	2	6	4	5	9	3	5	1	2	1	9	0	13	2	3	7
b_j	9	7	15	10	13	16	7	12	9	7	15	10	13	16	7	12	89

Besides, in Table 3 the average computing time on the t-GA is more short than that on the m-GA for each example. Obviously, the t-GA is more efficiency than the m-GA in the evolutionary process. It is because the tree-based encoding only requires $m + n - 2$ memory space for each chromosome but the matrix-based encoding requires $m \times n$ memory space for representing a chromosome. As a result, it takes more time spending on genetic operations for matrix-based encoding GA compared with the tree-based encoding GA. Therefore, for large-scale problem, the tree-based encoding GA proposed will be surely more time-saving than the matrix-based encoding GA.

Table 3: Comparison with m-GA and t-GA

problem size			memory for a solution		ACT(s)		parameter	
m	n	Q	m-GA	t-GA	m-GA	t-GA	pop_size	max_gen
3	4	2	12	5	10.42	5.45	30	500
7	8	2	56	13	1632.75	863.01	200	2000
8	9	3	72	15	6597.41	2375.53	200	2000
10	15	3	150	23	19411.08	11279.30	300	2000

m: number of origins
n: number of destinations
Q: number of objectives
m-GA: matrix-based encoding GA
t-GA: spanning tree-based encoding GA

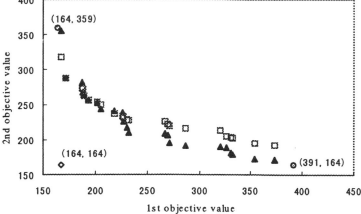

◇ ideal point

▲ obtained solutions by t-GA

□ obtained solutions by m-GA

◎ extreme points

Figure 6: Comparison between m-GA and t-GA on the obtained solutions for second tested problem

5   Conclusion

In this paper, we proposed a new approach by using spanning tree-based genetic algorithm to solve multi-objective transportation problem. Spanning tree-based encoding was implemented with Prüfer number and adopted to represent a balanced transportation solution. In order to keep the feasibility of chromosomes, a criterion of the feasibility was designed. Based on this criterion the crossover and mutation were operated and they can always generate the feasible chromosomes.

To demonstrate the effectiveness and efficiency of the proposed GA approach on the bicriteria transportation problem, we carried out the numerical experiments by both the matrix-based genetic algorithm and the spanning tree-based genetic algorithm. For the small-scale problem, there is no great difference on the computing results, which shows the effectiveness of the proposed GA approach. On the larger scale problem, the proposed spanning tree-based GA approach can get the Pareto solutions with more less CPU time than the matrix-based genetic algorithm approach and most of the results are not dominated by those obtained in the matrix-based genetic algorithm approach. Therefore, in the sense of Pareto optimality, the proposed spanning tree-based GA approach is more effective than the matrix-based genetic algorithm. So far as the computing time is concerned, the proposed spanning tree-based GA approach is much more efficient than the matrix-based genetic algorithm on the transportation problem.

Acknowledgment

This work was supported in part by a research grant from the Ministry of Education, Culture, & Sciences, the Japanese Government: Grant-in-Aid for Scientific Research, the International Scientific Research Program (No.07045032: 1995.4—1998.3).

References

[1] Dossey, J., A. Otto, L.spence and C. Eynden, 1993. *Discrete Mathematics.* Harper Collins.

[2] Gen, M., K. Ida, E. Kono and Y.Z. Li, 1994. Solving Bicriteria Solid Transportation Problem by Genetic Algorithm. *Proceedings of 16th International Conference on Computers & Industrial Engineering.* pp 572-575.

[3] Gen, M. and R. Cheng, 1997. *Genetic Algorithms and Engineering Design.* John Wiley & Sons, New York.

[4] Goldberg, D.E, 1989. *Genetic Algorithms in Search, Optimization and Machine Learning.* Addison Wesley, Reading, MA.

[5] Li, Y.Z. and M. Gen, 1997. Spanning Tree-Based Genetic Algorithm for Bicriteria Transportation Problem. *Proceedings of The Australia-Japan Joint Workshop on Intelligent and Evolutionary Systems.* Canberra. pp 112-119.

[6] Michalewicz, Z., 1992. *Genetic Algorithms + Data structure = Evolution Programs.* 1994. Second, extended ed. 1996. Third, revised and extended ed. Springer-Verlag, New York.

[7] Michalewicz, Z., G. A. Vignaux and M. Hobbs, 1991. A Non-Standard Genetic Algorithm for the Nonlinear Transportation Problem. *ORSA Journal on Computing, Vol.3, No.4.* pp 307-316.

[8] Palmer, C.C. and A. Kershenbaum, 1995. An Approach to a Problem in Network Design using Genetic Algorithms. *Networks, Vol.26.* pp 151-163.

[9] Vignaux, G.A. and Z. Michalewicz, 1991. A Genetic Algorithm for the Linear Transportation Problem. *IEEE Transactions on Systems, Man, and Cybernetics, Vol.21, No.3.* pp 445-452.

[10] Zhou, G. and M. Gen, 1997. Approach to Degree-Constrained Minimum Spanning Tree Problem Using Genetic Algorithm. *Engineering Design & Automation, Vol.3, No.2.* pp 157-165.

[11] Zhou, G. and M. Gen, 1997. Genetic Algorithm Approach on Multicriteria Minimum Spanning Tree Problem. *Accepted to European Journal of Operations Research.*

Chapter 3

Generic Issues: Model Representation, Multi-Objectives And Constraint Handling

Optimisation for Multilevel problems: A Comparison of Various Algorithms.
M.A. El-Beltagy, A.J. Keane

Evaluation of Injection Island GA Performance on Flywheel Design Optimization.
D. Elby, R.C. Averill, W.F. Punch III,
E.D. Goodman.

Evolutionary Mesh Numbering: Preliminary Results.
F. Sourd, M. Schoenauer

Two New Approaches to Multiobjective Optimization Using Genetic Algorithms.
C.A. Coello Coello

Mapping Based Constraint Handling for Evolutionary Search; Thurston's Circle Packing and Grid Generation
D.G. Kim, P. Husbands

Optimisation for Multilevel Problems: A Comparison of Various Algorithms

M. A. El-Beltagy   A.J. Keane

Department of Mechanical Engineering, University of Southampton

email: m.a.el-beltagy@soton.ac.uk

Abstract. The optimization of imprecisely specified functions is a common problem occurring in various applications. Models of physical systems can differ according to computational cost, accuracy and precision. In multilevel optimization, where different models of a system are used, there is a great benefit in understanding how many, fast evaluations of limited accuracy may be combined with a few accurate calculations to yield an optimum solution. The combination of different models or levels of representations can lead to an objective function surface characterized by multiple values at a single point. This paper compares various optimization methods with genetic algorithms using three different strategies of multilevel optimization. A modified 'bump' function is used as an example to compare the different methods and strategies. A sequential mixing strategy applied to a Genetic Algorithm with niche forming is shown to give best results. The paper highlights the need to develop a specialized optimization algorithm for this kind of problem.

Key Words: Optimization, Multilevel problems, Genetic Algorithms

1.   Introduction

In many optimization problems there may exist a number of different ways in which a particular problem is modelled. Some methods may be quite elaborate in their representation, while others involve a simplification of the problem, with the former being more accurate but at the same time more computationally expensive than the latter. It is therefore important to understand how a significant number of less accurate evaluations may be integrated with fewer accurate ones to arrive at an optimum design.

There are many ways in which the approximate and accurate representations can be integrated. In this paper three integration strategies are attempted, sequential multilevel optimization, gradually mixed multilevel optimization, and totally mixed multilevel optimization. These integration methods are explained in subsequent sections. Our main aim has been to see how different optimization methods work using these strategies paying particular attention to genetic algorithms (GAs).

This paper is arranged as follows: The next section describes the 'bump' function studied here. Section 3 gives an overview of niching as used in GA and its intuitive advantages in this type of problem. Sections 4, 5, and 6 detail experimental results using Sequential, Gradually Mixed, and Totally Mixed Multilevel optimization strategies. The paper closes with a brief conclusion and a discussion of future work.

2. The bump problem

The 'bump' problem, introduced by Keane[7], is very hard for most optimizers to deal with. It is quite smooth but contains many peaks, all of similar heights. Moreover, its optimal value is defined by the presence of a constraint boundary.

The problem is defined as

$$\text{maximize} \frac{\text{abs}(\sum_{i=1}^{n} \cos^4(x_i) - 2\prod_{i=1}^{n} \cos^2(x_i))}{\sqrt{\sum_{i=1}^{n} i x_i^2}} \qquad (1)$$

for

$$0 < x_i < 10 \quad i=1,...,n \qquad (2)$$

subject to

$$\prod_{i=1}^{n} x_i > 0.75 \quad \text{and} \quad \sum_{i=1}^{n} x_i < 15n/2 \qquad (3)$$

starting from

$$x_i = 5, \qquad i=1,..,n$$

An interesting feature of this function when $n=2$ is that the surface is nearly but not quite symmetrical in $x_1 = x_2$, so the peaks always occur in pairs but with one always bigger than its sibling. The global optimum is also defined by the product constraint. When the problem is generalized for n greater than two it becomes even more demanding with families of similar peaks occurring with a highly complex constraint surface. These properties of bump have made it suitable for the study of GA performance and optimizing GA control parameters[8] as well as the control parameters of other evolutionary optimization methods[9].

To investigate optimization with a distorted surface, equation (1) may be generalized to have a frequency shift parameter α and a spatial shift parameter β.

$$\text{maximize} \frac{\text{abs}(\sum_{i=1}^{n} \cos^4(\alpha(x_i + \beta)) - 2\prod_{i=1}^{n} \cos^2(\alpha(x_i + \beta)))}{\sqrt{\sum_{i=1}^{n} i(x_i + \beta)^2}} \qquad (4)$$

The α frequency shift parameter distorts bump, spreading out the peaks ($\alpha < 1$) or making them closer together ($\alpha > 1$). The spatial shift parameter β just shifts the peaks of bump in x_i. Hence the undistorted bump becomes one in which $\alpha=1$ and $\beta=0$.

3.   The advantages of niching

In natural terms a niche is viewed as an organism's environment, while a species is a collection of organisms with similar features. The subdivision of an environment based on an organism's role reduces interspecies competition for environmental resources, and this reduction in competition helps stable sub-populations to form around different niches in the environment[4,5]. This division of several sub-populations/species may be useful when there is sudden environmental change as one of the species could be very successful in the new environment while others may perish.

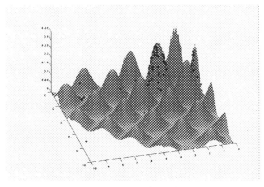

Figure 1-The distribution of the population over bump after the optimization has been carried out.

For many optimization problems there exist multiple peaks within the parameter search space: bump is an example of such a problem. A simple GA cannot readily maintain stable populations at the different optima of such functions. Figure 1 shows the distribution of the population over a 2D bump for a simple GA. It is clear here that the population has largely converged on a single peak which is not the global maximum (the global maximum is on the constraint boundary). On the other hand a GA with niching would force the population to be distributed over many peaks in the parameter space (Figure 2).

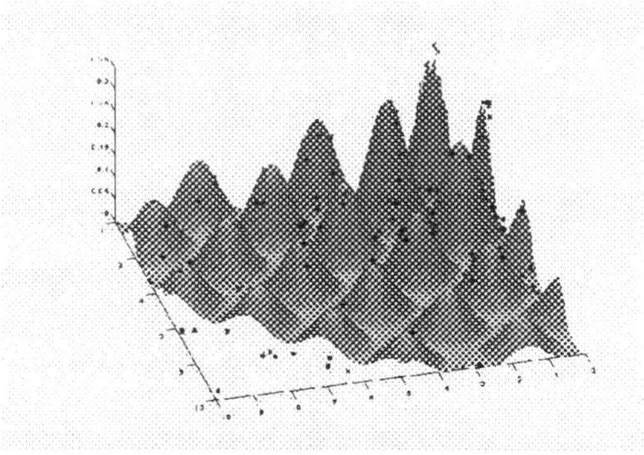

Figure 2-The distribution of the population over bump after the optimization has been carried out with niching.

In this paper we use a niche forming method based on MacQueen's adaptive KMEAN clustering algorithm. It is have been found effective at revealing unknown multimodal function structures and is able to maintain subpopulation diversity[12].

4.   Sequential Multilevel Optimization

In this approach, the optimization is started using the least accurate level of representation then, after a certain set number of function evaluations, the optimization on this level is stopped and the results used as starting points for the next more accurate level. This is carried on sequentially and the number of function evaluations is decreased from one level to the next until the most accurate level is reached where fewest function evaluations are carried out. The number of function evaluations carried out at each level would ordinarily be chosen to roughly equalize the computational effort expended at each level. For our bump problem the first less accurate level corresponds to maximum distortion (high values of α and/or β). The optimization then proceeds through an intermediate level ($\alpha \to 1$ and $\beta \to 0$) after which the final accurate representation is reached ($\alpha=1$ and $\beta=0$).

As has just been noted, when the maximum number of function evaluations for a certain level is reached its final parameter vector x_i becomes the initial one for the next level. In the case of population based methods the final population is used and re-evaluated using the more accurate function and becomes the initial population for the next level.

As has already been mentioned, the number of function evaluations per level decreases from the less accurate level to the more accurate one. This reduction mimics the situation where more refined models become more computationally expensive and hence only a more limited number of evaluations would be afforded. Details of the number of generations used here for each level are shown in Table 1. Assuming that an equal amount of effort were expended at each level, this supposes that the true function evaluation is 25 times more expensive than that for $\alpha=1.5$ and/or $\beta=0.5$ and 5 times more expensive than for $\alpha=1.1$ and/or $\beta=0.1$.

Optimization Level	1	2	3
Number of evaluations	12500	2500	500
α	1.5	1.1	1
β	0.5	0.1	0

Table 1 Number of evaluations at each level and corresponding distortion parameters

4.1. The Optimization methods

Seven optimization methods have been used here to compare their performance in this domain. All these methods are available in a design exploration system developed by Keane[10] called OPTIONS. The methods used here are

- A Genetic Algorithm based on clustering and sharing (GACS)[12].
- A Bit Climbing algorithm (BClimb) [3]
- A Dynamic Hill Climbing algorithm (DHClimb) [13]
- A population based Incremental learning algorithm (PBIL) [2]
- Simulated Annealing (SA)[11]
- Evolutionary Programming (EP)[6]
- Evolution Strategy based on the earlier work of Back, Hoffmeister and Schwefel (ES)[1]

4.2. Experiments and results

Experiments were performed for the 2-D and 20-D bump functions. The optimization was carried out for (1) varying α, (2) varying β, (3) varying both α and β using the values and limit on function evaluations for the levels as shown in Table 1. The results were averaged over thirty optimization runs. The initial population was randomly selected, but contains the point $x_i=5$ for $i=1..n$.

For the case of the 2D bump, the results in descending order of performance are as shown in Table 2

	Alpha (1)	Beta (2)	Alpha & Beta (3)	AVG	Avg No. Of steps
GACS	0.2767	0.2739	0.3082	0.2863	15500
SA	0.2907	0.2852	0.2736	0.2832	15500
DHClimb	0.2980	0.2674	0.2722	0.2792	15526
EP	0.2839	0.2615	0.2745	0.2733	15500
Bclimb	0.2626	0.2432	0.2679	0.2579	15500
PBIL	0.1756	0.1916	0.3464	0.2379	15500
ES	0.2206	0.2005	0.2179	0.2130	15500

Table 2. Sequential Multilevel Optimization of a 2D bump.

For a 20D bump, the results are as shown in Table 3.

	Alpha (1)	Beta (2)	Alpha & Beta (3)	AVG	Avg No. Of steps
GACS	0.5227	0.5951	0.5306	0.5495	15500
DHClimb	0.5379	0.4968	0.5468	0.5272	15933
PBIL	0.4417	0.5601	0.4416	0.4811	15500
EP	0.4222	0.4820	0.4209	0.4417	15500
SA	0.4004	0.4081	0.3880	0.3988	15500
Bclimb	0.4502	0.3343	0.3759	0.3868	15425
ES	0.2206	0.2005	0.2179	0.2130	15500

Table 3. Sequential Multilevel Optimization of a 20D bump

In Table 3 it is clear that the Dynamic Hill Climbing (DHClimb) algorithm performance is very comparable to that of the GACS and it performs better in two distortion sequences.

5. Gradually Mixed Multilevel Optimization

Here the optimisation procedure is carried out with three levels which are heterogeneously mixed throughout the optimization process. The probability of using a particular level varies with the number of function evaluations as shown in Figure 3.

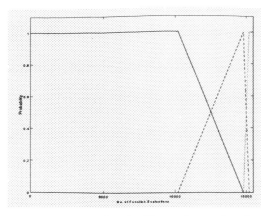

Figure 3 Probability of selecting a level after a given number of function Evaluations. The 1st level is solid, the 2nd Dashed, and the 3rd dotted.

Using this scheme the first 10200 evaluations are carried out using the 1st level. In next 4600 evaluations the 1st and 2nd levels are mixed gradually. This is followed by 400 evaluations, where the 2nd and 3rd levels are mixed, finally the last 300 evaluations are carried out solely at the third and most accurate level. The scheme is chosen such that the average overall computational cost is equal to that in the sequential mixing scheme. The values in Table 1 are used as an indicator of the relative cost between using different levels. The optimization was performed using some of the optimization methods discussed earlier in section 4.1.

Elitism was disabled in all methods. This was done to prevent an individual with a high objective function in the distorted representation from dominating and influencing the optimization in the latter stages where the more accurate representation is used. At these stages, this individual might have a lower objective function value. This contrasts with the sequential case where the optimization is carried over three distinct stages and hence elitism can be used at each stage.

As in the previous section, three distortion sequences were carried out. Results were again averaged over thirty runs, see Tables 4 and 5.

	Alpha	Beta	Alpha & Beta	AVG	Avg No. Of Trials
GACS	0.2807	0.2525	0.3251	0.2861	15500
SA	0.2572	0.2892	0.2032	0.2498	15500
PBIL	0.1492	0.1301	0.3413	0.2069	15500
DHClimb	0.0018	0.1211	0.3296	0.1508	15526
EP	0.0015	0.0668	0.3359	0.1348	15500
ES	0.1304	0.1191	0.1332	0.1276	15500
Bclimb	0.0076	0.1259	0.2019	0.1118	15500

Table 4 Gradually Mixed Multilevel Optimization of a 2D bump

Table 4 shows that, on average, GACS outperforms all the other methods on the 2D problem. Note that in the 2D case most methods do particularly well in the distortion sequence where α and β are being varied simultaneously, but they perform poorly on the other distortion sequences. This was found to be due to the 1st level peaks being very close to the 2nd and 3rd level peaks. This allows for easy migration between the peaks reached at one level on to the next. This behaviour, however, is not maintained in the 20D case where the peaks shift position in far more complex ways when α and β are varied simultaneously. Also, the search space is then too large for such convenient occurrences to have significant effect, see Table 5.

	Alpha	Beta	Alpha & Beta	AVG	Avg No. Of Trials
GACS	0.4788	0.5853	0.4765	0.5135	15500
SA	0.3995	0.4509	0.4023	0.4176	15500
PBIL	0.3367	0.5580	0.2732	0.3893	15500
DHClimb	0.2402	0.4313	0.1731	0.2815	15777
EP	0.2883	0.4292	0.0833	0.2669	15500
Bclimb	0.2553	0.2972	0.0808	0.2111	15425
ES	0.1374	0.1377	0.1342	0.1364	15500

Table 5 Gradually Mixed Multilevel Optimization of a 20D bump

It is also clear from the table that the other methods tried performed relatively better on the 20D problem than for the 2D case. This would appear to be because the changes in function value between levels are then less dramatic due to the increased complexity of the function.

6. Totally Mixed Multilevel optimization

In totally multilevel optimization the probability of using a particular level is constant throughout most of the optimization process. Towards the very end of the optimization only the most accurate level is used. In our case the 1st level has a probability of 82.22%, the 2nd has a probability of 16.44%, and the 3rd and most accurate level has a probability of 1.315%. In the last 300 evaluations the 3rd level probability becomes 100%, see Figure 4. This scheme is set up such that the computational cost on average is the same as in the last two schemes.

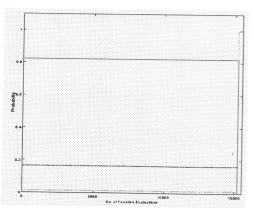

As in the previous section, elitism was disabled here for similar reasons.

Figure 4 Probability of selecting a level after a given number function Evaluations. The 1st level is solid, the 2nd Dashed, and the 3rd dotted

Again, three distortion sequences were carried out. The results were averaged over thirty runs and are as shown in Table 6 and Table 7.

	Alpha	Beta	Alpha & Beta	AVG	Avg No. Of Trials
GACS	0.1996	0.2070	0.3026	0.2364	15500
PBIL	0.1178	0.1261	0.3376	0.1938	15500
DHClimb	0.0016	0.1190	0.3284	0.1496	15535
EP	0.0017	0.1018	0.3342	0.1459	15500
ES	0.1335	0.1444	0.1187	0.1322	15500
Bclimb	0.0659	0.1337	0.1899	0.1298	15500
SA	0.0017	0.0802	0.2891	0.1237	15500

Table 6 Totally Mixed Multilevel Optimization of a 2D bump

Again GACS performed better then the other methods on average for the 2D case. And again all the other methods performed poorly on the α and β only distortion sequences.

	Alpha	Beta	Alpha & Beta	AVG	Avg No. Of Trials
PBIL	0.3838	0.5711	0.4822	0.4791	15500
GACS	0.3526	0.5440	0.2953	0.3973	15500
Bclimb	0.3332	0.2854	0.3273	0.3153	15981
EP	0.3055	0.4365	0.1079	0.2833	15500
DHClimb	0.2738	0.3796	0.1788	0.2774	15425
SA	0.2634	0.3771	0.1569	0.2658	15500
ES	0.1390	0.1365	0.1376	0.1377	15500

Table 7 Totally Mixed Multilevel Optimization of a 20D bump

As in the gradual mixed case, the relative performance of the other methods improves for the 20D case. Here, PBIL has overtaken GACS although the latter is no longer performing as well as in the previous two strategies.

7. Discussion

To gain a basis of comparison it is useful to consider the case where the entire computational time was dedicated to using the most accurate level (i.e. 1500 such evaluations assuming the ratios between the number of evaluation as in Table 1). Using this approach, the performance of the stochastic optimizers is as in Table 8. Shown are the averaged (AVG) and best ever obtained results (BST) over 30 runs for each optimizer.

	2 D AVG	20 D AVG	2 D BST	20 D BST
GACS	0.3118	0.4727	0.3638	0.5974
DHClimb	0.2812	0.3968	0.3576	0.5888
PBIL	0.2931	0.2243	0.3584	0.2962
EP	0.3111	0.3106	0.3630	0.3969
SA	0.3007	0.3935	0.3638	0.4821
Bclimb	0.2623	0.4508	0.3626	0.5908
ES	0.2734	0.1833	0.3576	0.2110

Table 8 Results for a single level Optimization using the accurate bump.

The comparative improvement of each method using any of the three proposed strategies may then be calculated, see Table 9. Here the results are normalised by diving the average performance of each method in Tables 2 to 7 by the values in Table 8. The results may also be normalised by dividing by the best ever results obtained for the accurate function evaluation (0.365 and 0.8035, respectively, for the 2D and 20D cases), see Table 10.

	Sequential Mixing 2 D	20 D	Gradual Mixing 2 D	20 D	Total Mixing 2 D	20 D	Average 2 D	20 D
GACS	0.9182	1.1624	0.9176	1.0864	0.7582	0.8405	0.8647	1.0298
DHClimb	0.9930	1.3286	0.5364	0.7095	0.5322	0.6991	0.6872	0.9124
PBIL	0.8116	2.1450	0.7058	1.7357	0.6613	2.1358	0.7262	2.0055
EP	0.8785	1.4221	0.4331	0.8594	0.4690	0.9121	0.5935	1.0645
SA	0.9417	1.0136	0.8308	1.0611	0.4113	0.6755	0.7279	0.9167
Bclimb	0.9831	0.8580	0.4262	0.4682	0.4950	0.6994	0.6348	0.6752
ES	0.7791	0.9541	0.4666	0.7443	0.4836	0.7513	0.5764	0.8165
Average	0.9007	1.2691	0.6167	0.9521	0.5444	0.9591	0.6872	1.0601

Table 9 Comparative improvements using the three mixing strategies

	Sequential Mixing 2 D	20 D	Gradual Mixing 2 D	20 D	Total Mixing 2 D	20 D	Average 2 D	20 D
GACS	0.7844	0.6838	0.7838	0.6391	0.6477	0.4944	0.7386	0.6058
DHClimb	0.7650	0.6561	0.4132	0.3504	0.4100	0.3452	0.5294	0.4505
PBIL	0.6517	0.5987	0.5668	0.4845	0.5311	0.5962	0.5832	0.5598
EP	0.7487	0.5497	0.3692	0.3322	0.3997	0.3526	0.5059	0.4115
SA	0.7758	0.4963	0.6845	0.5196	0.3389	0.3308	0.5997	0.4489
Bclimb	0.7065	0.4813	0.3063	0.2627	0.3557	0.3924	0.4562	0.3788
ES	0.5836	0.2176	0.3495	0.1698	0.3622	0.1714	0.4318	0.1863
Average	0.7165	0.5262	0.4962	0.3940	0.4350	0.3833	0.5492	0.4345

Table 10 Absolute performance using the three mixing strategies

It clear from the tables that:

- Sequential Mixing is a better strategy when averaged across all methods.
- None of the three proposed strategies provide any improvement for the 2D case.
- For the 20D case only GACS, PBIL, and EP were in general improved by adopting the mixing strategies proposed.
- The overall best approach for the 20D case was to use GACS with the sequential strategy. This gives an average final objective function on the three tests (α only, β only, α plus β) of 0.5495, see Table 3. However, this final value is only 16% better than a straight-forward use of GACS on the accurate function for 1500 steps.
- Although PBIL and EP showed improvements in the mixed methods (with a factor of 2 for PBIL and 1.0645 for EP), their objective function values were generally lower than GACS (the exception being PBIL on the 20D totally mixed case), also their performance was rather erratic between runs.

These observations lead to the conclusion that to work well in mixed method environments an optimizer must be specifically designed to cope with noisy functions. In the case of the GA this may perhaps point to the need for a diploid scheme as means for coping with this kind of environment, or perhaps a modified injection island scheme.

8. Conclusion and future work

We have presented a distorted bump function as being representative of a multimodal, distorted objective function that may be used as a tool for understanding how different optimizers behave in multilevel environments. Optimization of this function is characterized by the cost of evaluating the function being directly related to its accuracy. Hence, precise evaluations must be used sparingly for an efficient search. We have used different multilevel strategies and optimization methods to carry out this process. The use of a GA based on clustering and sharing (GACS) which distributes the population over many peaks in a changing fitness landscape, has been shown to give improved results using most of the different strategies. Though a number of methods came close to the GACS on some tests, none was as robust under all the different distortion sequences and different multilevel integration strategies. Overall, a sequential strategy using cheap but inaccurate solutions first, followed by a lesser number of intermediate solutions before finally using a few calls to the fully accurate but most expensive function proved to be the most effective approach. It was, however, only 16% more efficient than a simple used of only the most accurate function over an equivalent, limited number of trials.

Future work will be directed towards more efficient mixing strategies as well as more specialized optimization techniques specifically designed to be able to handle this kind of problem.

Acknowledgments

This work was supported under EPSRC grant no GR/L04733

References

1. Back T, Hoffmeister F, Schwefel H P. A survey of evolution strategies *Proceedings of The Fourth International Conference on Genetic Algorithms.* Morgan Kaufman Publishers, Inc,1991, 2-9.
2. Baluja S *Population-Based Incremental Learning: A Method for Integrating Genetic Search Based Function Optimization and Competitive Learning.* Carnegie Mellon University,1994.
3. Davis L. Bit-Climbing, representational bias, and the test suite design *Proceedings of The Fourth International Conference on Genetic Algorithms.* Morgan Kaufman,1991, 18-23.
4. Deb K and Goldberg D E . An Investigation of Niche and Species Formation in Genetic Function Optimization *Proceedings of The Fifth International Conference on Genetic Algorithms and Their Applications.* Morgan Kaufmann,1989, 42-50.
5. Deb. K *Genetic Algoritms in Multimodal Function Optimization.* M.S. Thesis, College of Engineering, Univ. of Alabama, Tuscaloosa AL,1989.
6. Fogel DB. Applying evolutionary programming to selected traveling salesman problems *Cybernetics and Systems, ,*1993, 24(1):27-36.
7. Keane A J. Experiences with Optimizers in Structural Design *Proceedings of the Conference on Adaptive Computing in Engineering Design and Control(ACEDC'94).* PEDC, University of Plymouth, UK,1994, , 14-27.
8. Keane A J. Genetic Algorithm Optimization of Multi-peak Problems: Studies in Convergence and Robustness *Artificial Intelligence in Engineering, ,*1995, **9(2),** 75-83.
9. Keane A J. A brief Comparison of Some Evolutionary Optimization Methods *Proceedings of the Conference on Applied Decision Technologies (Modern Heuristic Search Methods).* Uxbrigde,1995, 125-137.
10. Keane A J *The Options Design Exploration System: Reference Manual and User Guide,*1995.
11. Kirkpatrick S, Gelatt C D, and Vecchi M P. Optimization by simulated annealing *Science, ,*1983, 220(4598):671-680.
12. Yin X and Germay N *A Fast Genetic Algorithm with Sharing Scheme Using Cluster Methods in Multimodal Funtion Optimization, Proceedings of the International Conference on Artificial Neural Nets and Genetic Algorithms.* Springer-Verlag, Innsbruck,1993, 450-457.
13. Yuret D and de la Maza M. Dynamic Hill Climbing: Overcoming the limitations of optimization techniques *Proceedings of the 2nd Turkish Symposium of AI and ANN,*1993, 254-260.

Evaluation of Injection Island GA Performance on Flywheel Design Optimisation

David Eby, R. C. Averill
Department of Materials Science and Mechanics
William F. Punch III, Erik D. Goodman
Genetic Algorithms Research and Applications Group (GARAGe)
Michigan State University, East Lansing, MI 48824 USA
Phone (517)355-6453 Fax (517)432-0704
goodman@egr.msu.edu

Abstract. This paper first describes optimal design of elastic flywheels using an Injection Island Genetic Algorithm (iiGA). An iiGA in combination with a finite element code is used to search for shape variations to optimize the Specific Energy Density of flywheels (SED is the rotational energy stored per unit mass). iiGA's seek solutions simultaneously at different levels of refinement of the problem representation (and correspondingly different definitions of the fitness function) in separate subpopulations (islands). Solutions are sought first at low levels of refinement with an axisymmetric plane stress finite element code for high-speed exploration of the coarse design space. Next, individuals are injected into populations with a higher level of resolution that uses an axisymmetric three-dimensional finite element model to "fine-tune" the flywheel designs. Solutions found for these various "coarse" fitness functions on various nodes are injected into nodes that evaluate the ultimate fitness to be optimized. Allowing subpopulations to explore different regions of the fitness space simultaneously allows relatively robust and efficient exploration in problems for which fitness evaluations are costly. First the paper treats a greatly simplified case – one for which all two million possible solutions were enumerated, yielding a known global optimum. Then the success and speed of many methods, including several variations of an iiGA, in finding this known global optimum are compared. The iiGA methods always found the global optimum, and the other methods never did. Hybridizing the iiGA with a local search operator and a Threshold Accepting (TA) search at the end of each generation provided the fastest solutions, without sacrificing robustness. Finally, a problem with a large design space is presented and results are compared for a hybrid iiGA to a parallel GA that uses a topological "ring" structure. The hybrid iiGA greatly outperforms the topological "ring" GA in terms of fitness and search efficiency for this given problem.

1.0 Introduction

New optimization problems arise every day -- for instance, what is the quickest path to work? Where and how congested is the road construction? Am I better off riding my bike? If so, what is the shortest path? Sometimes these problems are easily solved, but many engineering problems cannot be handled satisfactorily using traditional optimization methods. Engineering involves a wide class of problems and optimization techniques. Many engineering design approaches such as "make-it-and-break-it" are simply out-of-date, and have been replaced by computer simulations that exploit various mathematical methods such as the finite element method to avoid costly design iterations. However, even with high-speed supercomputers, this design process can still be hindersome, producing designs that evolve slowly over a long period of time. The next step in the engineering of systems is the automation of optimization through computer simulation. If the desired performance factors for the system can be appropriately captured, then optimization over them is simply engineering on a grander scale.

Optimization approaches include hill climbing, stochastic search, directed stochastic search and hybrid methods. Hill climbing or gradient-based methods are single-point search methods that have been applied successfully to many shape optimization problems [1-3], and are extensible via neighborhood sampling even to cases in which derivatives are not analytically given. However, these methods are severely restricted in their application due to the likelihood of quickly converging to local exrema [4]. Random search methods simply evaluate randomly sampled designs in the search space, and are therefore generally limited to problems that have small search spaces, if practical search times are required. A directed random search method, such as a Genetic Algorithm (GA), is a multiple-point, directed stochastic search method that can be an effective optimization approach to a broad class of problems. The use of GA's for optimal design requires that a large number of possible designs be analyzed, even though this number generally still represents only a miniscule fraction of the total design space. When each evaluation is computationally intensive, a traditional simple or parallel GA can thus be difficult to apply. Injection Island Genetic Algorithms (iiGA's) can help reduce the computational intensity associated with typical GA's by searching at various levels of resolution within the search space using multiple analyses that can vary in levels of complexity, accuracy and computational efficiency.

Structural optimization via GA's is the main area of interest for this paper [5-13]. Recently, GA's have been successfully applied in the optimization of laminated composite materials [14-18]. The authors of the current paper have used an iiGA in the design of laminated composite structures [14,15]; others use different GA approaches [16-18]. iiGA's have also been applied to engineering problems such as [19]. [20-24] deal with the application of GA's to shape optimization problems. [20, 21] used GA's to find optimal shapes based on various polynomials while [22] presents the concept of fictitious domains to generate new shapes and [23] reduced computational costs associated with generating meshes for finite element evaluations by a point heat sink approach. [24] modeled flywheels as a series of concentric rings (see Figure 1) using a simple GA measuring fitness with a plane stress finite difference model. Although [24] has already performed optimization of flywheels using a simple GA, this paper differs in many respects: [24] *seeded* the initial population with flywheels that varied linearly in thickness from the inner to outer radii, allowing genetic operators to find new shapes, while this paper allows for ring thickness to be *randomly* chosen in the initial population; [24] searched for shapes using a simple GA while this paper will present various optimization approaches such as Threshold Accepting (TA), GA's, iiGA's and hybrid techniques; [24] based fitness on a *single objective* in each run while *multiple fitness* definitions where used *concurrently* in each iiGA run for this paper; [24] measures fitness only with a *plane stress evaluation* while the current paper presents techniques that *concurrently* use *multiple evaluations* that vary in levels of complexity, accuracy and computational efficiency.

Combining a GA with the finite element method is by now a familiar approach in the optimization of structures, but using an iiGA with multiple evaluation tools and with different fitness functions is a new approach aimed at decreasing computational time while increasing the robustness of a typical GA. Typically, a useful approximation to the overall response of most structures can be captured with a computationally efficient, simplified model, but often, these simplified models cannot capture all complex structural behaviors. If the model does not accurately capture the physics of the problem, then the results of any optimization technique will be an *artifact* of the simplified analysis, dooming the solution(s) to be incorrect. This forces the designer to use a more refined model, which can be computationally demanding, sometimes leading to evaluation times too long to be practical for use in GA search. These obstacles are nearly always present in interesting structural optimization problems. This paper will show how an efficient, simplified axisymmetric plane stress finite element model, when used to evaluate fitness in an optimization problem, produces solutions that are *artifacts* of the simplified analysis. The paper will also show that an ordinary parallel GA using the refined axisymmetric finite element model requires excessively long search times, in comparison to an iiGA approach which employs both types of FEA representation.

An eventual goal of this effort is to develop tools for multi-criterion optimization of large-scale, 3-dimensional composite structures, using an iiGA that searches at various levels of resolution and model realism. It incorporates several simultaneous and interconnected searches, including some that are faster (but often less accurate). This approach is designed to spend less time evaluating poor designs with computationally intensive fitness functions (this is to be done with the efficient, less accurate evaluations) and to spend more time evaluating potentially good designs with the computationally intensive fitness evaluation.

For the flywheel problem treated here, the lowest level of the iiGA searches with a simple axisymmetric plane stress finite element model (with a "sub-fitness" function), which quickly finds "building blocks" to inject into a series of GA populations using several more refined, axisymmetric, three-dimensional finite element models. An optimal annular composite flywheel shape is sought using both this iiGA approach and, for comparison, using a "ring" topology parallel GA (PGA). The flywheel is modeled as a series of concentric rings (see Figure 1). The thickness within each ring varies linearly in the radial direction. A diverse set of material choices is provided for each ring. Figure 2 shows a typical planar finite element model used to represent a flywheel, in which symmetry about the transverse normal direction and about the axis of rotation is used to increase computational efficiency. The overall fitness function for the genetic algorithm GALOPPS [1] was the specific energy density (SED) of the flywheel, which is defined as:

$$SED = \frac{\frac{1}{2}I\omega^2}{mass} \qquad 1.)$$

where ω is the angular velocity of the flywheel ("sub-fitness" function), I is the mass moment of inertia defined by:

$$I = \int_V \rho \cdot r^2 dV \qquad 2.)$$

and ρ is the density of the material.

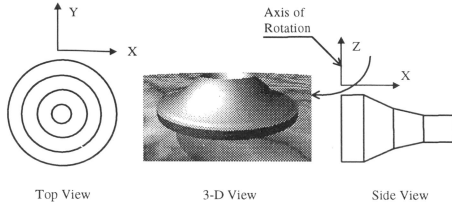

Top View   3-D View   Side View

Figure1. Visual Display of Flywheel

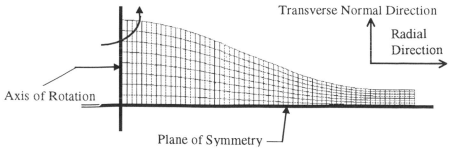

Figure 2. Typical Flywheel Model

1.1 Simulated Annealing and Threshold Accepting

Simulated Annealing (SA) is a combinatorial optimization technique that is based on the statistical mechanics of annealing of solids [25]. To understand how such an approach can be used as an optimization tool, one must consider how to coerce a solid into a low energy state. Annealing is a process typically applied to solid materials to force the atomic structure of the material into a highly ordered state. Atomic structures that maintain a highly ordered state are also at a low energy state. In an annealing process, a material is heated to a temperature that allows many atomic arrangements, then cooled slowly, minimizing energy, while statistically allowing an occasional increase in atomic energy. When the material is extremely hot, the probability of an increase in atomic energy is very high. As the cooling continues, the probability of an increase in atomic energy decreases. Similarly, SA methods use analogous set of parameters that simulate controlled cooling effects found in the annealing of materials.

SA methods begin with an initial solution that is often generated randomly, and try to perturb the solution to improve it. If the perturbation improves the solution then it is accepted and the process of perturbing continues. In this manner, SA methods are like iterative methods that climb hills. As with hill climbing methods, this process of searching just for a better solution tends to force the process to a local optimum. However, SA methods are different in this respect: annealing occasionally allows perturbations that are harmful to the solution to be accepted. This allows SA methods to "climb out" of local optima to search for a global optimum. In real physical systems, jumps to a higher ("worse") state of energy actually do occur. Probability of these jumps is reflected in the current temperature. As the annealing process (cooling) continues, the probability that only better solutions will be accepted increases. At the beginning of the annealing process (associated with a high temperature), the chance that a worse solution is accepted is, while later in the annealing process (at a lower temperature) the chance that a worse solution is accepted is small. This probability of accepting worse solutions is based on a Boltzman distribution:

$$\Pr[Accept] = e^{-\frac{\Delta E}{T}} \qquad\qquad 3.)$$

By successively lowering the temperature T, the simulation of material coming into equilibrium at each newly reduced temperature can effectively simulate physical annealing.

Threshold Accepting (TA) is a simplified version of Simulated Annealing. The probability of accepting a worse solution is governed by the Boltzmann distribution for SA applications and the TA algorithm, but the TA algorithm is not dependent upon a specified temperature. Instead, the TA algorithm rate of cooling is based on a specified percentage of the current solution fitness. This percentage decreases over the set of generations. This causes the TA in earlier generations to have a higher probability of accepting a worse individual, while later generations in the optimization are less likely to accept a worse solution.

1.2 Parallel Genetic Algorithms

Two problems associated with GA's are their need for many fitness evaluations and their propensity to converge prematurely. An approach that ameliorates both of these problems is a parallel GA (PGA), which also produces a more realistic model of nature than a single large population. PGA's typically decrease processing time to a given solution quality, even when executed on a single processor, and better explore the search space. If they are executed using parallel processors, an additional speedup (in wall clock time) nearly linear with processor number may be achieved.

Unlike some specialized sequential GA's which may pay a nontrivial computational cost for maintaining a structured population (demes, etc.) based on similarity comparisons (niching techniques, etc.), PGA's maintain multiple, separate subpopulations which are allowed to evolve nearly independently. This allows each subpopulation to explore different parts of the search space, each maintaining its own high-fitness individuals and each controlling how mixing occurs with other subpopulations, if at all.

1.3 Injection Island GA's

iiGA's represent an extension to the usual notion of parallel GA's, allowing each subpopulation to search at a different levels of resolution within a given space, or to search using representations or fitness functions which differ in some other way among subpopulations. This includes searching at low levels of resolution on some nodes (islands) and injecting their highest-performance individuals into islands of higher resolution for "fine-tuning". This injection occurs while all islands continue to search simultaneously, although it is also possible to stop or re-assign low-resolution islands once they have converged. The parallel GA environment in which the iiGA is run is based on the GALOPPS toolkit [26] developed by one of the authors. The software can be run on one or multiple workstations [26] (a single processor was used for all runs reported here). Figure 3a shows the iiGA used for this paper to perform multiple refinements in the geometric representation by increasing the number of rings in the flywheels. For composite analysis, material properties can vary from ring to ring of the flywheel. Islands with different levels of resolution evaluate fitness using either a simplified analysis that is computationally cheaper or a refined, computationally expensive analysis. Different GA parameters can be used for each population. The rates of crossover, mutation, and island interaction can all vary from island to island. For example, an island can exploit a simplified evaluation tool that is computationally cheap by increasing the island's population size. Also, islands using a computationally cheap evaluation function can be allowed to evaluate more generations before injecting their results into other islands.

Many engineering problems require satisfying multiple fitness criteria in some sort of weighted overall fitness function to find an optimal design, if not actually requiring multicriterion optimization. Each individual fitness measure may have its own optimal or suboptimal solutions. In an iiGA, it may be useful to use each individual criterion as the fitness function for some subpopulations, allowing them to seek "good" designs with respect to each individual criterion, as potential building blocks for the more difficult weighted fitness function, or as useful points for assessment of Pareto optimality. iiGA's take advantage of the low communication required to migrate individuals from island to island. Often, only the best individual in a population migrates to allow "good" ideas (building blocks) to be combined with other "good" ideas to find "better" ideas amongst islands using different "sub-fitness" functions. Finally, for weighted fitness evaluation, individuals may be injected into a set of nodes where the evaluation of an overall weighted fitness function is employed. This search method facilitates robust exploration of the search space for all aspects of the overall fitness. Of course, many variations on these injection island architectures can be custom tailored for specific problems.

iiGA's using islands of different resolutions have the following advantages over other PGA's:

(i) Building blocks of lower resolution can be directly found by search at that resolution. After receiving lower resolution solutions from its parent island(s), an island of higher resolution can "fine-tune" these solutions, but may also reject those inferior to better solution regions already located.

(ii) The search space in islands with lower resolution is proportionally smaller. This typically results in finding "fit" solutions more quickly, which are injected into higher resolution islands for refinement.

(iii) Islands connected in the hierarchy (islands with a parent-child relationship) share portions of the same search space, since the search space of the parent is typically contained in the search space of the child. Fast search at low resolution by the parent can potentially help the child find fitter individuals.

(iv) iiGA's embody a divide-and-conquer and partitioning strategy which has been successfully applied to many problems. In iiGA's, the search space is usually fundamentally divided into hierarchical levels with well-defined overlap (the search space of the parent is contained in the search space of the child).

(v) In iiGA's, nodes with smaller block size can find the solutions with higher resolution. Although Dynamic Parameter Encoding (DPE) [27] and ARGOT [28] also deal with the resolution problem, using a zoom or inverse zoom operator, they are different from iiGA's. First, they are working at the phenotype level and only for real-valued parameters. iiGA's typically divide the string into small blocks regardless of the meaning of each bit. Second, it is difficult to establish a well-founded, general trigger criterion for zoom or inverse zoom operators in PDE and ARGOT. Furthermore, the sampling error can fool them into prematurely converging on suboptimal regions. Unlike PDE and ARGOT, iiGA's search different resolution levels in parallel and may reduce the risk of zooming into the wrong target interval, although there remains, of course, a risk that search will prematurely converge on a suboptimal region.

2.0 Finite Element Models of Flywheels

Two axisymmetric finite element models were developed to predict planar and three-dimensional stresses that occur in flywheels composed of orthotropic materials undergoing a constant angular velocity. Both finite element models were developed applying the principle of minimum potential energy. The finite element model that assumes a plane stress state is truly a one-dimensional finite element model, and is accurate when the gradient of the flywheel thickness is small. The finite element model that yields a three-dimensional stress state is truly a two-dimensional finite element model, and is accurate for all shapes. An automated mesh generator was written to allow for mesh refinement through the transverse normal and the radial directions. Therefore, the finite element code that predicts three-dimensional stresses can have various levels of refinement. A coarse mesh with a small number of degrees of freedom will be less accurate but more efficient than a refined mesh that contains more degrees of freedom. The mesh was also generated to minimize the time required to solve the set of linear equations created by the finite element code. By first assuming an initial angular velocity, the stresses and strains were calculated. Next, the initial angular velocity was scaled to the maximum failure angular velocity. The maximum stress failure criterion was used to predict the maximum failure angular velocity in the analysis of isotropic flywheels, while the maximum strain criterion was used for composite flywheels.

3.0 Global Optima for a Simplified Flywheel

In order to explore how effective the iiGA search is in finding the global optimum for this sort of problem, and to compare the speed of finding it using iiGA's with various enhancements, a simplified flywheel problem was posed. A flywheel that contains 6 concentric rings (i.e., 7 heights) with 8 possible values for each height (see Figure 3b) created a design space of 8^7 or about 2 million possible designs. Using a coarse (962 DOF), three-dimensional finite element model, it was possible to calculate the SED of all of these designs, in about 50 hours on a SPARC Ultra processor. With the global optimum design known from exhaustive search, other search methods could be judged as to robustness and efficiency.

The TA algorithm alone began its search with a randomly initiated design. All hybrid algorithms that incorporated the TA algorithm were initiated with the best individual of the current generation, performing at most 10 TA operations, with the resulting solution always replacing the worst in the population. The local search method took the best individual of each generation and varied the thickness profile of whichever ring the FEA code found to fail first. The inner and outer thicknesses were increased and decreased independently, so a total of four evaluations occurred. When incorporating the local search method in any algorithm, the worst solution in the population was replaced only when a better solution was found by the local search. All multipoint search methods used the same total population size, 2,200 individuals. All GA runs used elitism (guaranteed survival of best individual) and one-point crossover. Typically, for larger, computationally expensive problems, each island would be located on a separate processor, but for this problem, only a single Sun Sparc Ultra workstation was used. All runs used a 1% mutation rate.

The motivation for the particular iiGA topology used here requires some explanation. The search space for the plane stress finite element model evaluation contains good building blocks for the iiGA. Also, the plane stress evaluation (0.001seconds per evaluation) is up to 1000 times faster than the three-dimensional evaluation of stress (for this analysis). To make the iiGA search less computationally intensive and more robust, the iiGA shown in Figure 3a was designed to exploit these facts. A full cycle in an iiGA consists of evaluating a specified number of generations (which varies from island to island) in each island. Genetic operations can also be varied from island to island. Islands 0 through 1 had a 75% rate of crossover, population size of 300, and completed 12 generations per cycle before migrating 3 individuals in accordance with Figure 3a. Islands 0 and 1 measured fitness with plane stress finite element code, basing fitness on the sub-fitness function (angular velocity alone). Islands 0 and 1 contained designs with 3 and 6 rings with 7 and 13 DOF, respectively. The choice of a high crossover rate was chosen to motivate those particular islands to discover new designs. A large population size and high number of generations per cycle was used due to the computational efficiency of the plane stress evaluation and to force the islands to converge quickly to potentially productive regions of the design space, presumably containing useful building blocks. Islands 2 and 3 had a crossover rate of 70%, population size of 200, and completed 6 generations per cycle before migrating 3 individuals, evaluating fitness with the three-dimensional axisymmetric finite element code basing fitness on SED (130 DOF). Islands 4 and 5 had a 65% crossover rate, population size of 200 and completed 4 generations before migrating individuals, measuring fitness with the three-dimensional axisymmetric finite element code basing fitness on SED (430 DOF). Islands 6 through 8 had a crossover rate of 60%, population size of 100, and received migrated individuals every 2 generations, measuring fitness with the three-dimensional axisymmetric finite element code basing fitness on SED (962 DOF). Islands 6 through 8 have a lower population size and number of generations per cycle to explore the space more slowly and to avoid a large number of costly evaluations. Islands 6 through 8 should fine tune potentially good designs (building blocks) received from the islands at a lower resolution. Figure 3a also displays a hybrid iiGA design that groups the islands according to the method by which they perform their specialized heuristic search (if any) at the end of each generation. Of course, many variations on these hybrid iiGA designs can be custom tailored for specific problems.

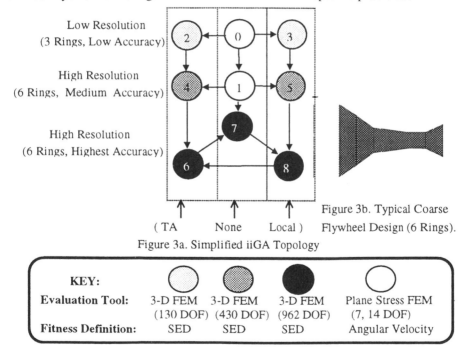

Figure 3b. Typical Coarse Flywheel Design (6 Rings).

Figure 3a. Simplified iiGA Topology

Figure 3. Simplified Injection Island GA Topology with Coarse Flywheel Representation.

4.0 Results of Global Optimization Study

Table 1 shows the results of the various methods. Each run lasted 6000 seconds on the same processor. In five runs of each method, the simple GA, with and without TA and local search heuristics, and the ring topology parallel GA, never found the global optimum. Figure 4 displays the fitness (reevaluated with the most accurate FEM evaluation) as a function of time of a typical run for a TA algorithm, simple GA and a simple GA that incorporated either a TA algorithm or a local search method. Elitism was used in all GA runs, so solutions are only plotted when better solutions are found (leading to the appearance of different run lengths).

Other hybrid iiGA topologies were tested that incorporated either Threshold Accepting or local search methods. Without the local search or TA heuristics, the iiGA took an average of 768 seconds to find the global optimum. The hybrid iiGA that also used local search found the global optimum in 715 seconds (average) while the iiGA that incorporated the TA found the global solution in 674 seconds (average). Figures 5 and 7 display the fitness as a function of time for the iiGA (same topology as Figure 3a) and hybrid iiGA (Figure 3a, TA/None/local), respectively. The iiGA alone found the global solution in 768 seconds (average), while the hybrid iiGA (Figure 3a, TA/None/Local) found the global optimum in 417 seconds (average). The hybrid iiGA that used the TA algorithm and local search method evaluated less than 5% of the entire search space, taking less than 0.5% of the time needed to enumerate the entire search space, measuring more than half of the evaluations with the plane stress finite element model to find the global optimum. Examination of Figure 4, shows that the local search and the TA help the simple GA find better solutions. Also, the TA alone quickly climbs to a suboptimal solution. Figure 5 shows the iiGA quickly finding "building blocks" at low levels of resolution that are injected into islands of higher resolution. Figure 6 displays the hybrid iiGA (Figure 3a, TA/none/Local) benefiting from the combination of TA and local search heuristics. Figures 4-6 only display the first 1000 seconds because no better solutions were ever found thereafter.

Table 1. Comparison of Optimization Approaches.

Optimization Technique	Average Time to Find Global Solution (5 Runs)
TA	Never Found
Simple GA	Never Found
Simple GA with Local Search	Never Found
Simple GA with TA	Never Found
Ring Topology GA	Never Found
iiGA	Always Found, 768 Seconds
Hybrid iiGA with Local Search	Always Found, 715 Seconds
Hybrid iiGA with TA	Always Found, 674 Seconds
Hybrid iiGA with Local Search and TA	Always Found, 417 Seconds

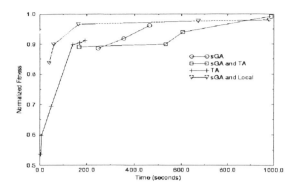

Figure 4. Fitness as a Function of Time on a Single Processor for a Typical Run of a Simple GA, GA with TA, and Simple GA with Local Search Method.

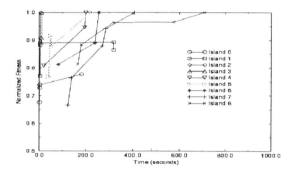

Figure 5. Fitness as Function of Time on a Single Processor for Typical iiGA Run.

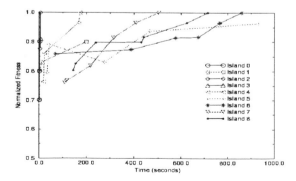

Figure 6. Fitness as a Function of Time on a Single Processor for a Typical Hybrid iiGA that Incorporated TA and Local Search Methods.

5.0 Searching Larger Design Spaces using GA's and PGA's

A larger search domain was created to compare the iiGA to a topological "ring" GA. A 24-ring flywheel with 1024 heights per thickness created a huge design space. Again, to make the GA search less computationally intensive and more robust, an iiGA as shown in Figure 7a was designed. Islands 0 through 2 evaluate fitness based on angular velocity with a simplified plane stress finite element model with varying geometric resolutions (3, 6 and 12 rings). Islands 0 through 2 have 7, 13 and 25 computational degrees of freedom, respectively. Islands 3 through 11 measure fitness based on SED using the three-dimensional axisymmetric finite element model. Islands 3 and 4 are low in geometric resolution (3 rings), but have 160 degrees of freedom. Islands 5 and 6 are medium in geometric resolution (6 rings), containing 558 degrees of freedom. Islands 7 and 8 are high in geometric resolution (12 rings), having 1,952 degrees of freedom. Islands 9 through 11 are the highest in geometric resolution (24 rings) with 9,982 degrees of freedom.

A full cycle consists of evaluating a specified number of generations (which varies from island to island) in the injection island topology. Islands 0 through 2 had a 75% rate of crossover, population size of 300, and completed 15 generations per cycle before migrating the island's best individual in accordance with Figure 7a. Islands 3 and 4 had a crossover rate of 70%, population size of 100, and completed 7 generations per cycle before migrating 3 individuals. Islands 5 and 6 had a 65% crossover rate, population size of 130 and completed 4 generations before migrating individuals. Islands 7 and 8 had a crossover rate of 65%, population size of 130 and migrated individuals after evaluating 3 generations. Islands 9 through 11 had a crossover rate of 60%, population size of 100 and received migrated individuals every 3 generations. Islands 0 through 2 can converge much faster to "good" building blocks when compared to the rest of the islands due to the simplification of the plane stress evaluation and the level of resolution. The iiGA topology design in Figure 7a uses this as an advantage because the topology injects building blocks from the simplified plane stress evaluation based on angular velocity into two isolated islands that evolve independently, searching separate spaces efficiently using the axisymmetric three-dimensional finite element model to evaluate SED.

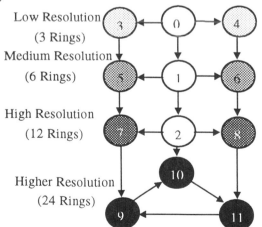

Figure 7a. Injection Island Topology

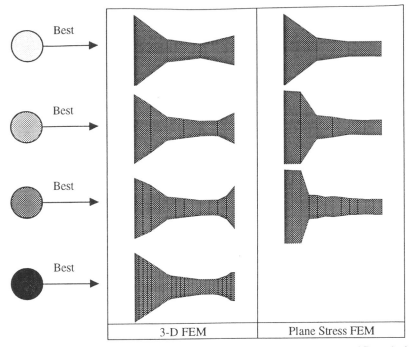

Figure 7b. Best Flywheel Found for Each Level of Resolution

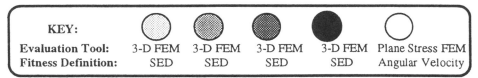

Figure 7. Injection Island GA Topology with Annular Composite Flywheel Results.

A "ring" PGA topology was also considered. This topology shown in Figure 8a, contains 20 islands with the same total number of individuals (divided equally among the islands) as the iiGA topology shown in Figure 7a. The flywheel geometric resolution was 24 rings, containing 9,982 computational degrees of freedom. In all studies, the composite flywheel material was E-glass.

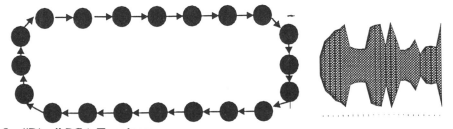

8a. "Ring" PGA Topology.

8b. Best Ever Flywheel from "Ring" PGA within the time used by iiGA.

Figure 8. 20-Ring Island GA Topology with Annular Composite Flywheel Results.

6.0 Results for Composite Annular Flywheels

Figure 7b displays the "best ever" annular composite flywheel at all the levels of geometric resolution. Also, Figure 7b compares the three-dimensional to the plane stress axisymmetric results. The plane stress results based on angular velocity are exaggerated shapes that are *artifacts* of the analysis. However, the plane stress results cannot be dismissed because they are the building blocks that helped to form the final "finely tuned" flywheels rapidly. Figure 8b displays the best ever flywheel found in the "ring" topology, which ran for the same amount of time (3 hours). Clearly the iiGA is more efficient in its search technique than the "ring" PGA. Also, the iiGA annular composite flywheel has a 35% increase in SED when compared to the PGA results.

7.0 Discussion and Conclusions

The iiGA offers some new tools for approaching difficult optimization problems. For many problems, the iiGA can be used to break down a complex fitness function into "sub-fitness" functions, which represent "good" aspects of the overall fitness. The iiGA can build solutions in a sequence of increasingly refined representations, spatially or according to some other metric. The iiGA can also use differing evaluation tools, even with the same representation. A simplified analysis tool can be used to quickly search for good building blocks. This, in combination with searching at various levels of resolution, makes the iiGA efficient and robust. Mimicking a smart engineer, the iiGA can first quickly evaluate the overall response of a structure with a coarse representation of the design and finish the job off by slowly increasing the levels of refinement until a "finely tuned" structure has been evolved. This approach allows the iiGA to decrease computational time and increase robustness in comparison with a typical GA, or even a typical parallel GA. This was illustrated when the iiGA found a flywheel with a 35% increase in SED when compared to the best flywheel found in the same amount of time by the "ring" topology parallel GA. It was demonstrated much more strongly with the results for the simpler problem with the known global optimum, in which all variants of iiGA found the solution unerringly and rapidly, and all variants of the SGA with local search and threshold accepting heuristics, and the parallel ring GA, never found the solution. Of course, finding the global optimum for a problem with a reduced search space does not guarantee that the iiGA will find the global optimum for more complex cases, but it at least lends plausibility to the idea that the iiGA methods are helpful in searching such spaces relatively efficiently for near-optimal solutions. In many engineering domains in which each design evaluation may take many minutes (or hours), the availability of such a method, parallelizable with minimal communication workload, could make good solutions attainable for problems not previously addressable.

References:

1. C. Soto and R. Diaz, 1993, "Optimum Layout and Shape of Plate Structures using Homogenization," *Topology Design of Structures*, pp. 407-420.
2. K. Suzuki and N. Kikuchi, 1990, "Shape and Topology Optimization by a Homogenization Method," *Sensitivity Analysis and Optimization with Numerical Methods*, 115:15-30.

3. K. Suzuki and N. Kikuchi, 1991, "A Homogenization Method for Shape and Topology Optimization," *Computational Methods in Applied Mechanics in Engineering*, 93:291-318.
4. E. Sangren, E. Jensen and J. Welton, 1990, "Topological Design of Structural Components using Genetic Optimization Methods," *Sensitivity Analysis and Optimization with Numerical Methods,* 115:31-43.
5. P. Hajela and E. Lee, 1997, "Topological Optimization of Rotorcraft Subfloor Structures for Crashworthiness Considerations," *Computers and Structures*, 64: 65-76.
6. C. Mares and C. Surace, 1996, "An Application of Genetic Algorithms to Identify Damage in Elastic Structures" *Journal of Sound and Vibration*, 195:195-215.
7. C. Chapman and M. Jakiela, 1996, "Genetic Algorithm-Based Structural Topology design with Compliance and Topology Simplification Considerations," *Journal of Mechanical Design*, 118:89-98.
8. S. Rajan, 1995, "Sizing, Shape and Topology Design Optimization of Trusses using Genetic Algorithms," *Journal of Structural Engineering*, 121:1480-1487.
9. A. Keane, 1995, "Passive Vibration Control via Unusual Geometeries: The Application of Genetic Algorithm Optimization to Structural Design," *Journal of Sound and Vibration*, 185: 441-453.
10. Y. Nakaishi and S. Nakagiri, 1996, "Optimization of Frame Topology Using Boundary Cycle and Genetic Algorithm," *JSME International Journal*, 39: 279-285.
11. N. Queipo, R. Devarakonda and J. Humphery, 1994, "Genetic Algorithms for Theromosciences research: Application to the Optimized Cooling of Electronic Components," *International Journal of Heat and Mass Transfer*, 37:893-908.
12. R. Flynn and P. Sherman, 1995, "Multi-Criteria Optimization of Aircraft Panels: Determining Viable Genetic Algorithm Configurations," *International J of Intelligent Systems*, 10: 987-999.
13. H. Furuya and R. Haftka, 1995, "Placing Actuators on Space Structures by Genetic Algorithms and Effective Indicies," *Structural Optimization*, 9: 69-75.
14. W. F. Punch III, R.C. Averill, E.D. Goodman, S. C. Lin, and Y. Ding, February 1995,"Design Using Genetic Algorithms - Some Results for Laminated Composite Structures," *IEEE Expert*,10:42-49.
15. W. F. Punch III, R.C. Averill, E.D. Goodman, S. C. Lin, Y. Ding, and Y.C. Yip, 1994, "Optimal Design of Laminated Composite Structures using Coarse-Grain Parallel Genetic Algorithms," *Computing Systems in Eng*, 5:414-423.
16. N. Kosigo, L. T. Watson, Z. Gural, and R. T. Haftka, 1993, "Genetic Algorithms with Local Improvement for Composite Laminate Design," *Structures & Controls Optimization*, ASME, Aerosp. Div., 38:13-28.
17. T. Le Riche and R. T. Haftka, 1993, "Optimization of Laminate Stacking Sequence for Buckling Load Maximization by Genetic Algorithm," *AIAA J* 31:951-956.
18. A. Todoroki, K. Watanabe and H. Kobayashi, 1995, " Application of Genetic Algorithms to Stiffness Optimization of Laminated Composite Plates with Stress-Concentrated Open Holes," *JSME Internat. Journal*, 38(4), 458-464.

19. I. Parmee and H. Vekeria, 1997, "Co-Operative Evolutionary Strategies for Single Component Design," *Proceedings of the Seventh International Conference on Genetic Algorithms*, Michigan State University, pp. 529-536.

20. G. Fabbri, 1997, "A Genetic Algorithm for Fin Profile Optimization," *International Journal of Heat and Mass Transfer*, 40:2165-2172.

21. N. Foster and G. Dulikravich, 1997, "Three-Dimensional Aerodynamic Shape Optimization Using Genetic and Gradient Search Algorithms," *Journal of Spacecraft and Rockets*, 34:pp. 36-42.

22. J. Haslinger and D. Jedelsky, 1996, "Genetic Algorithms and Fictitious Domain Based Approached in Shape Optimization," *Structural Optimization*, 12:257-264.

23. J. Wolfersdorf, E. Achermann and B. Weigand, May 1997, "Shape Optimization of Cooling Channels Using Genetic Algorithms," *Journal of Heat Transfer*, 119: 380-388.

24. G. Genta and D. Bassani, 1995, "Use of Genetic Algorithms for the Design of Rotors," *Meccanica*, 30: 707-717.

25. R. Ruthenbar, 1989, "Simulated Annealing Algorithms: An Overview," *IEEE Circuits and Devices Magazine*, 5: 19-26.

26. E. D. Goodman, 1996, "GALOPPS, The Genetic Algorithm Optimized for Portability and Parallelism System, User's Guide," Technical Report, Genetic Algorithms Research and Applications Group (GARAGe), Michigan State University, East Lansing, July, 1996, pp. 100.

27. E. D. Goodman, R. C. Averill, W. F. Punch III and D. J. Eby, 1997, "Parallel Genetic Algorithms in the Optimization of Composite Structures," WSC2 Proceedings , World Wide Web Conference, http://garage.cps.msu.edu/papers/GARAGe97-05-02.ps.

28. N. N. Schraudolph and R. K. Belew, "Dynamic Parameter Encoding for Genetic Algorithms," Technical Report LAUR 90-2795 (revised), Los Alamos National Laboratories, 1991.

Evolutionary Mesh Numbering: Preliminary Results

Francis Sourd and Marc Schoenauer
CMAP – URA CNRS 756
Ecole Polytechnique
Palaiseau 91128, France
marc.schoenauer@polytechnique.fr

No Institute Given

Abstract. Mesh numbering is a critical issue in Finite Element Methods, as the computational cost of one analysis is highly dependent on the order of the nodes of the mesh. This paper presents some preliminary investigations on the problem of mesh numbering using Evolutionary Algorithms. Three conclusions can be drawn from these experiments. First, the results of the up-to-date method used in all FEM softwares (Gibb's method) can be consistently improved; second, none of the crossover operators tried so far (either general or problem specific) proved useful; third, though the general tendency in Evolutionary Computation seems to be the hybridization with other methods (deterministic or heuristic), none of the presented attempt did encounter any success yet. The good news, however, is that this algorithm allows an improvement over the standard heuristic method between 12% and 20% for both the 1545 and 5453-nodes meshes used as test-bed. Finally, some strange interaction between the selection scheme and the use of problem specific mutation operator was observed, which appeals for further investigation.

1 Introduction

Most Design Problems in engineering make an intensive use of numerical simulations of some physical process in order to predict the actual behavior of the target part. When the mathematical model supporting the numerical simulation involves Partial Differential Equations, Finite Element Methods (FEM) are today one of the most widely used method by engineers to actually obtain an approximate numerical solution to their theoretical model. However, the computational cost of up-to-date numerical methods used in FEM directly depends on the way the nodes of the underlying mesh (i.e. the discretization) are numbered. Solving the Mesh Numbering Problem (MNP) amounts to find the permutation of the order of the nodes that minimizes that computational cost.

Numerical engineers have developed a powerful heuristic technique (the so-called *Gibb's method*) that gives reasonably good results, thus providing a clear reference to any a new mesh numbering algorithm.

The goal of this paper is to use Evolutionary Algorithms (EAs) to tackle the MNP. EAs have been widely applied to other combinatorial optimization problems, among which the popular TSP [10, 4]. However, as far as we know, this is the first attempt to solve the MNP using EAs. Unfortunately, though both problem look for a solution in the space of permutations of $[0, n]$, the specificity of the MNP might make inefficient the simple transposition of TSP techniques to the MNP. Indeed, looking at the history of Evolutionary TSP, it seems clear that the key of success is hybridization with problem-specific techniques: from the Grefenstette's early incorporation of domain knowledge [10] to the most recent works [4, 12, 6] where evolutionary results can – at last – be compared to the best Operational Research results, even for large size TSP instances. So the path to follow here seems rather clear: design some NMP-specific operators, and compare their performances to either "blind" problem independent operators or TSP-specific operators.

The paper is organized as follows: Section 2 recalls the basics of Finite Element Methods, and precisely defines the objective function. The state-of-the-art "Gibbs method' is also briefly described. Section 3 presents the design of the particular Evolutionary Algorithm used thereafter, discussing in turn the representation issue, specific crossover and mutation operators, and the initialization procedure. This algorithm is then experimented in section 4, with emphasis on the tuning of the probabilities of application of all operators at hand (both problem-specific and problem independent). First, the crossover operators all rapidly appear harmful. Second, surprising interactions between the selection scheme and the different mutation operators seem to indicate that, though at the moment domain knowledge did not increase the performances of the algorithm, there is still some room for improvement in that direction. Finally, the usefulness of evolutionary mesh numbering with respect to Gibbs' method is discussed: the results are indeed better, but the cost is also several orders of magnitude greater.

2   Mesh Numbering

2.1   Theoretical Background

Many models for physical, mechanical, chemical and biological phenomenon end up in Partial Differential Equations (PDE) where the unknown are functions defined on some domain Ω of $\mathbb{R}^n$.

A popular way to transform a system of PDEs into a finite linear system is the Finite Element Method (FEM) [18, 3]. The domain Ω is discretized into small *elements*, who build up a *mesh*. A typical mesh – on a non-typical domain – is given in Figure 1. The solution of the initial PDEs is sought in spaces of functions that are polynomial on each element. Such an approximate solution is completely determined by its values at some points of each element, called the *nodes* of the mesh. Those values are computed by writing the original PDEs locally on each element, resulting in a linear system of size the number of nodes times the number of *degrees of freedom*, or unknown values, at each node). Usual sizes for such systems range from a few hundreds (e.g. in two-dimensional structural mechanics) to millions (e.g. for three-dimensional problems in aerodynamics).

However, as a consequence of the local discretization, the equation at each node only involves values at a few neighboring nodes: the resulting matrix is hence very sparse. And specific methods exist for sparse system [9], whose complexity is proportional to the square of the *bandwidth* of the matrix, i.e. the average size of the *profile* of the matrix, given by the sum over all lines of the maximal distance to the diagonal of non-zero terms. For full matrices, the bandwidth is $n(n-1)/2$ while it is n for tridiagonal matrices (for $n \times n$ matrices).

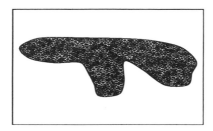

Fig. 1. *Sample mesh with 1545 nodes.*

2.2   Computing the bandwidth

The contribution of each single line to the total bandwidth of the matrix is highly dependent on the order of the nodes of the mesh: the equation for node number i only involves the neighboring nodes; hence the only non-zero terms of the corresponding equation will appear in the matrix in the column equal to the number of the node in the mesh. Depending on the order of the nodes in the mesh, the bandwidth can range from a few units to almost the size of the matrix.

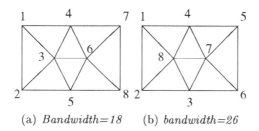

(a) *Bandwidth=18*   (b) *bandwidth=26*

Fig. 2. *A simple example of bandwidth with respect to nodes order.*

A simple example of a mesh is given in Figure 2 for a two-dimensional domain discretized into triangles. The nodes of that mesh are the summits of the triangles (note that, while common, this situation is not the rule, and the middle of the edges, the center of gravity of the triangles, ... can also be nodes). The effect of numbering is demonstrated in Figures 2-a and 2-b, where the same mesh can give a bandwidth of 18 or 26, depending on the order in which the nodes are considered.

The only useful data for mesh numbering is, for each node, the list of all neighbors. For instance, the mesh of Figure 2-a can hence be viewed as
(2 3 4) (1 3 5) (1 2 4 5 6) (1 3 6 7) (2 3 6 8) (3 4 5 7 8) (4 6 8) (5 6 7)
Once an order for the nodes is chosen, the profile of the matrix can be constructed: an upper bound [1] of the fitness function is given by the following equation (1), where N is the number of nodes, and, for each i in $[1, N]$, $\mathcal{N}(i)$ is the set of neighbors of node i:

$$B = \sum_{i=1}^{i=N} \min_{(j>i);(j\in\mathcal{N}(i))} (j - i) \tag{1}$$

Note that due to symmetry, only the contribution to the bandwidth of the upper part of the matrix is considered.

For the simple case of the mesh of figure 2, the bandwidth is $3 + 3 + 3 + 3 + 3 + 2 + 1 = 18$ for the order (a) and $7 + 6 + 5 + 4 + 2 + 1 + 1 = 26$ for the order (b).

The goal of mesh renumbering is to find the order of the nodes (i.e. a permutation on $[1, N]$) that minimizes the bandwidth B.

2.3 State of the art

The problem of mesh numbering is clearly a $NP - complete$ combinatorial problem, as the search space is the space of permutations in $[1, n]$. Hence, no exact deterministic method can be hoped for. Numerical scientists have paid much attention to that problem, developing heuristic methods. The favorite method nowadays, used in most FEM software packages, is the so-called *Gibbs method* presented in detail in [5]). It performs three successive steps, and use the graph G representing the "neighbor to" relationship between nodes.

- Find both ends of the numbering. They are chosen such that, first, their distance (in term of minimal path in G joining them) is maximal, and second, each one has the lowest possible number of neighbors.
- Make a partition of graph G into layers containing nodes at the same distance from the origins.
- Number the nodes breadth-first, starting from one origin (i.e. numbering nodes layer by layer).

---

[1] Some of these candidates to be non-zero can actually be null, depending on the actual formulation of the PDEs. But this rare eventuality will not be considered in the general framework developed here.

As Gibbs method is used in all FEM packages, the minimum requirement for any other algorithm to be considered interesting by numerical scientists is to obtain better results than those of Gibbs. Hence, in the following, all results of performance will be given relatively to those of Gibbs.

3 The evolutionary algorithm

3.1 MNP is not a TSP

Combinatorial optimization is a domain where evolutionary algorithms have encountered some successes, when compared to state-of-the-art heuristic methods. Probably the most studied combinatorial optimization problem is the Traveling Salesman Problem (TSP). In both cases (MNP and TSP), the search space is that of all permutations of N elements.

However, a first obvious difference is that both the starting point and the "direction" of numbering are discriminant in the MNP while they are not in the TSP. As a consequence, the size of the search space in the MNP is $n!$ while it is $(n-1)!/2$ for the TSP. On the other hand, no *degeneracy* (see [14]) is present in the *permutation* representation for the MNP (see below).

Second, the MNP is not so easily *decomposable*. A whole part of a solution of the TSP can be modified without modifying the remaining of the tour. In the MNP on the other hand, the propagation of any modification has consequences on all geometrical neighbors of all modified nodes.

As a consequence, the useful notion of "neighbor" is totally different from one problem to another: In the TSP, two towns are usually called neighbors if they are visited one after the other on the tour at hand; In the MNP, the "neighbor of" relationship is absolute geometrical data, independent of any order.

Practically, a major difference between both problems is the absence of any local well known optimization algorithm for the MNP: as quoted in the introduction, the best results obtained so far by evolutionary algorithms on the TSP are due to hybrid "memetic" algorithms searching the space of local optima with respect to a local optimization procedures (e.g. the 2-opt or 4-opt procedures in [4, 12, 6]). Such strategy cannot be reproduced for the MNP.

3.2 Representation

One important consequence of these differences is that the concepts of "edges" or "corners" [14], known to be of utter importance for the TSP, do not play any role in the MNP.

Two representations will be experimented with in that paper:

- In the *permutation* representation, a permutation is represented by the sequential list of the numbers of all nodes. Note that this representation relies on a predefined order of the nodes (corresponding to the identity permutation $(1, 2, 3, \ldots, N)$). Some consequences are discussed in section 3.5.

— All permutations can be decomposed in a sequence of transpositions. More-over, all sequences of transpositions make a unique valid permutation. The *transposition* representation describes a permutation as an ordered list of transpositions. Note that this representation is highly degenerate: many genotypes correspond to the same phenotype.

3.3 Crossover operators

Four crossover operators have been tested, from general-purpose operators to MNP-specific crossovers.

— The transposition crossover is a straightforward crossover for the transposi-tion representation: It exchanges portions of the transposition lists of both parents. As any combination of transpositions make a valid permutation, it directly generates valid permutations.

— The *edge crossover* is a general-purpose operator for permutation problems, designed and tested on the TSP problem [17]. No specific knowledge about the problem is used, but the underlying assumption is that edges are the important features in the permutations. All experiments using the edge crossover gave lousy results, thus confirming that edges do not play for the MNP the important role they have in the TSP. That crossover will not be mentioned any more here.

— The *breadth-first* crossover is based on the heuristic technique described in section 2.3, but uses additional information from both parents to generate one offspring.

A starting point is randomly chosen, is given the number it has in parent A, and becomes the current node. All neighbors of the current node are numbered in turn, being awarded an unoccupied number, before being put in a FIFO stack. To number the neighbor N of a node M already numbered i_M, the differences Δ_A and Δ_B of the numbers of nodes M and N respectively in parent A and parent B are computed. If the number $i_M + min(\Delta_A, \Delta_B)$ is free, it is given to node N. Otherwise, if the number $i_M + Max(\Delta_A, \Delta_B)$ is free, it is given to node N. Otherwise, the number closest to i_M which is not yet used is given to node N.

All nodes are processed once, and are given a yet-unattributed number, generating a valid permutation. This crossover tries as much as possible to reproduce around each node the best local numbering among those of both parents.

— The *difference* crossover was designed to both preserve the diversity in the population and try to locally minimize the fitness function. From both per-mutation representations of the parents, all nodes having the same number are given that common number in the offspring permutation. Further, all remaining nodes in turn (in a random order) are given the number which is the closest possible from all its neighbors numbers.

3.4 Mutation operators

Here again, both general-purpose and problem specific mutation operators were tested.

- The minimal mutation for the permutation representation is the exchange of numbers between two randomly chosen nodes (i.e. the application of a transposition), thereafter termed *transposition mutation*. Its strength can be adjusted by repeated application during the same mutation operation.
- The *neighbor transposition mutation* is a slight modification of the above operator: after the first node has been chosen randomly, it is exchanged with one of its neighbors.
- The *neighbor permutation mutation* is a further step in the direction of using neighbor information to perform mutation: a node is randomly chosen, and a random permutation among the numbers of all its neighbors is performed.
- The *inversion mutation* reverts some part of the permutation. A special case of the inversion mutation is when the whole permutation is inverted. This operator was found useful in Gibbs method (section 2.3).
- The choice of the origin of the numbering is important in the MNP. The *origin mutation* was designed to handle this issue. An integer i is randomly chosen, and all numbers j are replaced by either $i + j$ mod n either $n - (i + j)$ mod n, on a local minimization argument (the brute translation only gives birth to usually very bad numbering when the old origins met).

3.5 Initialization procedure

The standard way to initialize a population is to perform a uniform sampling of the genotype space. However, alternative specific ways have been proposed: for instance, the greedy heuristic for the TSP (from a random initial town, chose the nearest town not yet visited) constructs individuals with a tour length of about 20% more than the optimal value on average [4]. In the same line, three different initialization procedures have been tested for the MNP.

- The standard uniform sampling of the standard representation (section 3.2) was the first obvious choice. It is termed *random initialization*.
- As the state-of-the-art solution is given by the output of the Gibbs method (section 2.3), the *Gibbs initialization* performs only slight perturbations of the Gibbs solution to generate the initial permutations. This is achieved using the transposition representation (section 3.2) taking the Gibbs order as reference, and allowing only a small number of transpositions. Note that in this case, the original Gibbs result is included as the first individual of the population.
- The trouble with the above Gibbs initialization is a very strong bias toward solution quite similar to Gibbs. If higher optima are located in very different regions of the permutation space, they will probably not be found. To address that issue, the *point initialization* was designed, based on some breadth-first heuristic similar to Gibbs', but using a random starting point (Gibbs process is very sensitive to the choice of the initial point).

The average bandwidth of permutation drawn using the uniform initialization is of course quite large (more than 6 times that of Gibbs method). The point initialization gives much better individuals: their bandwidth range from 20% to 100% above Gibbs results while Gibbs initialization stays between 0% and 15% above Gibbs order.

4 Experimental Results

4.1 The meshes

Three meshes have been used to test the algorithm described above: the *small* mesh has 164 nodes, the *medium* one has 1544 nodes (see Figure 1) and the *large* mesh has 5453 nodes. The computational cost of one fitness function evaluation increases linearly with the size of the mesh. Hence, most initial experiments were performed in the *small* mesh. The validation of clear tendencies was then carried on for confirmation on the *medium* mesh. The best combination of parameters were finally tested on the *large* mesh, as its size is becoming to be of some interest for real world application.

4.2 Experimental settings

Two basic evolution schemes were used: a standard GA, with linear ranking and elitist generational replacement; a $(\mu+\lambda)$-ES scheme, in which all μ parents give birth to λ offspring, the best μ of the $\mu + \lambda$ parents + offspring becoming the parents of the next generation. The first series of experiments were performed on the mesh, using a population sizes of 50 for the GA scheme, and a (7+50)-ES. Both schemes were tested with the same combinations of P_c-P_m, (crossover rate - mutation rate). If an individual undergoes crossover, a mate is selected and only one offspring is generated. One single type of crossover was made possible at each run. The resulting offspring (or the initial individual if no crossover occurred) then undergoes mutation with probability P_m. If mutation happens, one mutation operator is chosen according to user-defined weights, and, in the case of transposition mutations, one single transposition is performed.

4.3 First tuning on the small mesh

As said above, intensive experiments were performed on the small mesh. During those experiments, all values for P_c and P_m between 0 and 1 by 0.1 steps were tried independently, for all possible crossover operators. Those runs were allowed 150000 fitness evaluations (unless otherwise mentioned).

Preliminary runs were performed to tune the mutation weights ([16]). The weights for the 5 mutations were first set equal. Then a close look at the types of mutations that the best individual in the population was submitted to, along different runs, allowed to eliminate both the *inversion mutation* and the *origin mutation.* Only the *random transposition mutation,* the *neighbor transposition mutation* and the *neighbor permutation mutation* proved useful, and their weights *PmRand, PmNeighbor* and *PmAround* were set to values between 0 and 1, their sum being equal to 1.

Initialization procedures As could be predicted, the best on-line results were obtained for the *Gibb's initialization,* as its starting point was rather better than both other. But whereas the *point initialization* almost caught up (as will be seen in forthcoming section 4.4), the *random initialization* stayed far beyond, even when allowed ten times the number of fitness evaluations. So only the *Gibb's* and the *point* initializations will be considered in the following.

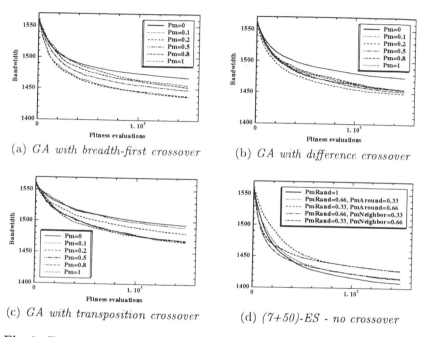

(a) *GA with breadth-first crossover*   (b) *GA with difference crossover*

(c) *GA with transposition crossover*   (d) *(7+50)-ES - no crossover*

Fig. 3. *Typical on-line results (averages over 21 runs) on the small mesh: for the three GA runs, population size is 50, P_c is 0.6 and P_m is as indicated; for the (7+50)-ES runs, $P_m = 1$; in all cases, whenever an individual is mutated, PmRand=0.66 and PmNeighbor=0.33.*

Crossover operators and evolution schemes The first experiments aimed at comparing the crossover operators, and adjusting the crossover rate P_c. A common feature could be observed for all three crossovers: when using the GA scheme, and except for high values of P_m, for which it made no significant difference, the results decreased when P_c decreased from 1 to 0.6 or 0.5. Moreover, and almost independently of the settings of the mutation weights, the best results were obtained with the (7+50)-ES scheme, with $P_c = 0$ (higher values of

P_c for the ES scheme performed rather poorly - but this might be because of the rather small population size of 7).

When it comes to compare the crossover operators, the results of Figure 3 (a), (b) and (c), are what could be expected: the "blind" *transposition crossover* (c) performs rather poorly – and gets its best results for the highest values of P_m which seems to indicate that it is really not helping much. On the opposite, both other operators, that do incorporate some domain knowledge, get their highest performances for $P_m = 0.2$ and $P_m = 0.1$. Moreover, the *breadth-first crossover* (a) performs better than the *difference crossover* (b) (and the difference is statistically significant with 99% confidence T-test after 300000 fitness computations).

A last argument favoring the abandon of crossover operators is the extra cost they require, as based on local optimization heuristics of complexity $o(N)$. For instance the total CPU time is increased by a factor around 4 between runs with $P_c = 0$ and $P_c = 1$ (from 3 to 13mn on average for 300000 evaluation runs on a Pentium P200).

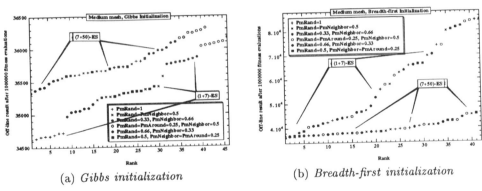

(a) *Gibbs initialization*　　　　　(b) *Breadth-first initialization*

Fig. 4. *Off-line results of both (7+50)-ES and (1+7)-ES after 1000000 fitness evaluations, for different settings of the mutation weights.*

4.4　Mutations and population size

This section presents some results obtained on the *medium* and *large* meshes, using mutation operators only inside some *ES* evolution scheme. At first, the goal of these experiments was to sort out the usefulness of problem-specific knowledge, in the initialization procedure and in the mutations operators. Bur it rapidly turned out that the population size also was a very important parameter.

Figure 4 witnesses the surprising results obtained on the medium mesh: each dot indicates the best fitness reached after 1 million fitness evaluations of a single run of the evolutionary algorithm. The different shapes of the dots represent different settings of the relative mutation rates $PmRand$, $PmNeighbor$ and $PmAround$. Figure 4-a shows the runs that used the Gibbs initialization procedure while Figure 4-b those who used the breadth-first initialization procedure.

On each Figure, the two distinct sets of points correspond to the $(1+7)$- and the $(7+50)$-ES schemes, as indicated. Note that all trials with larger population sizes were unsuccessful.

Some clear conclusions can be drawn from these results. First, the overall best results are obtained for the $(1+7)$-ES scheme starting from the Gibbs initialization and using the "blind" *transposition mutation* only (Figure 4-a) (see section 4.5 for a detailed comparison with the results of Gibbs method). But comparing the results between Gibbs and breadth-first initialization procedures on the one hand, and $(1+7)$- and $(7+50)$-ES schemes on the other hand gave some unexpected results:

- Whereas the $(1+7)$-ES scheme consistently outperformed the the $(7+50)$-ES scheme when using Gibbs initialization, the reverse is true for the breadth-first initialization;
- Whereas different settings of the mutation rates gave rather different results for the Gibbs runs, this does not seem to be so clear for the breadth-first runs;
- whatever the initialization, the $(7+50)$ results are more stable than the $(1+7)$ results with respect to mutation rates; This is striking on the breadth-first plot, but also true on the other plot.
- the worst results are obtained for the lowest values of $PmRand$ (the results with even lower values of $PmRand$ are not presented here, but were very poor). However, domain specific mutation operators (see the black circles and squares, compared to the "+" dots) are more efficient with the $(7+50)$ scheme than with the $(1+7)$ scheme. This is specially visible on the Gibbs runs.

Note that these tendancies were confirmed when the runs were allowed more fitness evaluations (e.g. 10 millions, see forthcoming section 4.5).

Some tentative explanations can however be proposed, after noticing that the $(1+7)$ scheme can be viewed more like a *depth-first* algorithm while the $(7+50)$ scheme searches in a more *breadth-first* manner.

So, using Gibbs initialization probably tights the population in a very limited area of the search space from which it is useless to try to escape: this favors the performance of depth-first search, as breadth oriented search does not have the possibility to jump to other promising regions. Moreover, it seems that the transpositions of neighbor nodes in a depth-first search does not allow large enough moves to easily escape local optima, resulting in the best results for the pure random mutation for the $(1+7)$.

On the other hand, successive local moves have greater chances of survival in the $(7+50)$ breadth search, and give the best results in that case (though some random mutation are still needed). And when it comes to a more widely spread population after the breadth-first initialization, the breadth-first search demonstrates better results by being able to use more efficiently in that case the domain neighboring information.

This situation suggests further directions of research: First, the $(1 + 7)$-ES scheme with pure random mutation resembles some sort of Tabu search [7, 8], and so might be greatly improved by adding some memory to it, either deterministically, like in standard Tabu search, or stochastically, as proposed in [15, 13].

Second, more than one neighbor transposition seems necessary to generate improvements. Hence, the number of transpositions should not be forced to 1, and can be made either self-adaptive, with the same problems than in the case of integer variables in Evolution Strategies [1], or exogenously adaptive, as proposed in [2] where some hyperbolic decreasing law is used for the mutation rate along generations.

4.5 Evolutionary mesh numbering vs Gibbs method

But apart from optimizing the evolutionary algorithm itself on the MNP, a critical point is whether evolutionary mesh numbering can compete with the Gibbs method. Of course, if the Gibbs initialization is used, as the original Gibbs numbering is included in the initial population, any improvement is in fact giving a better result than the Gibbs method. But how interesting is that improvement, especially when the computational cost is taken into account?

On the medium mesh (1545 nodes), the bandwidth using Gibbs method is 39055. As can be seen on Figure 4, the best result of the $(1 + 7)$-ES with pure random mutation after 1 millions evaluations is 34604, i.e. an improvement of 11.4%. The computational cost of one run is 15-20mn (on a Pentium 200Mhz Linux workstation), to be compared to the 20s seconds of Gibbs method!

If the maximum number of function evaluations is set to 10 millions, the best result in the same conditions is 32905 (i.e. 15.75% improvement), with an average over 21 runs of 33152 (15.11%). Of course, in that latter case, the computational cost is 10 times larger . . .

The results on the large mesh (5453 nodes) follow the same lines, though only the combinations of parameters found optimal on the medium mesh were experimented with, due to the computational cost (around 12-14 hours for 10 Millions evaluations).

From a Gibbs bandwidth of 287925 (obtained in about one minute), the best result for the $(1 + 7)$-ES with only random mutation was 257623 (10.52%) while the average of 6 runs was 258113 (10.35%). On the other hand, the best for $(7 + 50)$-ES was 262800 (8.73%), the average being 263963.43 (8,32%), with $PmRand = 0.66$ and $PmVois = 0.33$ (best parameters of Figure 4-a).

The first a priori conclusion is that the computational cost of the evolutionary algorithm makes it useless in practical situations: an improvement of between 10 and 15% requires a computational power of several order of magnitude larger. This quick conclusion must however be moderated: First, due to the quadratic dependency of the computational cost of the matrix inversion in term of the bandwidth, the actual gain in computing time is around 35% for a 15% bandwidth decrease. And second, many meshes used nowadays in industry require a few months of manpower to be built and validated. So 24 more hours of computation for a better numbering is relatively low increase in cost. And if the mesh is then put in an exploitation environment, and is used in several thousands of different Finite Element Analyses, then the overall gain might be in favor of getting a really better numbering, even at what a priori seems a high computational cost. But of course this means that meshes of up to a few thousands nodes can be handled by evolutionary algorithms.

It is nevertheless important to notice that the computation of the bandwidth can be greatly optimized. The present algorithm was designed to be very general, handling any possible operator. Hence it always computes the fitness from scratch. However, in the case where only a small number of transpositions are performed, the variation of the fitness could be computed, by examining only the neighbors of the transposed nodes.

5 Conclusion

We have presented feasibility results for the application of Evolutionary Computation to the problem of mesh numbering. Our best results outperform the state-of-the-art Gibbs method by 10 to 15% on the two test meshes used (with 1545 and 5453 nodes respectively). Whereas these sizes would appear fairly high for TSP problems for instance, they are still small figures with respect to real-world problems, where hundreds of thousands of nodes are frequent.

From the Evolutionary point of view, two issues should be highlighted. First, though both general-purpose and domain-specific crossover operators were tried, none proved efficient. A possible further trial could be to use more global geometrical information (e.g. divide the mesh into some connected components, and exchange the relative orders of such blocks, in the line of [11]).

Second, the overall best results were obtained by a $(1 + 7)$-ES using pure random transposition mutation and starting from an initial population made of slightly perturbed Gibbs meshes. This which might be an indication that other heuristic local search methods (e.g. Tabu search) might be better suited to the MNP. However, as discussed in section 4, some hints make us believe that there is still a large room for improvement using evolutionary ideas: on the one hand, the problem-specific mutations proved useful for the $(7 + 50)$-ES, indicating that we might not have make good usage of the domain knowledge; on the other hand, the $(7 + 50)$-ES (with problem-specific mutation) outperformed the $(1 + 7)$-ES when the initial population was not limited to modified Gibbs meshes: our hope is that starting from totally different parts of the search space could provide much better results in some particular situations ... which still remain to be identified. But in those yet hypothetical cases, the relevance of the evolutionary approach for the MNP would be clear.

References

1. T. Bäck and M. Schütz. Evolution strategies for mixed-integer optimization of optical multilayer systems. In J. R. McDonnell, R. G. Reynolds, and D. B. Fogel, editors, *Proc. of the 4th Annual Conf. on Evolutionary Programming*. MIT Press, March 1995.
2. T. Bäck and M. Schütz. Intelligent mutation rate control in canonical GAs. In Z. W. Ras and M. Michalewicz, editors, *Foundation of Intelligent Systems 9th Intl Symposium, ISMIS '96*, pages 158–167. Springer Verlag, 1996.
3. P. G. Ciarlet. *Mathematical Elasticity, Vol I : Three-Dimensional Elasticity*. North-Holland, Amsterdam, 1978.
4. B. Freisleben and P. Merz. New genetic local search operators for the TSP. In H.-M. Voigt, W. Ebeling, I. Rechenberg, and H.-P. Schwefel, editors, *Proc. of the 4th Conf. on Parallel Problems Solving from Nature*, LNCS 1141, pages 890–899. Springer Verlag, 1996.
5. P.L. George. *Automatic mesh generation, application to Finite Element Methods*. Wiley & Sons, 1991.
6. M. George-Schleuter. Asparagos96 and the travelling salesman problem. In T. Bäck, Z. Michalewicz, and X. Yao, editors, *Proc. of the 4th IEEE Intl Conf. on Evolutionary Computation*, pages 171–174. IEEE Press, 1997.
7. F. Glover. Heuristics for integer programming using surrogate constraints. *Decision Sciences*, 8(1):156–166, 1977.
8. F. Glover and G. Kochenberger. Critical event tabu search for multidimensional knapsack problems. In *Proc. of the Intl Conf. on Metaheuristics for Optimization*, pages 113–133. Kluwer Publishing, 1995.
9. G.H. Golub and C. F. van Loan. *Matrix Computations*. John Hopkins University Press, 1996 - Third edition.
10. J. J. Grefenstette. Incorporating problem specific knowledge in genetic algorithms. In Davis L., editor, *Genetic Algorithms and Simulated Annealing*, pages 42–60. Morgan Kaufmann, 1987.
11. C. Kane and M. Schoenauer. Genetic operators for two-dimensional shape optimization. In J.-M. Alliot, E. Lutton, E. Ronald, M. Schoenauer, and D. Snyers, editors, *Artificial Evolution*, LNCS 1063. Springer Verlag, Septembre 1995.
12. P. Merz and B. Freisleben. Genetic local search for the TSP: New results. In T. Bäck, Z. Michalewicz, and X. Yao, editors, *Proc. of the 4th IEEE Intl Conf. on Evolutionary Computation*, pages 159–164. IEEE Press, 1997.
13. M. Peyral, A. Ducoulombier, C. Ravisé, M. Schoenauer, and M. Sebag. Mimetic evolution. In *Artificial Evolution'97*, Springer Verlag. To appear.
14. N. J. Radcliffe and P. D. Surry. Fitness variance of formae and performance prediction. In L. D. Whitley and M. D. Vose, editors, *Foundations of Genetic Algorithms 3*, pages 51–72. Morgan Kaufmann, 1995.
15. M. Sebag, M. Schoenauer, and C. Ravisé. Toward civilized evolution: Developping inhibitions. In Th. Bäeck, editor, *Proc. of the 7th Intl Conf. on Genetic Algorithms*. Morgan Kaufmann, 1997.
16. F. Sourd. Renumérotation de maillages par algorithmes génétiques, June 1996. Stage d'option de l'Ecole Polytechnique.
17. D. Whitley, T. Starkweather, and D. Fuquay. Scheduling problems and travelling salesman: The genetic edge recombination operator. In J. D. Schaffer, editor, *Proc. of the 3rd Intl Conf. on Genetic Algorithms*. Morgan Kaufmann, 1989.
18. O. C. Zienkiewicz. *The Finite Element Method in Engineering Science*. McGraw-Hill, New-York, 1st edition, 1967, 3rd edition, 1977.

Two New Approaches to Multiobjective Optimisation Using Genetic Algorithms

Carlos A. Coello Coello

Departamento de Sistemas, Universidad Autónoma Metropolitana
email: cacc@hp9000a1.uam.mx

Abstract. In this paper, two new multiobjective optimization techniques based on the genetic algorithm (GA) are introduced. These methods are based in the concept of min-max optimum, and can produce the Pareto set and the best trade-off among the objectives. The results produced by these approaches are compared to those produced with mathematical programming techniques and other GA-based approaches using a multiobjective optimization tool called MOSES. This tool, developed by the author, is a convenient testbed for analyzing the performance of new and existing multicriteria optimization techniques, and it is an effective engineering design tool.

1. Introduction

Much work has been done in engineering optimization during the last few years, but the trend has been to deal with ideal and unrealistic problems, rather than with real-world applications. One of the reasons for this has been that for many years only single-objective functions were considered. As we know, this is not a realistic assumption, since most real-world problems have several (possibly conflicting) objectives. This situation has led designers to make decisions and trade-offs based on their experience, instead of using some well-defined optimality criterion. Over the years, more than 20 mathematical programming techniques have been developed to deal with multiple objectives. However, the main focus of these approaches is to produce a *single* trade-off based on some notion of optimality, rather than producing several possible alternatives from which the designer may choose. More recently, the genetic algorithm (GA) has been found to be effective on some scalar optimization problems. In order to extend the GA to deal with multiple objectives, the structure of the GA has been modified to handle a vector fitness function. This paper will introduce two new multiobjective optimization techniques, proposed by the author, as well as MOSES (Multiobjective Optimization of Systems in the Engineering Sciences), a system developed as a testbed for new and existing multiobjective optimization techniques [2]. The new approaches, based on the notion of min-max optimum, are able to generate the Pareto set and better trade-offs than any of the other techniques included in MOSES.

1.1. Statement of the Problem

Multiobjective optimization (also called multicriteria optimization, multiperformance or vector optimization) can be defined as the problem of finding [8]:

> a vector of decision variables which satisfies constraints and optimizes a vector function whose elements represent the objective functions. These functions form a mathematical description of performance criteria which are usually in conflict with each other. Hence, the term "optimize" means finding such a solution which would give the values of all the objective functions acceptable to the designer.

Formally, we can state it as follows: Find the vector $\bar{x}^* = \left[x_1^*, x_2^*, \ldots, x_n^*\right]^T$ which will satisfy the m inequality constraints: $\quad g_i(\bar{x}) \geq 0 \quad i = 1, 2, \ldots, m$ (1)

the p equality constraints $h_i(\bar{x}) = 0 \quad i = 1, 2, \ldots, p$ (2)

and optimize the vector function $\bar{f}(\bar{x}) = \left[f_1(\bar{x}), f_2(\bar{x}), \ldots, f_k(\bar{x})\right]^T$ (3)

where $\bar{x} = \left[x_1, x_2, \ldots, x_n\right]^T$ is the vector of decision variables.

1.2. Min-Max Optimum

The idea of stating the *min-max optimum* and applying it to multiobjective optimization problems, was taken from game theory, which deals with solving conflicting situations. The min-max approach to a linear model was proposed by Jutler and Solich [7]. It has been further developed by Osyczka [6], Rao [9] and Tseng and Lu [11].

The min-max optimum compares relative deviations from the separately attainable minima. Consider the ith objective function for which the relative deviations can be calculated from

$$z_i'(\bar{x}) = \frac{|f_i(\bar{x}) - f_i^0|}{|f_i^0|} \quad (4) \quad \text{or from} \quad z_i''(\bar{x}) = \frac{|f_i(\bar{x}) - f_i^0|}{|f_i(\bar{x})|} \quad (5)$$

It should be clear that for (4) and (5) we have to assume that for every $i \in I$ and for every $\bar{x} \in X, f_i(\bar{x}) \neq 0$. If all the objective functions are going to be minimized, then equation (4) defines function relative increments, whereas if all of them are going to be maximized, it defines relative decrements. Equation (5) works conversely. Let $\bar{z}(\bar{x}) = \left[z_1(\bar{x}), \ldots, z_i(\bar{x}), \ldots, z_k(\bar{x})\right]^T$ be a vector of the relative increments which are defined in $\mathfrak{R}^k$. The components of the vector $z(\bar{x})$ will be evaluated from the formula $\forall_{i \in I}(z_i(\bar{x})) = max\{z_i'(\bar{x}), z_i^{\cdot}(\bar{x})\}$ (6)

2. Multiobjective Optimization using GAs

Genetic algorithms require scalar fitness information to work, which means that when approaching multicriteria problems, we need to perform a scalarization of the objective vectors. One problem is that it is not always possible to derive a global criterion based on the formulation of the problem. In the absence of information, objectives tend to be given equivalent importance, and when we have some understanding of the problem, we can combine them according to the information available, probably assigning more importance to some objectives. Optimizing a combination of the objectives has the advantage of producing a single compromise solution, requiring no further interaction with the decision maker [4]. The problem is, that if the optimal solution cannot be accepted, either because the function used excluded aspects of the problem which were unknown prior to optimization or because we chose an inappropriate setting of the coefficients of the combining function, additional runs may be required until a suitable solution is found. To test our 2 methods, we used MOSES (Multiobjective Optimization of Systems in the Engineering Sciences), a system developed to compare several multiobjective optimization techniques that contains the following methods: MOGA, NSGA, NPGA, Hajela's method, VEGA, Lexicographic Ordering, two Monte Carlo methods, and Osyczka's Multiobjective Optimization System (this system contains 5 mathematical programming techniques). For a complete description of MOSES and the techniques that it implements, see [2].

2.1. A new GA-based approach based on a weighted min-max strategy

1. The initial population is generated in such a way that all its individuals constitute feasible solutions. This can be ensured by checking that none of the constraints is violated by the solution vector encoded by the corresponding chromosome.
2. The user should provide a vector of weights, which are used to spawn as many processes as weight combinations are provided (normally this number will be reasonably small). Each process is really a separate genetic algorithm in which the given weight combination is used in conjunction with a min-max approach to generate a single solution.
3. After the n processes finish (n=number of weight combinations), a final file is generated containing the Pareto set, which is formed by picking up the best solution from each of the processes spawned in the previous step.
4. Since this approach requires knowing the ideal vector, the user is given the opportunity to provide such values directly (in case he/she knows them) or to use

another genetic algorithm to genete it. This additional program works in a similar manner, spawning k processes (k=number of objective functions), where each process corresponds to a genetic algorithm responsible for a single objective function. When all the processes terminate, there will be a file containing the ideal vector, which turns out to be simply the best values produced by each of the spawned processes.

5. The crossover and mutation operators were modified to ensure that they produced only feasible solutions. Whenever a child encodes an infeasible solution, it is replaced by one of its parents.

6. Notice that the Pareto solutions produced by this method are guaranteed to be feasible, as opposed to the other GA-based methods in which there could be convergence towards a non-feasible solution.

2.2. A new GA-based approach based on min-max selection with sharing

This is another approach in which a Min-Max selection strategy replaces the Pareto ranking selection scheme previously reported in the literature, and sharing is used to avoid that the GA converges to a single solution. The basic algorithm is the following:

1. The initial population is generated as in the previous approach.

2. By exploring the population at each generation, the local ideal vector is produced. This is done by comparing the values of each objective function in the entire population.

3. The binary tournament selection algorithm is modified, so that instead of comparing the fitnesses of two individuals, their maximal deviations with respect to the local ideal vector are compared. If one dominates the other, then it wins the tournament, and if there is a tie, then sharing is used to decide who is the winner, in a way similar to the NPGA. This means that we count the number of individuals within each niche of each of the competitors, and the individual with a lower count wins.

4. The crossover and mutation operators were modifed as in the previous algorithm to ensure that they produced only feasible solutions. When a child encodes an infeasible solution, it is replaced by one of its parents.

5. Any solutions produced with this method are also guaranteed to be feasible as before.

So far, much better results have been found using an *ad-hoc* floating point representation proposed by the author in which a floating point number is represented using a string of digits. Previous work in this area shows that this representation turns out to be very useful in numerical optimization problems [1] when combined with a technique, developed by the author, that performs a dynamic adjustment of parameters [2].

Method	x_1	x_2	x_3	x_4	f_1	f_2
Monte Carlo 1	30.84	28.26	3.79	4.06	**188.65**	0.06175
Monte Carlo 1	52.97	44.08	1.99	0.99	555.22	**0.00849**
Min-Max (OS)	74.97	44.97	1.97	1.97	**316.85**	0.01697
Min-Max (OS)	74.99	44.99	1.99	2.06	326.49	**0.01636**
GA (Binary)	66.39	38.63	0.90	0.91	**128.27**	0.05241
GA (Binary)	80.00	50.00	4.99	4.99	848.41	**0.00591**
GA (FP)	61.14	41.14	0.90	0.90	**127.46**	0.06034
GA (FP)	80.00	50.00	5.00	5.00	850.00	**0.00590**
Literature	60.70	49.90	0.90	0.90	**128.47**	0.06000
Literature	80.00	50.00	5.00	5.00	850.00	**0.00590**

Table 4.1 Comparison of results computing the ideal vector of the design of an I-beam. For each method the best results for optimum f_1 and f_2 are shown in **boldface**. OS stands for Osyczka's Multiobjective Optimization System. Every objective is being minimized.

3. Example

To illustrate the use of MOSES and the efficiency of the two new techniques proposed, we selected one engineering design example from the literature [2]. More examples may be found somewhere else [2]. Since it is generally intractable to obtain an analytical representation of the Pareto front, it is usually very difficult to measure the performance of a multiobjective optimization technique. For the purposes of this paper, we compared the results only in terms of the best trade-offs that could be achieved. For that sake, we used the expression

$$L_p(f) = \sum_{i=1}^{k} w_i \left| \frac{f_i^0 - f_i(x)}{\rho_i} \right|$$

where k is the number of objectives, $\rho_i = f_i^0$ or $f_i(x)$, depending on which gives the maximum value for $L_p(f)$ and w_i refers to the weight assigned to each objective (if not known, equal weights are assigned to all the objectives). This should favor all the mathematical programming techniques incorporated in MOSES, while disfavoring all the GA-based approaches, including our two new techniques. It is also important to mention that all the GA-based techniques developed so far have assumed only unconstrained problems, which is unrealistic, and does not apply to most engineering problems. That is why several constrained problems were chosen from the literature for our experiments [2]. Highly constrained search spaces remain to be a problem even for single objective optimization, because it could be quite difficult for the GA to even generate a single feasible solution to a problem, and some sort of hybrid with another technique (e.g., a mathematical programming algorithm) may be necessary to help the GA generate feasible solutions at generation zero. This sort of situation has been faced by the author before (see for example [1]), but the approaches presented in this paper have been successful at finding feasible designs since the earliest stages of the search.

3.1. Design of an I-beam

The multiobjective optimization problem is formulated as follows [8]:

Find the dimensions of an I-beam which satisfy the geometric and strength constraints and which optimize the following criteria:

1. cross section area of the beam which for the given length minimizes its volume; and
2. static deflection of the beam for the displacement under the force P.

Both criteria are to be minimized. It should be noted that these criteria are contrary to one another (i.e., the best solution for the first objective function gives the worst solution for the second one and viceversa). It is assumed that: the permissible bending stress of the beam material $k_g = 16kN/cm^2$, Young's Modulus of Elasticity $E = 2 \times 10^4 kN/cm^2$, and Maximal bending forces $P = 600kN$ and $Q = 50kN$. The vector of decision variables is $x = [x_1, x_2, x_3, x_4]^T$. Their values will be given in centimeters. The geometric constraints are: $10 \le x_1 \le 80$, $10 \le x_2 \le 50$, $0.9 \le x_3 \le 5$, $0.9 \le x_4 \le 5$
The strength constraint is

$$\frac{M_y}{W_y} + \frac{M_z}{W_z} \le k_g$$

where M_y and M_z are maximal bending moments in Y and Z directions respectively; W_y and W_z are section moduli in Y and Z directions respectively. For the forces acting the values of M_y and M_z are $30,000kN-cm$ and $2,500kN-cm$ respectively. The section moduli can be expressed as follows:

$$W_y = \frac{x_3(x_1-2x_4)^3 + 2x_2x_4\left[4x_4^2 + 3x_1(x_1-2x_4)\right]}{6x_1}, \quad W_z = \frac{(x_1-2x_4)x_3^3 + 2x_4x_2^3}{6x_2}$$

Thus the strength constraint is:

$$16 - \frac{180,000x_1}{x_3(x_1-2x_4)^3 + 2x_2x_4\left[4x_4^2 + 3x_1(x_1-2x_4)\right]} - \frac{15,000x_2}{(x_1-2x_4)x_3^3 + 2x_4x_2^3} \ge 0$$

The objective functions can be expressed as follows

1. Cross-section area $f_1(x) = 2x_2x_4 + x_3(x_1 - 2x_4)cm^2$

2. Static deflection $f_2(x) = \dfrac{Pl^3}{48EI} cm$

where I is the moment of inertia which can be calculated from

$$I = \frac{x_3(x_1 - 2x_4)^3 + 2x_2x_4\left[4x_4^2 + 3x_1(x_1 - 2x_4)\right]}{12}$$

4. Comparison of Results

The ideal vector that each method incorporated in MOSES generates for the example presented in the previous section will be compared against the best results reported in the literature. Such methods include two Monte Carlo techniques, Osyczka's multiobjective optimization system and several GA-based approaches [2]. For the GA-based approaches, the same parameters were used (i.e., population size, crossover and mutation rate) and, if niching was required, the niche size was computed according to the methodology suggested by the developers of the method (see [2] for details).

4.1. Design of an I-beam

The ideal vector of this problem was computed using Monte Carlos methods 1 and 2 (generating 100 points), Osyzcka's multiobjective optimization system and a GA (with a population of 100 chromosomes running during 50 generations) using binary and floating point representation. The corresponding results are shown in Table 4.1, including the best results reported in the literature [8]. The results for Monte Carlo Method 2 are the same as for Method 1, and the results presented for the Min-max method are also the basis for computing the best trade-off for all the methods in Osyczka's system. As can be seen from the results, the GA provided the best ideal vector, combining the results produced with both binary and floating point representation, although the second representation scheme provides better results in general [2]. As can be seen in Table 4.2, the two new GA-based approaches proposed by the author, named *GAminmax1* and *GAminmax2* respectively, provide the best overall results when a floating point representation is used. The second method slightly improves the results obtained using the first method, although it should be mentioned that the first technique does not require any sort of niching parameters as the second approach. The results obtained for this problem show how easily the mathematical programming techniques can be surpassed by a GA-based approach, using the same number of points, though the GA starts with a completely random population. Although the same random numbers generator was used for both the Monte Carlo and the GA-based approaches, the results are quite different. For those who think that a simple linear combination of the objectives should be good enough to deal with multiobjective optimization problems, the results for GALC

158

(see Table 4.2) show the contrary even for this simple bi-objective problem. Finally, it is interesting to notice how some simple approaches, like Lexicographic ordering (in which an objective is randomly selected at each turn) work remarkably well with problems that have few objectives, like this. Such behavior, however, is not as good when several objectives are used [2]. For the example in this paper, 10 weight combinations were used (all combinations of two, from 0.1 to 0.9) for the first of the new methods proposed. Although the Pareto front is not included in this paper, the two techniques introduced produced a very good contour, with less dominated solutions than any of the other approaches included in MOSES (see [2] for details).

Method	x_1	x_2	x_3	x_4	f_1	f_2	$L_n(f)$
Ideal Vector					127.46	0.0059	0.000000
Monte Carlo 1	77.57	20.59	3.47	3.88	401.77	0.0159	3.837581
Monte Carlo 2	75.01	31.02	1.76	2.48	277.09	0.0198	3.525181
Min-Max (OS)	75.06	44.99	1.99	1.99	320.55	0.0167	3.350946
GCM (OS)	75.06	44.99	1.99	1.99	320.55	0.0167	3.350946
WMM (OS)	75.06	44.99	1.99	1.99	320.55	0.0167	3.350946
PMM (OS)	74.97	44.97	1.97	1.97	316.85	0.0170	3.360191
NMM (OS)	74.99	44.99	1.99	2.06	326.49	0.0164	3.332501
GALC (B)	80.00	50.00	0.92	3.98	463.99	0.0083	3.042494
GALC (FP)	80.00	50.00	0.90	3.80	445.55	0.0086	2.953417
Lexicographic (B)	80.00	45.25	0.98	2.73	319.95	0.0124	2.614093
Lexicographic (FP)	80.00	50.00	0.90	2.26	293.74	0.0134	2.572228
VEGA (B)	80.00	50.00	0.94	2.24	295.59	0.0134	2.589958
VEGA (FP)	80.00	23.33	3.52	5.00	479.49	0.0116	3.736026
NSGA (B)	80.00	44.28	2.39	4.35	555.19	0.0080	3.714160
NSGA (FP)	80.00	50.00	5.00	1.18	506.56	0.0132	4.210141
MOGA (B)	80.00	46.48	1.29	2.70	347.27	0.0119	2.743380
MOGA (FP)	80.00	30.38	0.90	3.53	279.95	0.0146	2.668388
NPGA (B)	78.75	36.39	1.40	3.71	372.17	0.0117	2.909137
NPGA (FP)	78.52	29.36	2.51	2.74	344.44	0.0160	3.409632
Hajela (B)	80.00	50.00	0.90	4.72	535.48	0.0072	3.418376
Hajela (FP)	80.00	50.00	1.92	5.00	634.05	0.0066	4.090622
GAminmax1 (B)	80.00	40.58	0.92	3.02	312.77	0.0127	2.603628
GAminmax1 (FP)	80.00	50.00	0.90	2.43	310.33	0.0126	2.568096
Gaminmax2 (B)	80.00	49.59	1.12	2.33	315.36	0.0129	2.653479
GAminmax2 (FP)	80.00	50.00	0.90	2.35	303.06	0.0129	2.567664

Table 4.2 Comparison of the best overall solution found by each one of the methods included in MOSES for the design of an I-beam. GA-based methods were tried with binary (B) and floating point (FP) representations. The following abbreviations were used: OS = Osyczka's Systems, GCM=Global Criterion Method (exponent = 2.0), WMM =Weighting Min-max, PWM=Pure Weighting Method, NWM=Normalized Weighting Method, GALC=Genetic Algorithm with a linear combination of objectives using scaling. In all cases, weights were assumed equal to 0.5 (equal weight for both objectives). Each objective is being minimized.

5. Conclusions

Two new GA-based multiobjective optimization techniques based on the min-max optimization approach have been proposed. The first approach is very robust because it transforms the multiobjective optimization problem into several single objective optimization problems, and it works very well independently of the representation scheme used. However, a floating point representation seems to work better for numerical optimization applications with any of the two approaches proposed. The main drawbacks of the first min-max approach proposed is that it requires the ideal vector and a set of weights to delineate the Pareto set. Nevertheless, when the ideal vector is not known, a set of target (desirable) values for each objective can be provided instead. Regarding the issue of finding the proper weights, from the personal experience of the author, it can be said that it is relatively easy to find a set of weights that can produce a considerable fraction of the Pareto front. However, it may be harder in certain problems in which there are too many points of the Pareto front clustered nearby an specific region, since the use of inappropriate weights could bias the GA to find repeatedly the same point. The second technique proposed does not require the ideal vector, since it is able to compute it based on the local populations generated. However, to avoid convergence to a single solution, a form of sharing similar to that employed by the NPGA (Niched Pareto Genetic Algorithm) was implemented, but the problem of finding an optimum tournament size was eliminated by using a min-max tournament selection strategy. Nevertheless, the niche sharing factor still has to be provided by the user. For that purpose, the author has proposed the use of sharing on the parameter values by computing the optimum niche size using the guidelines provided by Goldberg and Deb [3]. The second method proposed is very fast and reliable except in cases in which it is possible to find solutions that highly favor more than one objective (namely when some elements of the ideal vector are easily achievable) but that highly disfavors other objectives [2]. This behavior is normally present in highly convex search spaces which are unfortunately very common in engineering optimization problems and more work is required to extend this method to deal with such situations. The two techniques developed by the author ensure that only feasible points are produced at generation zero, and the crossover and mutation operators were modified in such a way that infeasible solutions are never generated by the algorithms. This property makes these approaches unique, since none of the other GA-based techniques considered this important issue. This is mainly because most of the previous work with multiobjective optimization techniques dealt only with unconstrained problems. Finally, the importance of MOSES as a benchmark for new multiobjective optimization methods should be obvious, since no other similar tools, combining GA-based approaches with mathematical programming techniques, were previously available. Additional details of the methods that it contains and their implementation details may be found in [2].

6. Future Work

Much additional work remains to be done, since this is a very broad area of research. For example, it is desirable to do more theoretical work on niches and population sizes for multiobjective optimization problems to verify some empirical results previously reported [2]. In that sense, MOSES is expected to be useful as an experimentation tool for those interested in this area. To talk about convergence in this context seems a rather difficult task, since there is no common agreement on what optimum really means. However, if we use concepts from Operations Research such as the min-max optimum, it should be possible to develop such a theory of convergence for these kinds of problems. Also, it is highly desirable to be able to find more ways of incorporating knowledge about the domain into the GA, as long as it can be automatically assimilated by the algorithm during its execution and does not have to be provided directly by the user (to preserve its generality).

References

1. Coello Carlos A, Christiansen, Alan D & Hernández Aguirre Arturo, 1996. Using genetic algorithms to design combinational logic circuits. In: Dagli Cihan H, Akay Metin, Chen C. L. Philip, Fernández Benito R. and Ghosh Joydeep (eds), 1996. *Intelligent Engineering Systems Through Artificial Neural Networks, Volume 6. Fuzzy Logic and Evolutionary Programming*, ASME Press, St. Louis, Missouri, USA, pp 391-396.
2. Coello Carlos A, 1996. *An Empirical Study of Evolutionary Techniques for Multiobjective Optimization in Engineering Design*. PhD thesis, Department of Computer Science, Tulane University, New Orleans, Louisiana, USA.
3. Deb Kalyanmoy and Goldberg David E, 1989. An investigation of niche and species formation in genetic function optimization. In Schaffer J. David (editor), 1989. *Proceedings of the Third International Conference on Genetic Algorithms*. San Mateo, California, George Mason University, Morgan Kaufmann Publishers, pp 42-50.
4. Fonseca Carlos M. and Fleming Peter J, 1994. *An overview of evolutionary algorithms in multiobjective optimization*. Technical report, Department of Automatic Control and Systems Engineering, University of Sheffield, Sheffield, U K.
5. Goldberg David E, 1989. *Genetic Algorithms in Search, Optimization and Machine Learning*. Reading, Mass.:Addison-Wesley Publishing Co.
6. Osyczka A, 1978. An approach to multicriterion optimization problems for engineering design. *Computer Methods in Applied Mechanics and Engineering*, 15:309-333.
7. Osyczka A, 1984. *Multicriterion Optimization in Engineering with FORTRAN Programs*. Ellis Horwood Limited.
8. Osyczka A, 1985. Multicriteria optimization for engineering design. In Gero John S (editor). *Design Optimization*, 1985. Academic Press, pp 193-227.
9. Rao S, 1986. Game theory approach for multiobjective structural optimization. *Computers and Structures*, 25(1):119-127.
10. Rosenberg R S, 1967. *Simulation of genetic populations with biochemical properties*. PhD thesis, University of Michigan, Ann Harbor, Michigan.
11. Tseng C H and Lu T W, 1990. Minimax multiobjective optimization in structural design. *International Journal for Numerical Methods in Engineering*, 30:1213-1228.

Mapping Based Constraint Handling for Evolutionary Search; Thurston's Circle Packing and Grid Generation

Dae Gyu Kim   Phil Husbands

School of Cognitive and Computing Sciences, University of Sussex
Falmer, Brighton, United Kingdom, BN1 9QH
E-mail: {dgkim, philh}@cogs.susx.ac.uk

Abstract

This paper presents a mapping based constraint handling method for optimisation problems to be used with evolutionary search techniques. Two different mapping techniques are used; Thurston's circle packing algorithm and TTM (Thompson-Thames-Martin) grid generator. Both methods are used to find approximate Riemann mappings between arbitrarily shaped 2D feasible solution spaces and a simple rectangular domain in which evolutionary search can be applied in a straightforward manner.

The mapping based constraint handling method is compared, on a number of problems, with penalty-based and repair-based constraint handling methods. Discussions on the method's limitations and ways to overcome these, and extension to higher dimensional problems are given.

1 Introduction

Most practical optimization problems have constraints of many types: linear, nonlinear, or some combination of these. These constraints define the topology of the feasible solution space of a given problem. Any evolutionary search technique needs to be aware of the existence of infeasible solutions to solve such a problem.

In general, when evolutionary search (ES) methods are applied to solve optimisation problems, each of the variables in an n-dimensional space are encoded to lie between predefined upper and lower limits. Constraints of the given problem impose complicatedly shaped feasible solution domain(s) within the search space which is a simple n-dimensional hypercube.

Most of the work on evolutionary search for constrained problems uses penalty functions or repairing heuristics to steer the search away from infeasible regions. The complexity of the feasible solution domain often makes it difficult both to initialise the populations and to reproduce feasible offspring from feasible parents. Therefore, ES may work easily and efficiently if it does not have to worry about infeasible solutions. With some problems a judiciously chosen encoding scheme may be all that is required [5]. Another, more general, way of achieving this would be if arbitrarily shaped feasible solution domains could be mapped into a simple hypercube. In this case, ES can be applied in a straightforward manner inside the mapped domain where all solutions are legal. An inverse mapping can be used to allow fitness evaluation of each solution.

Riemann's Mapping Theorem says that "any simply connected domain (not the entire space) can be mapped one-to-one and conformally into any other simply connected domain (not the entire space) ([4])". This means that any two dimensional feasible solution domain could be mapped into a simple hypercube (ractangle) in which evolutionary search methods can be applied easily.

There are several methods in the literature for doing Riemann mapping for a given set of domains. Among such methods, the simplest is *Bilinear Mapping* which unfortunately has lots of restrictions ([8]), especially on the shape of the original domain. Thurston proposed *Circle Packing*, Appendix 2 in [15], an algorithmic approach to generate mappings from any simply connected two dimensional domain to a unit circle, which has been proved to converge to the Riemann Mapping [15]. Structured grid generation methods which are a usual tool for solving partial differential equations of fluid dynamics also can be used to find mappings between feasible solution domain(s) and a hypercube.

The mapping methods mentioned above are expected to be useful in handling constraints for evolutionary search methods like Genetic Algorithms. Such a mapping can map a complicated feasible solution domain into a rectangular domain. A unit circle from Riemann Mapping can be understood as a rectangle in a polar coordinate system. Once such a simple feasible solution domain is obtained, evolutionary search can use straightforward encodings and its reproduction operators do not explicitly have to handle infeasible solutions. Test results presented in this paper show that the mapping based constraint handling method is a promising technique.

The structure of this paper is as follows. In Section 2, some basic concepts on constraints and search space, and constraint handling methods for Evolutionary Search will be discussed.In Section 3, a mapping based constraint handling method using Circle Packing and grid generation as the approximate methods to find Riemann Mappings will be discussed. In Section 4, three example problems with non-linear constraints will be discussed to show how constrains can be handled according to the proposed constraint handling method. The test results are compared to other constraint handling methods. A summary and some future research directions conclude this paper.

2   Constraint Handling in Evolutionary Search

Evolutionary Search Methods generally rely on parameter values to encode solutions. The parameter values are initially randomly distributed within their bounds. When the feasible solution domain is not in a hypercubic shape, it is difficult to avoid producing infeasible solutions during the initialization phase. Modification of valid solutions through the reproduction process, crossover and mutation in genetic algorithms for example, does not guarantee the production of feasible solutions only. Heuristics are needed to handle either infeasible solutions or the reproduction process itself. These heuristics are usually referred to as *constraint handling* methods.

2.1   Constraints, Search Domain, and Feasible Solution Domain

Most search methods explore the search space defined by the domains of variables, which usually is a hypercube if there is no constraints. The feasible solution space is a region formed by all valid solutions within the search space.

Linear equality constraints can be eliminated to reduce the number of independent variables. Linear inequality constraints trim the search space, which usually produces *convex* feasible solution domains. Non-linear inequality constraints define more complicated feasible domains which may be either convex or concave. A concave domain does not guarantee feasible solutions when offspring are made by linear combination of their parents, which complicates evolutionary search. Non-linear equality constrains are the most difficult type of constraints for any optimisation technique, [14]. A way of handling non-linear equality constraints using the mapping based method will be discussed in Section 5.

2.2   Penalty Based Constraint Handling

Infeasible solutions are discouraged by receiving a penalty during the evaluation process, and feasible solutions are indirectly encouraged in this way. There are various ways of defining actual penalty values to be given to each infeasible solution. Penalty methods are usually problem dependent for better performance. [10] has a good summary of those used in the literature.

2.3   Repair Based Constraint Handling

An infeasible solution can be *repaired* to make a feasible solution, or another feasible solution can be used just for the fitness evaluation. Repairing can be done in many different ways: create a feasible solution by modifying from an infeasible solution, or take a feasible solution for the repaired one. Repaired feasible solutions can be used as parents for reproduction or un-repaired infeasible solution can be used. Most repair heuristics are problem dependent[1] as they should consider the problem specific constraints.

2.4 Constraint Handling with Special Encoding

Some elaborate representation methods, such as [12], which guarantee that all individuals remain feasible, can be another constraint handling method. Such a representation will usually be highly problem dependent and can be thought of as mapping the feasible solution domain to another domain defined by the specific encoding.

2.5 Mapping Based Constraint Handling

If feasible solution domains can be mapped into a simple domain like a hypercube, then simple binary or float encodings can be used and make application of ES straightforward. "*Mapping Based Method*" in this paper refers to this approach.

One important factor when using such a mapping based method is that the mapping should be one-to-one and preserve locality properties. Some conditions that such a mapping should meet are:

> "(1) for each solution $s \in \mathcal{F}$, $\mathcal{F}$ is the feasible solution domain, there is a decoded solution, $d \in \mathcal{S}$, (2) each decoded solution d corresponds to a feasible solution s, and (3) all solutions in $\mathcal{F}$ should be represented by the same number of decodings d. ... (4) the transformation T is computationally fast and (5) it has locality feature. ... [10]"

3 Mapping Based Constraint Handling

One of the important factors of the mapping based constraint handling is how to find appropriate mappings for each problem. Two different mapping techniques are used for this purpose: Thurston's circle packing, Appendix 2 in [15], and TTM (Thompson-Thames-Martin) grid generator, [7].

These mapping can be applied to handle multiple disjoint domains as well as a decomposition of a single but difficult domain. Figure 1 is an example of this procedure for the case of circle packing.

Domains are mapped into unit circles, and their radii are adjusted so that the circles keep the same area ratio as they had in the original domain, see Figure 1.c. They will be stacked together to make one domain, Figure 1.e, where the evolutionary search can work, and an inverse mapping to the original domains, Figure 1.a, will be used for fitness calculation of each individual.

---

[1]The repairing process used in *GENOCOP III*, [11], is an example of a problem independent repairing method, which is similar to the one from *complex* method, [14].

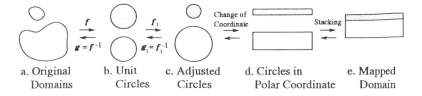

a. Original   b. Unit   c. Adjusted   d. Circles in   e. Mapped
Domains   Circles   Circles   Polar Coordinate   Domain

Figure 1: Handling multiple disjoint regions

3.1 Circle Packing Algorithm

Thurston's circle packing is a way to find the Riemann mapping between a two dimensional arbitrary domain and a unit circle. Figure 2 shows an example of this mapping. There are three major steps to find the circle packing from a given domain to a unit circle.

1. Hexagonal circle packing for approximation, Figure 2.a

2. Radii iteration and Thurston packing, Figure 2.b

3. Transformation into a unit circle, Figure 2.c

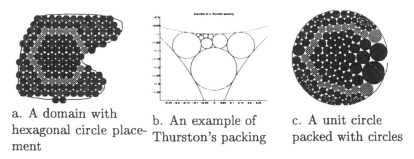

a. A domain with hexagonal circle placement   b. An example of Thurston's packing   c. A unit circle packed with circles

Figure 2: Example of a circle packing (riemannmap by [1] was used to generate a, and c.

Hexagonal Circle Packing

As can be seen in Figure 2.a, the domain is filled with circles of the same radius. Each circle can have up to six neighbouring circles in two dimensional cases. Minimum number of neighbouring circles are set to be three for convergence purpose. The number of neighbours is also called the "kissing number", [17].

The neighbouring relationship between circles is important and holds throughout the algorithm. The centres of every three neighbouring circles form triangles which are the basic structure of the approximation and a key to the inverse mapping from the unit circle to the original domain.

An approximation of the original domain is then defined by the Jordan curve surrounding the circles. This approximation has difficulty in describing sharp corners when a finite number of circles are involved. This problem can be solved mostly by using Thurston's boundary modification, [2], and this will be discussed later in Section 5.

Radii Iteration, Thurston's Packing

This is a step to find a set of circle radii to squeeze all the circles into a small region while all the neighbouring circles are tangential to each other. Suppose that there are n circles in the hexagonal circle packing. The radius of any two neighbouring circles are set to be 1 to become tangential to a unit circle, and the region between the two circles and the unit circle is where all the other circles are going to be packed. For the remaining $n - 2$ circles a measure called *curvature* is defined respectively, and the iteration process tries to make all the curvature values vanish. Details of the formulation can be found in Appendix 2 in [15]. The implementation in this research uses the Newton-Rapson method to find the set of radii values.

When all the curvatures values vanish, the circles can be laid down on a plane without any overlapping while all the neighbours are tangential to each other. One example of the layout is shown in Figure 2.b.

Transformation of circles into a unit circle

All the circles in Thurston's packing, Figure 2.b, will then be transformed into the unit circle as follows. An *inverse mapping* and some appropriate adjustment such as *scaling*. Inversion puts all the outer circles into the unit circle, scaling makes the circles fit into the unit circle. Circle centres may need further transformations such as *rotation* and *flipping*. Detailed formulations about these transformations can be found in [6].

3.2   TTM (Thompson-Thames-Martin) Grid Generation

Numerical grid generation is a common tool to solve the partial differential equations of fluid dynamics with complex geometry. This transformation makes a complex physical domain into a simple region. One disadvantage is that the equations usually become more complex after the transformation. It is important to have an inverse mapping if the mapping is to be used with an evolutionary search techniques. A non-zero *Jacobian* of a mapping is a condition to be satisfied to guarantee the existence of an inverse mapping.

Among various kind of grid generation algorithms available, we chose the *homogeneous Thompson Thames Martin (TTM)* grid generator, [7], for this research. The mapping is simple and relatively robust against folding[2]. Detailed formulation for two and three dimensional TTM grid generator can be found in [7]. Figure 3 shows some examples of generated grids.

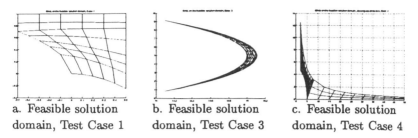

| a. Feasible solution | b. Feasible solution | c. Feasible solution |
| domain, Test Case 1 | domain, Test Case 3 | domain, Test Case 4 |

Figure 3: Grids generated on feasible solution domains. Examples are taken from [10]

The sample grids shown in Figure 3 are taken from test problems suggested in [10]. Considering the TTM grid generator does not guarantee unfolded grids although it is robust, decomposition of domains can be a way to tackle some tricky cases. Figure 3.c is a case when the feasible solution domain is decomposed into two, then the procedure suggested at the start of Section 3, Figure 1.

3.3 Mapping between domains

Once a mapping between domains is obtained, a simple interpolation method can be used to find mappings between specific points in both domains. Triangular elements are the basis of the mapping and interpolation.

A point in a domain can be mapped into another domain keeping its relation to the three corners points of a triangle in which it lies.

For example[3], suppose there are three points A, B, C making a triangle in the mapped domain and their matching points in the original domain are A', B', C'. There is a point, P, in the mapped domain and its matching point, P', in the original domain is to be found. Three coefficients, a, b, c, can be determined for this interpolation.

$$
\begin{aligned}
a + b + c &= 1 \\
aA.x + bB.x + cC.x &= P.x \\
aA.y + bB.y + cC.y &= P.y
\end{aligned}
\quad\Longrightarrow\quad
\begin{aligned}
P'.x &= aA'.x + bB'.x + cC'.x \\
P'.y &= aA'.y + bB'.y + cC'.y
\end{aligned}
$$

---

[2]When a mapping has zero Jacobian, then some part of the original domain are folded into the mapped domain, or there may be a region in the original domain that is not covered by the mapped domain. Inverse mapping is not defined in both cases.

[3].x and .y represent x and y coordinate of a point. For example, $A.x$ means x coordinate of point A.

4 Mapping Based Constraint Handling: Examples

Two different optimisation problems with non-linear constraints are tested. The first example shows the detailed procedure of handling non-linear constraints with the mapping based constraint handling method and how to apply Genetic Algorithm as the search technique.

4.1 Test Problem 1

This example[4] has two non-linear constraints and bounds are $0 \leq x \leq 3$ and $0 \leq y \leq 4$. The resulting feasible solution domains are shown in Figure 4.a.

$$\begin{array}{ll} \text{Minimize} & \quad \text{subject to,} \quad y \leq 2x^4 - 8x^3 + 8x^2 + 2, \\ f(x,y) = -x - y, & \qquad\qquad\qquad\quad y \leq 4x^4 - 32x^3 + 88x^2 - 96x + 36, \end{array}$$

Finding Mappings Between Domains

Figure 4.b is a hexagonal circle packing with circle radius 0.07. This approximation does not cover regions near corners. Although not seen here, decomposition of the domains and using smaller radius values can improve the approximation as can be found in [6]. As another approximation method, TTM grid generators also has been used as in Figure 4.c.

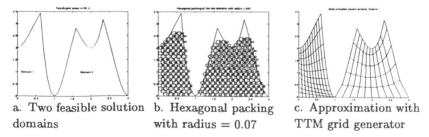

a. Two feasible solution domains   b. Hexagonal packing with radius = 0.07   c. Approximation with TTM grid generator

Figure 4: Feasible solution domains and approximation with circle packing and with TTM grids, Problem 1

Applying A Genetic Algorithm

Once mappings between the original domain and a rectangular domain is found, then a Genetic Algorithm can be started within the mapped rectangular domain. Disjoint domains as is in this problem are handled with a procedure

---

[4]This test problem is originally from Test Problem 6, Page 30, [3]

shown in Figure 1. Inverse mappings from the mapped domain to the original domain are used for the fitness evaluation for every individuals.

Table 1 shows the test result of the mapping based constraint handling method together with results from other techniques. There are four test cases: Mapping based constraint handling using circle packing (MBC-C), MBC with grid generation (MBC-G), Genocop II [10], and a penalty based method (PE)[5].

Parameter settings for the GAs using MBC and PE are: Number of populations is 5, 10 individuals in each population. Result in Table 1 are averaged out of 100 runs except the one from Genocop II which are taken from [10]. The number of iterations of Genocop II is calculated based on 4 iterations and each iteration has 1,000 trials. Initial trials of Genocop II to locate the starting point are not counted.

Known global solution for this problem is -5.5080 when $(x, y) = (2.3295, 3.1785)$. The best solutions found with MBC-C is -5.4951, MBC-G is -5.5079, However, Genocop II is -5.5085, and PE found -5.4657. Genocop II slightly violates both constraints to produce this result. Obviously the MBC methods do not violate any constraints.

Result	Number of Evals		Solutions			Note
	Mean	Std. Dev.	Mean	Best	Worst	
MBC-C	1810	251	-5.1641	-5.4951	-5.1604	No
MBC-G	1625	217	-5.5079	-5.5079	-5.5079	No
Genocop II	4000	NA	NA	-5.5085	NA	Yes
PE	1800	385	-5.0875	-5.4657	-4.4999	No

Table 1: Number of iterations and solutions for Problem 2. MBC-C represents GA using circle packing approximation, MBC-G represents GA using the approximation with the grid generation. Note in the last column mentions about constraint vilation of the final solution.

MBC-C needs to move a bit further toward the corner to get a better result, but this was not possible for the current approximation accuracy. MBC-C has problems in approximating regions near boundaries and corners. Some possible improvement are discussed in Section 5. MBC-G shows the best performance out of the four test cases. The MBC methods reliably produce very good solutions (ie. the standard deviations of the number of evaluations to best solutions are lower than other approaches.)

---

[5]The penalty based method here is implemented on PGA, [16], and applies a fixed amount of penalty value to all infeasible solutions. The penalty value in this problem has been chosen as 100, which is expected to be large enough to discourage the infeasible solutions.

170

4.2 Test Problem 2

Test problem 2^6 is to design a rectangular structure to enclose a plant unit. Mathematical simplifications leave two linear inequality constraints and one non-linear constraint and the bounds are $0 \leq x_1 \leq 60$ and $0 \leq x_2 \leq 70$.

$$\text{Minimize} \quad f(x,y) = 30x_1x_2 + \frac{480000}{x_1} + \frac{960000}{x_2} \quad \text{subject to,} \quad \begin{aligned} g_1(x) &= 110 - x_1 - x_2 \geq 0 \\ g_2(x) &= 3x_1 - x_2 \geq 0 \\ g_3(x) &= \frac{2}{3}x_2 - \frac{16000}{x_1x_2} \geq 0 \end{aligned}$$

The feasible solution domain is shown in Figure 5.a. Approximation of the domain with circle packing and the grid generation method are shown in Figure 5.b and Figure 5.c respectively.

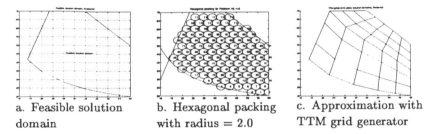

| a. Feasible solution domain | b. Hexagonal packing with radius = 2.0 | c. Approximation with TTM grid generator |

Figure 5: Feasible solution domains and approximation with circle packing and with TTM grids, Problem 2

Result	Number of Evals		Solutions		
	Mean	Std. Dev.	Mean	Best	Worst
MBC-C	2392	251	72002	72000	72012
MBC-G	1978	418	72004	72000	72012
Genocop III	10219		72000	72000	72000
Complex	23*		72319	72319	72319

Table 2: Number of iterations and solutions for Problem 2. MBC-C represents GA using circle packing approximation, MBC-G represents GA using the approximation with the grid generation. (* 23 replacements, Table 7.1 in [14])

Table 2 summarises the test result of four different search techniques. All methods except the *Complex method*, [14], find the global solution easily. Their performances can be compared using the number of evaluations taken to find

---

[6]This example is taken from Example 7.5 of [14], Page 271.

the solutions. MBC-G converged quicker than the other approaches. MBC-C converges slower than MBC-G but faster than Genocop III. [14] reports a difficulty[7] with the Complex method for this problem, and the method applied upto the 23-th replacement of the solution. The test result in [14] shows that the solution just oscillates after a few replacement.

Although there is not room to present them here, several other test problems have been investigated. Results were similar those reported here, the MBC outperformes other technique in nearly all cases, [6].

5 Future Work and Conclusion

Future Work

The two example problems of Section 4 show that the mapping based constraint handling method is a promising approach for handling constraints in evolutionary search. Some aspects that require further research can be listed as follows,

1. Solving higher dimensional cases.

2. Improving approximation of the circle packing.

3. Handling non-linear equality constraints.

A key to extending the circle packing to higher dimensional cases is how to find the curvature measure as in the two dimensional cases. Thw basis of the measure in the two dimensional case is circumference, and its extension to the three dimensional cases is the surface area of a sphere. Likewise, in n-dimensional cases the measure can have its basis in the surface area of a n-dimensional sphere. Formula to calculate such area are readily available from various literature, [9]. The TTM grid generator can be extended to higher dimensional case as the basis is an interpolation method, the *Transfinite interpolation*. Formula for its three dimensional version is available from [7]. A practical obstacle to handling higher dimensional cases is the growth of required memory size to store the lookup table for the mapping.

Some of the problems raised on using the circle packing method are convergence speed, convergence tolerance, and better approximations of boundaries and regions near sharp corners. Convergence speed can be improved by using the N-dimensional Newton Method, Chap. 9.6 in [13], for the radii iteration process. A the convergence speed also depends on the convergence tolerance, adequate value of the convergence tolerance depending on the problem size is needed.

The inherent problem of circle packing that it has difficulty in approximating boundaries and near sharp corner regions can be improved somewhat by

---

[7] "vertices tend to remain flattened against the constraint, and the search slows down considerably [14]"

applying Thurston's boundary modification, [2]. Figure 6 illustrates such an improvement. For the uncovered regions in Figure 6.b, a local search can be applied when necessary.

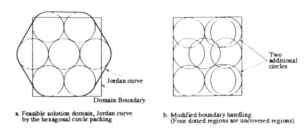

a. Feasible solution domain, Jordan curve
   by the hexagonal circle packing

b. Modified boundary handling
   (Four dotted regions are uncovered regions)

Figure 6: Approximation of the region near domain boundary. a. without boundary modification, b. with boundary modification

As with any other constraint handling methods, non-linear equality constraints are not easy to handle with the mapping based constraint handling method. An idea for handling these constraints is to consider the constraints as linear equality constraints that reduce the dimensionality of the problem. Consider a three dimensional case where a non-linear equality constraint defines a plane, then the feasible solution domain is defined on the plane. Likewise, when a problem has n independent variables, m non-linear equality constraints can be considered to lower the dimensionality to $n - m$ dimensions. This idea will be extended to the general cases.

Conclusion

A new mapping based constraint handling method for evolutionary search methods is suggested and tested in this paper. The method converts feasible solution domains into a rectangular domain so that the search can be made easily without considering about infeasible solutions.

Two different mapping methods, Thurston's circle packing algorithm and TTM grid generation, are used as the tools to find mappings from the original feasible solution domain(s) to a simple rectangular domain. The mapping based constraint handling method has been shown to work better than other constraint handling methods.

The two test cases in this paper show that the approximation using the grid generation method looks more promising than the approximation with Thurston's circle packing. Although it hasn't been shown in this paper, some cases where the circle packing approximation may work better than the approximation with grids are expected.

Current research concerns two dimensional cases. Work to extend this method to higher dimensional cases is underway. This extension is expected to make the mapping based constraint handling method more useful in handling most of the constraints for evolutionary search methods.

References

[1] Ithiel Carter. riemannmap, 1988. ftp://geom.umn.edu/pub.

[2] Ithiel Carter. *Circle Packing and Conformal Mapping*. PhD thesis, University of California, San Diego, 1989.

[3] C. A. Floudas and P. M. Pardalos. *A Collection of Test Problems for Constrained Global Optimization Algorithms*. Number 455 in Lecture Notes in Computer Science. Springer-Verlag, 1990.

[4] Michael D. Greenberg. *Advanced Engineering Mathematics*. Prentice-Hall International Inc., 1988.

[5] Phil Husbands. An ecosystem model for integrated production planning. *International Journal of Computer Integrated Manufacturing*, 6(1 & 2):74 – 86, 1993.

[6] Daegyu Kim and Phil Husband. Riemann mapping constraint handling method for genetic algorithms. Technical Report CSRP 469, COGS, University of Sussex, 1997.

[7] P. Knupp and S. Steinberg. *Fundamentals of Grid Generation*. CRC Press Inc., 1993.

[8] H. Kober. *Dictionary of Conformal Representations : 2 ed.* Dover, 1957.

[9] Alex Lopez. Frequently askes questions in mathematics. http://daisy. uwaterloo.ca/ alopez-o/math-faq/, 1997.

[10] Zbigniew Michalewicz. *Genetic Algorithms + Data Structures = Evolutionary Programs*. Springer-Verlag, third edition, 1996.

[11] Zbigniew Michalewicz and Girish Nazhiyath. Genocop III: A co-evolutionary algorithm for numerical optimization problems with nonlinear constraints. In *Proceedings of the second IEEE ICEC*, Perth, Australia, 1995.

[12] Charles C. Palmer and Aaron Kershenbaum. Representing trees in genetic algorithms. In *Proceedings of the First IEEE International Conference on Evolutionary Computation*, volume 1, pages 379–384, 1994.

[13] W. Press, S. Teukolsky, W. Vetterling, and B. Flannery. *Numerical Recipies in C*. Cambridge University Press, second edition, 1995. WWW Edition.

[14] G. V. Reklaitis, A. Ravindran, and K. M. Ragsdell. *Engineering Optimization: Methods and Applications*. John Wiley & Sons, Inc.

[15] Burt Rodin and Dennis Sullivan. The convergence of circle packing to the riemann mapping. *Journal of Differention Geometry*, 26:349–360, 1987.

[16] Peter Ross and Geoffrey H. Ballinger. PGA - parallel genetic algorithm testbed. Technical report, Department of Artificial Intelligence, University of Edinburgh, UK, 1996. Available from ftp.dai.ed.ac.uk/pub/pga-2.9/pga-2.9.1.tar.gz.

[17] Ian Stewart. Kissing number. Scientific American, February 1992.

Chapter 4

Structured Representations

Evolutionary Design of Analog Electrical Circuits using Genetic
Programming.
J.R. Koza, F.H. Bennett III, D. Andre, M.A. Keane.

Improving Engineering Design Models using an Alternative Genetic
Programming Approach.
A.H. Watson, I.C. Parmee

From Mondrian to Frank Lloyd Wright: Transforming Evolving
Representations.
T. Schnier, J.S. Gero

A Comparison of Evolutionary-Based Strategies for Mixed Discrete
Multilevel Design Problems.
K. Chen, I.C. Parmee

Evolutionary Design of Analog Electrical Circuits Using Genetic Programming

John R. Koza
Computer Science Dept.
Stanford University
Stanford, California 94305
koza@cs.stanford.edu
http://www-cs-faculty.stanford.edu/~koza/

Forrest H Bennett III
Visiting Scholar
Computer Science Dept.
Stanford University
Stanford, California 94305
fhb3@slip.net

David Andre
Computer Science Division
University of California
Berkeley, California
dandre@cs.berkeley.edu

Martin A. Keane
Martin Keane Inc.
5733 West Grover
Chicago, Illinois 60630
makeane@ix.netcom.com

Abstract. The design (synthesis) of analog electrical circuits entails the creation of both the topology and sizing (numerical values) of all of the circuit's components. There has previously been no general automated technique for automatically designing an analog electrical circuit from a high-level statement of the circuit's desired behavior. This paper shows how genetic programming can be used to automate the design of both the topology and sizing of a suite of five prototypical analog circuits, including a lowpass filter, a tri-state frequency discriminator circuit, a 60 dB amplifier, a computational circuit for the square root, and a time-optimal robot controller circuit. All five of these genetically evolved circuits constitute instances of an evolutionary computation technique solving a problem that is usually thought to require human intelligence.

1. Introduction

The design process entails creation of a complex structure to satisfy user-defined requirements. The design of analog electrical circuits is particularly challenging because it is generally viewed as requiring human intelligence and because it is a major activity of practicing analog electrical engineers.

The design process for analog circuits begins with a high-level description of the circuit's desired behavior and entails creation of both the topology and the sizing of a satisfactory circuit. The topology comprises the gross number of components in the circuit, the type of each component (e.g., a resistor), and a list of all connections between the components. The sizing involves specifying the values (typically numerical) of each of the circuit's component.

Considerable progress has been made in automating the design of certain categories of purely digital circuits; however, the design of analog circuits and mixed analog-digital circuits has not proved as amenable to automation. Describing "the analog dilemma," Aaserud and Nielsen (1995) noted

"Analog designers are few and far between. In contrast to digital design, most of the analog circuits are still handcrafted by the experts or so-called 'zahs' of analog design. The design process is characterized by a combination of experience and intuition and requires a thorough knowledge of the process characteristics and the detailed specifications of the actual product.

"Analog circuit design is known to be a knowledge-intensive, multiphase, iterative task, which usually stretches over a significant period of time and is performed by designers with a large portfolio of skills. It is therefore considered by many to be a form of art rather than a science."

There has been extensive previous work on the problem of circuit design using simulated annealing, artificial intelligence, and other techniques as outlined in Koza, Bennett, Andre, Keane, and Dunlap 1997, including work using genetic algorithms (Kruiskamp and Leenaerts 1995; Grimbleby 1995; Thompson 1996). However, there has previously been no general automated technique for synthesizing an analog electrical circuit from a high-level statement of the desired behavior of the circuit. This paper presents a uniform approach to the automatic design of both the topology and sizing of analog electrical circuits.

2. Five Problems of Analog Design

This paper applies genetic programming to a suite of five problems of analog circuit design. The circuits contain a variety of types of components, including transistors, diodes, resistors, inductors, and capacitors. The circuits have varying numbers of inputs and outputs.

(1) Design a lowpass filter having one-input and one-output composed of capacitors and inductors and that passes all frequencies below 1,000 Hz and suppresses all frequencies above 2,000 Hz.

(2) Design a tri-state frequency discriminator (source identification) circuit having one input and one output that is composed of resistors, capacitors, and inductors and that produces an output of 1/2 volt and 1 volt for incoming signals whose frequencies are within 10% of 256 Hz and within 10% of 2,560 Hz, respectively, but produces an output of 0 volts otherwise.

(3) Design a computational circuit having one input and one output that is composed of transistors, diodes, resistors, and capacitors and that produces an output voltage equal to the square root of its input voltage.

(4) Design a time-optimal robot controller circuit having two inputs and one output that is composed of the above components and that navigates a constant-

speed autonomous mobile robot with nonzero turning radius to an arbitrary destination in minimal time.

(5) Design an amplifier composed of the above components and that delivers amplification of 60 dB (i.e., 1,000 to 1) with low distortion and low bias.

3. Design by Genetic Programming

The circuits are developed using genetic programming (Koza 1992; Koza and Rice 1992), an extension of the genetic algorithm (Holland 1975) in which the population consists of computer programs. Multipart programs consisting of a main program and one or more reusable, parametrized, hierarchically-called subprograms can be evolved using automatically defined functions (Koza 1994a, 1994b). Architecture-altering operations (Koza 1995) automatically determine the number of such subprograms, the number of arguments that each possesses, and the nature of the hierarchical references, if any, among such automatically defined functions. For current research in genetic programming, see Kinnear 1994, Angeline and Kinnear 1996, Koza, Goldberg, Fogel, and Riolo 1996, Koza et al. 1997, and Banzhaf, Nordin, Keller, and Francone 1998.

A computer program is not a circuit. Genetic programming can be applied to designing circuits if a mapping is established between the program trees (rooted, point-labeled trees – that is, acyclic graphs – with ordered branches) used in genetic programming and the line-labeled cyclic graphs germane to electrical circuits. The principles of developmental biology, the creative work of Kitano (1990) on using genetic algorithms to evolve neural networks, and the innovative work of Gruau (1992) on using genetic programming to evolve neural networks provide motivation for a technique for mapping trees into circuits by means of a growth process that begins with an embryo. See also Brave 1996 on using genetic programming to evolve finite automata. For circuits, the embryo typically includes the inputs and outputs of the particular circuit being designed and a test harness of fixed components (such as source and load resistors). The embryo also contains modifiable wires. The embryo is a valid electrical circuit; however, until these wires are modified, the circuit does not produce interesting output. An electrical circuit is developed by progressively applying the functions in a circuit-constructing program tree to the modifiable wires of the embryo (and, during the developmental process, to new components and modifiable wires).

The functions in the circuit-constructing program trees are divided into four categories: (1) topology-modifying functions that alter the circuit topology, (2) component-creating functions that insert components into the circuit, (3) arithmetic-performing functions that appear in subtrees as argument(s) to the component-creating functions and specify the numerical value of the component, and (4) automatically defined functions that appear in the function-defining branches and potentially enable certain substructures of the circuit to be reused (with parameterization). Every program tree translates into a valid electrical circuit.

Each branch of the program tree is created in accordance with a constrained syntactic structure. Branches are composed of construction-continuing subtrees that continue the developmental process and arithmetic-performing subtrees that determine the numerical value of components. Topology-modifying functions have one or more construction-continuing subtrees, but no arithmetic-performing subtree. Component-creating functions have one or more construction-continuing subtrees

180

and typically have one arithmetic-performing subtree. This constrained syntactic structure is preserved using structure-preserving crossover with point typing (see Koza 1994a).

3.1.   The Embryonic Circuit

The starting point for the development of an electrical circuit is an embryo. The embryo is a valid (albeit useless) electrical circuit that contains at least one modifiable wire. The embryo will typically also include the input(s) and output(s) of the circuit and certain additional electrically sensible fixed components, such as a source resistor for each input and a load resistor for each output point. The specific embryo used depends on the number of inputs and outputs.

Figure 1 shows a one-input, one-output embryonic circuit in which **USOURCE** is the input signal and **UOUT** is the output signal (the probe point). The circuit is driven by an incoming alternating circuit source **USOURCE**. There is a fixed load resistor **RLOAD** and a fixed source resistor **RSOURCE** in the embryo. In addition to the fixed components, there is a modifiable wire **Z0** between nodes 2 and 3. All development originates from this modifiable wire.

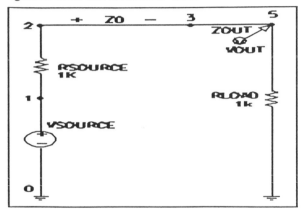

Figure 1   One-input, one-output embryo.

An electrical circuit is created by executing a circuit-constructing program tree that contains various component-creating and topology-modifying functions. The population of individuals being bred by genetic programming consists of such program trees. Each tree in the population creates one circuit. The modifiable wire(s) in the embryonic circuit are transformed into various circuit components and connections as the embryo is developed into a fully developed electrical circuit by a circuit-constructing program tree. The electrical circuit is progressively developed by progressively applying the functions of the circuit-constructing program tree, in a specified orderly way, to the embryonic circuit (and its successor circuits). After all the functions of the program tree have been executed, the developmental process is completed and the result is a fully developed electrical circuit.

3.2.   Component-Creating Functions

Each program tree contains component-creating functions and topology-modifying functions. The component-creating functions insert a component into the developing circuit and assign component value(s) to the component.

Each component-creating function has a writing head that points to an associated component in the developing circuit and modifies that component in a specified manner. The construction-continuing subtree of each component-creating functions points to a successor function or terminal in the circuit-constructing program tree.

The arithmetic-performing subtree of a component-creating functions consists of a composition of arithmetic functions (addition and subtraction) and random constants (in the range -1.000 to +1.000). The arithmetic-performing subtree specifies the numerical value of a component by returning a floating-point value that is interpreted on a logarithmic scale as the value for the component in a range of 10 orders of magnitude (using a unit of measure that is appropriate for the particular type of component.

The two-argument resistor-creating R function causes the highlighted component to be changed into a resistor. The value of the resistor in kilo Ohms specified by its arithmetic-performing subtree.

The left part of figure 2 shows a modifiable wire $Z0$ connecting nodes 1 and 2 of a partial circuit containing four capacitors ($C2, C3, C4$, and $C5$). The circle indicates that $Z0$ has a writing head (i.e., is the highlighted component and that $Z0$ is subject to subsequent modification). The right part of the figure shows the result of applying the R function to $Z0$. The circle indicates that the newly created $R1$ has a writing head $R1$ and thus remains subject to subsequent modification.

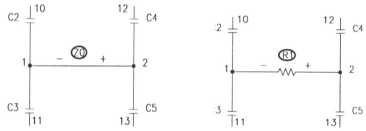

Figure 2 (a) **Modifiable wire** $Z0$. (b) **Result of** R **function.**

Similarly, the two-argument capacitor-creating C function causes the highlighted component to be changed into a capacitor whose value in micro Farads is specified by the arithmetic-performing subtrees.

The one-argument Q_D_PNP diode-creating function causes a diode to be inserted in lieu of the highlighted component. This function has only one argument because there is no numerical value associated with a diode and thus no arithmetic-performing subtree. In practice, the diode is implemented here using a pnp transistor whose collector and base are connected to each other. The Q_D_NPN function inserts a diode using an npn transistor in a similar manner.

There are also six one-argument transistor-creating functions (Q_POS_COLL_NPN, Q_GND_EMIT_NPN, Q_NEG_EMIT_NPN, Q_GND_EMIT_PNP, Q_POS_EMIT_PNP, Q_NEG_COLL_PNP) that insert a bipolar junction transistor in lieu of the highlighted component and that directly connect the collector or emitter of the newly created transistor to a fixed point of the circuit (the positive power supply, ground, or the negative power supply). For example, the Q_POS_COLL_NPN function inserts a bipolar junction transistor whose collector is connected to the positive power supply.

Each of the functions in the family of six different three-argument transistor-creating Q_3_NPN functions causes an npn bipolar junction transistor to be inserted in place of the highlighted component and one of the nodes to which the highlighted component is connected. The Q_3_NPN function creates five new nodes and three modifiable wires. There is no writing head on the new transistor, but there is a writing head on each of the three new modifiable wires. There are 12 members (called Q_3_NPN0, ..., Q_3_NPN11) in this family of functions because there are two choices of nodes (1 and 2) to be bifurcated and then there are six ways of attaching the transistor's base, collector, and emitter after the bifurcation. Similarly the family of 12 Q_3_PNP functions inserts a pnp bipolar junction transistor.

3.3.   Topology-Modifying Functions

Each topology-modifying function in a program tree points to an associated highlighted component and modifies the topology of the developing circuit.

The three-argument SERIES division function creates a series composition of the highlighted component (with a writing head), a copy of it (with a writing head), one new modifiable wire (with a writing head), and two new nodes.

The four-argument PSS parallel division function creates a parallel composition consisting of the original highlighted component (with a writing head), a copy of it (with a writing head), two new modifiable wires (each with a writing head), and two new nodes. Figure 3 shows the result of applying PSS to resistor **R1** of figure 2a.

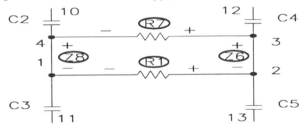

Figure 3   Result of the PSS function.

The one-argument polarity-reversing FLIP function reverses the polarity of the highlighted component.

There are six three-argument functions (T_GND_0, T_GND_1, T_POS_0, T_POS_1, T_NEG_0, T_NEG_1) that insert two new nodes and two new modifiable wires, and then make a connection to ground, positive power supply, or negative power supply, respectively.

There are two three-argument functions (PAIR_CONNECT_0 and PAIR_CONNECT_1) that enable distant parts of a circuit to be connected together. The first PAIR_CONNECT to occur in the development of a circuit creates two new wires, two new nodes, and one temporary port. The next PAIR_CONNECT creates two new wires and one new node, connects the temporary port to the end of one of these new wires, and then removes the temporary port.

The one-argument NOOP function has no effect on the highlighted component; however, it delays activity on the developmental path on which it appears in relation to other developmental paths in the overall program tree.

The zero-argument END function causes the highlighted component to lose its writing head, thereby ending that particular developmental path.

The zero-argument SAFE_CUT function causes the highlighted component to be removed from the circuit provided that the degree of the nodes at both ends of the highlighted component is three (i.e., no dangling components or wires are created).

4. Preparatory Steps

Before applying genetic programming to a problem of circuit design, seven major preparatory steps are required: (1) identify the suitable embryonic circuit, (2) determine the architecture of the overall circuit-constructing program trees, (3) identify the terminals of the program trees, (4) identify the primitive functions contained in these program trees, (5) create the fitness measure, (6) choose parameters, and (7) determine the termination criterion and method of result designation.

4.1. Embryonic Circuit

The embryonic circuit used on a particular problem depends on the circuit's number of inputs and outputs. For example, the one-input, one-output embryo (figure 1) was used for the lowpass filter. However, the robot controller circuit has two inputs and both inputs need their own separate source resistors. Moreover, the embryo has three modifiable wires in order to provide full connectivity between the two inputs and the one output. In some problems, such as the amplifier, the embryo contains additional fixed components because of additional problem-specific functionality of the test harness.

4.2. Program Architecture

Since there is one result-producing branch in the program tree for each modifiable wire in the embryo, the architecture of each circuit-constructing program tree depends on the embryonic circuit. One result-producing branch was used for the frequency discriminator and the computational circuit; two were used for lowpass filter problem; and three were used for the robot controller and amplifier. The architecture of each circuit-constructing program tree also depends on the use, if any, of automatically defined functions. Automatically defined functions and architecture-altering operations were used in the frequency discriminator, robot controller, and amplifier. For these problems, each program tree in the initial random population of programs had a uniform architecture with no automatically defined functions. In later generations, the number of automatically defined functions, if any, emerged as a consequence of the architecture-altering operations.

4.3. Function and Terminal Sets

The function set for each design problem depended on the type of electrical components that were used to construct the circuit. Capacitors, diodes, and transistors were used for the computational circuit, the robot controller, and the amplifier. Resistors were used for the frequency discriminator. When transistors were used, functions to provide connectivity to the positive and negative power supplies were also included,.

For the computational circuit, the robot controller, and the amplifier, the function set, $\mathcal{F}_{ccs\text{-}initial}$, for each construction-continuing subtree was

$\mathcal{F}_{ccs\text{-}initial} = \{$R, C, SERIES, PSS, PSL, FLIP, NOOP, T_GND_0, T_GND_1,
T_POS_0, T_POS_1, T_NEG_0, T_NEG_1, PAIR_CONNECT_0,
PAIR_CONNECT_1, Q_D_NPN, Q_D_PNP, Q_3_NPN0, ...,
Q_3_NPN11, Q_3_PNP0, ..., Q_3_PNP11, Q_POS_COLL_NPN,
Q_GND_EMIT_NPN, Q_NEG_EMIT_NPN, Q_GND_EMIT_PNP,
Q_POS_EMIT_PNP, Q_NEG_COLL_PNP$\}$.

For the npn transistors, the Q2N3904 model was used. For pnp transistors, the Q2N3906 model was used.

The initial terminal set, $T_{\text{ccs-initial}}$, for each construction-continuing subtree was

$T_{\text{ccs-initial}} = \{\text{END, SAFE\_CUT}\}$.

The initial terminal set, $T_{\text{aps-initial}}$, for each arithmetic-performing subtree consisted of

$T_{\text{aps-initial}} = \{\Re\}$,

where $\Re$ represents floating-point random constants from -1.0 to $+1.0$.

The function set, F_{aps}, for each arithmetic-performing subtree was,

$F_{\text{aps}} = \{+, -\}$.

The terminal and function sets were identical for all result-producing branches for a particular problem.

For the lowpass filter and frequency discriminator, there was no need for functions to provide connectivity to the positive and negative power supplies.

For the frequency discriminator, the robot controller, and the amplifier, the architecture-altering operations were used and the set of potential new functions, $F_{\text{potential}}$, was

$F_{\text{potential}} = \{\text{ADF0, ADF1, ...}\}$.

The set of potential new terminals, $T_{\text{potential}}$, for the automatically defined functions was

$T_{\text{potential}} = \{\text{ARG0}\}$.

The architecture-altering operations changed the function set, F_{ccs} for each construction-continuing subtree of all three result-producing branches and the function-defining branches, so that

$F_{\text{ccs}} = F_{\text{ccs-initial}} \cup F_{\text{potential}}$.

The architecture-altering operations generally changed the terminal set for automatically defined functions, $T_{\text{aps-adf}}$, for each arithmetic-performing subtree, so that

$T_{\text{aps-adf}} = T_{\text{aps-initial}} \cup T_{\text{potential}}$.

4.4. Fitness Measure

The fitness measure varies for each problem. The high-level statement of desired circuit behavior is translated into a well-defined measurable quantity that can be used by genetic programming to guide the evolutionary process. The evaluation of each individual circuit-constructing program tree in the population begins with its execution. This execution progressively applies the functions in each program tree to an embryonic circuit, thereby creating a fully developed circuit. A netlist is created that identifies each component of the developed circuit, the nodes to which each component is connected, and the value of each component. The netlist becomes the input to the 217,000-line SPICE (Simulation Program with Integrated Circuit Emphasis) simulation program (Quarles, Newton, Pederson, and Sangiovanni-Vincentelli 1994). SPICE then determines the behavior of the circuit. It was necessary to make considerable modifications in SPICE so that it could run as a submodule within the genetic programming system.

4.4.1 Lowpass Filter

A simple *filter* is a one-input, one-output electronic circuit that receives a signal as its input and passes the frequency components of the incoming signal that lie in a specified range (called the *passband*) while suppressing the frequency components that lie in all other frequency ranges (the *stopband*).

The desired lowpass LC filter should have a passband below 1,000 Hz and a stopband above 2,000 Hz. The circuit is driven by an incoming AC voltage source with a 2 volt amplitude. If the source (internal) resistance **RSOURCE** and the load resistance **RLOAD** in the embryonic circuit are each 1 kilo Ohm, the incoming 2 volt signal is divided in half.

The *attenuation* of the filter is defined in terms of the maximum signal in its stopband relative to the reference voltage (half of 2 volt here). A *decibel* is a unitless measure of relative voltage that is defined as 20 times the common (base 10) logarithm of the ratio between the voltage at a particular probe point and a reference voltage.

A voltage in the passband of exactly 1 volt and a voltage in the stopband of exactly 0 volts are considered ideal. A voltage in the passband of between 970 millivolts and 1 volt and a voltage in the stopband of between 0 volts and 1 millivolt are regarded as acceptable. The (preferably small) shortfall from 1 volt in the passband is called the *passband attenuation*. Similarly, the (preferably small) signal in the stopband is called the *stopband attenuation*. Any voltage lower than 970 millivolts in the passband and any voltage above 1 millivolts in the stopband is regarded as unacceptable. A fifth-order *elliptic (Cauer) filter* with a modular angle Θ of 30 degrees (i.e., the arcsin of the ratio of the boundaries of the passband and stopband) and a reflection coefficient ρ of 20% satisifes these design goals.

Since the high-level statement of behavior for the desired circuit is expressed in terms of frequencies, the voltage **UOUT** is measured in the frequency domain. SPICE performs an AC small signal analysis and report the circuit's behavior over five decades (between 1 Hz and 100,000 Hz) with each decade being divided into 20 parts (using a logarithmic scale),so that there are a total of 101 fitness cases.

Fitness is measured in terms of the sum over these cases of the absolute weighted deviation between the actual value of the voltage that is produced by the circuit at the probe point **UOUT** and the target value for voltage. The smaller the value of fitness, the better. A fitness of zero represents an (unattainable) ideal filter.

Specifically, the standardized fitness is

$$F(t) = \sum_{i=0}^{100} \left(W(d(f_i), f_i) d(f_i) \right).$$

where f_i is the frequency of fitness case i; $d(x)$ is the absolute value of the difference between the target and observed values at frequency x; and $W(y,x)$ is the weighting for difference y at frequency x.

The fitness measure is designed to not penalize ideal values, to slightly penalize every acceptable deviation, and to heavily penalize every unacceptable deviation. Specifically, the procedure for each of the 61 points in the 3-decade interval between 1 Hz and 1,000 Hz is as follows: If the voltage equals the ideal value of 1.0 volt in this interval, the deviation is 0.0. If the voltage is between 970 millivolts and 1 volt, the absolute value of the deviation from 1 volt is weighted by a factor of 1.0. If the voltage is less than 970 millivolts, the absolute value of the deviation from 1 volt is weighted by a factor of 10.0. The acceptable and unacceptable deviations for each of the 35 points from 2,000 Hz to 100,000 Hz are similarly weighed (by 1.0 or 10.0).

For each of the five "don't care" points between 1,000 and 2,000 Hz, the deviation is deemed to be zero.

The number of "hits" for this problem (and all other problems herein) is defined as the number of fitness cases for which the voltage is acceptable or ideal or that lie in the "don't care" band (for a filter).

Many of the random initial circuits and many that are created by the crossover and mutation operations in subsequent generations cannot be simulated by SPICE. These circuits receive a high penalty value of fitness (10^8) and become the worst-of-generation programs for each generation.

4.4.2   Tri-state Frequency Discriminator

Fitness is the sum, over 101 fitness cases, of the absolute weighted deviation between the actual value of the voltage that is produced by the circuit and the target value.

The three points that are closest to the band located within 10% of 256 Hz are 229.1 Hz, 251.2 Hz, and 275.4 Hz. The procedure for each of these three points is as follows: If the voltage equals the ideal value of 1/2 volts in this interval, the deviation is 0.0. If the voltage is more than 240 millivolts from 1/2 volts, the absolute value of the deviation from 1/2 volts is weighted by a factor of 20. If the voltage is more than 240 millivolts of 1/2 volts, the absolute value of the deviation from 1/2 volts is weighted by a factor of 200. This arrangement reflects the fact that the ideal output voltage for this range of frequencies is 1/2 volts, the fact that a 240 millivolts discrepancy is acceptable, and the fact that a larger discrepancy is not acceptable.

Similar weighting was used for the three points (2,291 Hz, 2,512 Hz, and 2,754 Hz) that are closest to the band located within 10% of 2,560 ,Hz.

The procedure for each of the remaining 95 points is as follows: If the voltage equals the ideal value of 0 volts, the deviation is 0.0. If the voltage is within 240 millivolts of 0 volts, the absolute value of the deviation from 0 volts is weighted by a factor of 1.0. If the voltage is more than 240 millivolts from 0 volts, the absolute value of the deviation from 0 volts is weighted by a factor of 10. For details, see Koza, Bennett, Lohn, Dunlap, Andre, and Keane 1997b.

4.4.3   Computational Circuit

SPICE is called to perform a DC sweep analysis at 21 equidistant voltages between –250 millivolts and +250 millivolts. Fitness is the sum, over these 21 fitness cases, of the absolute weighted deviation between the actual value of the voltage that is produced by the circuit and the target value for voltage. For details, see Koza, Bennett, Lohn, Dunlap, Andre, and Keane 1997a.

4.4.4   Robot Controller Circuit

The fitness of a robot controller was evaluated using 72 randomly chosen fitness cases each representing a different target point. Fitness is the sum, over the 72 fitness cases, of the travel times. If the robot came within a capture radius of 0.28 meters of its target point before the end of the 80 time steps allowed for a particular fitness case, the contribution to fitness for that fitness case was the actual time. However, if the robot failed to come within the capture radius during the 80 time steps, the contribution to fitness was 0.160 hours (i.e., double the worst possible time).

SPICE performs a nested DC sweep, which provides a way to simulate the DC behavior of a circuit with two inputs. It resembles a nested pair of FOR loops in a computer program in that both of the loops have a starting value for the voltage, an increment, and an ending value for the voltage. For each voltage value in the outer loop, the inner loop simulates the behavior of the circuit by stepping through its range of voltages. Specifically, the starting value for voltage is –4 volt, the step size is 0.2 volt, and the ending value is +4 volt. These values correspond to the dimensions of the robot's world of 64 square meters extending 4 meters in each of the four directions from the origin of a coordinate system (i.e., 1 volt equals 1 meter). For details, see Koza, Bennett, Keane, and Andre 1997.

4.4.5   60 dB Amplifier

SPICE was requested to perform a DC sweep analysis to determine the circuit's response for several different DC input voltages. An ideal inverting amplifier circuit would receive the DC input, invert it, and multiply it by the amplification factor. A circuit is flawed to the extent that it does not achieve the desired amplification, the output signal is not perfectly centered on 0 volts(i.e., it is biased), or the DC response is not linear. Fitness is calculated by summing an amplification penalty, a bias penalty, and two non-linearity penalties – each derived from these five DC outputs. For details, see Bennett, Koza, Andre, and Keane 1996.

4.5.   Control Parameters

The population size, M, was 640,000 for all problems. Other parameters were substantially the same for each of the five problems and can be found in the references cited above.

4.6.   Implementation on Parallel Computer

Each problem was run on a medium-grained parallel Parsytec computer system (Andre and Koza 1996) consisting of 64 80-MHz PowerPC 601 processors arranged in an 8 by 8 toroidal mesh with a host PC Pentium type computer. The distributed genetic algorithm was used with a population size of $Q = 10,000$ at each of the $D = 64$ demes (semi-isolated subpopulations). On each generation, four boatloads of emigrants, each consisting of $B = 2\%$ (the migration rate) of the node's subpopulation (selected on the basis of fitness) were dispatched to each of the four adjacent processing nodes.

5. Results

In all five problems, fitness was observed to improve over successive generations. Satisfactory results were generated in every case on the first or second run. When two runs were required, the first produced an almost satisfactory result.

5.1.   Lowpass Filter

Many of the runs produced lowpass filters having a topology similar to that employed by human engineers. For example, in generation 32 of one run, a circuit (figure 4) was evolved with a near-zero fitness of 0.00781. The circuit was 100% compliant with the design requirements in that it scored 101 hits (out of 101). This circuit had the recognizable ladder topology (46) of a Butterworth or Chebychev filter (i.e., a composition of series inductors horizontally with capacitors as vertical shunts).

Figure 5 shows the behavior in the frequency domain of this evolved lowpass filter. As can be seen, the evolved circuit delivers about 1 volt for all frequencies up to 1,000 Hz and about 0 volts for all frequencies above 2,000 Hz.

In another run, a 100% compliant recognizable "bridged T" arrangement was evolved. In yet another run using automatically defined functions, a 100% compliant circuit emerged with the recognizable elliptic topology that was invented and patented by Cauer. When invented, the Cauer filter was a significant advance (both theoretically and commercially) over the Butterworth and Chebychev filters.

Thus, genetic programming rediscovered the ladder topology of the Butterworth and Chebychev filters, the "bridged T" topology, and the elliptic topology.

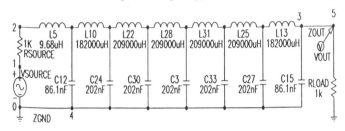

Figure 4   Evolved 7-rung ladder filter.

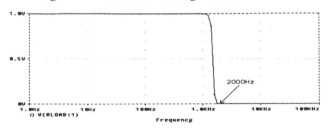

Figure   5   Frequency domain behavior of evolved 7-rung ladder filter.

5.2.   Tri-state Frequency Discriminator

The evolved three-way tri-state frequency discriminator circuit from generation 106 scores 101 hits (out of 101). Figure 6 shows this circuit (after expansion of its automatically defined functions). The circuit produces the desired outputs of 1 volt and 1/2 volts (each within the allowable tolerance) for the two specified bands of frequencies and the desired near-zero signal for all other frequencies.

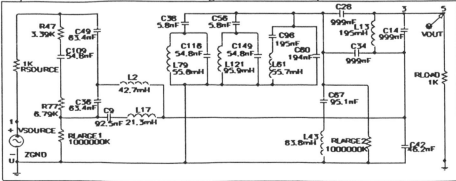

Figure 6   Evolved frequency discriminator.

5.3. Computational Circuit

The genetically evolved computational circuit for the square root from generation 60 (figure 7), achieves a fitness of 1.68, and has 36 transistors, two diodes, no capacitors, and 12 resistors (in addition to the source and load resistors in the embryo). The output voltages produced by this best-of-run circuit are almost exactly the required values.

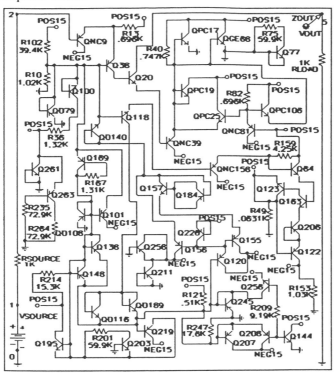

Figure 7 Evolved square root circuit.

5.4. Robot Controller Circuit

The best-of-run time-optimal robot controller circuit (figure 8) appeared in generation 31, scores 72 hits, and achieves a near-optimal fitness of 1.541 hours. In comparison, the optimal value of fitness for this problem is known to be 1.518 hours. This best-of-run circuit has 10 transistors and 4 resistors. The program has one automatically defined function that is called twice (incorporated into the figure).

5.5. 60 dB Amplifier

The best circuit from generation 109 (figure 9) achieves a fitness of 0.178. Based on a DC sweep, the amplification is 60 dB here (i.e., 1,000-to-1 ratio) and the bias is 0.2 volt. Based on a transient analysis at 1,000 Hz, the amplification is 59.7 dB; the bias is 0.18 volts; and the distortion is very low (0.17%). Based on an AC sweep, the amplification at 1,000 Hz is 59.7 dB; the flatband gain is 60 dB; and the 3 dB bandwidth is 79, 333 Hz. Thus, a high-gain amplifier with low distortion and acceptable bias has been evolved.

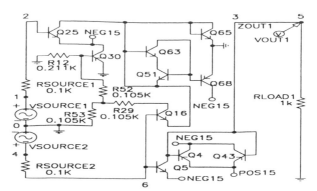

Figure 8   Evolved robot controller.

6. Other Circuits

Numerous other circuits have been similarly designed, including an asymmetric bandpass filter, crossover filter, comb filter, amplifier, temperature-sensing circuit, and voltage reference circui (Koza, Bennett, Andre, and Keane 1996; Koza, Andre, Bennett, and Keane 1996; and Koza, Bennett, Andre, Keane, and Dunlap 1997; Bennett 1997).

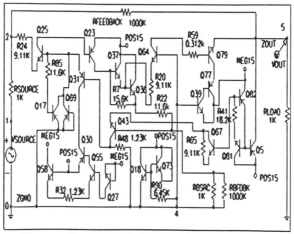

Figure 9   Genetically evolved amplifier.

References

Aaserud, O. and Nielsen, I. Ring. 1995. Trends in current analog design: A panel debate. *Analog Integrated Circuits and Signal Processing*. 7(1) 5-9.

Andre, David and Koza, John R. 1996. Parallel genetic programming: A scalable implementation using the transputer architecture. In Angeline, P. J. and Kinnear, K. E. Jr. (editors). 1996. *Advances in Genetic Programming 2*. Cambridge: MIT Press.

Angeline, Peter J. and Kinnear, Kenneth E. Jr. (editors). 1996. *Advances in Genetic Programming 2.* Cambridge, MA: The MIT Press.

Banzhaf, Wolfgang, Nordin, Peter, Keller, Robert E., and Francone, Frank D. 1998. *Genetic Programming – An Introduction.* San Francisco, CA: Morgan Kaufmann and Heidelberg: dpunkt.

Bennett III, Forrest H, Koza, John R., Andre, David, and Keane, Martin A. 1996. Evolution of a 60 Decibel op amp using genetic programming. In Higuchi, Tetsuya, Iwata, Masaya, and Lui, Weixin (editors). *Proceedings of International Conference on Evolvable Systems: From Biology to Hardware.* Lecture Notes in Computer Science, Volume 1259. Berlin: Springer-Verlag. Pages 455-469.

Bennett III, Forrest H. 1997. Programming computers by means of natural selection: Applicaton to analog circuit synthesis. In Hara, F. and Yoshida, K. (editors). *Proceedings of International Symposium on System Life, July 21 – 22, 1997.* Tokyo: The Japan Society of Mechanical Engineers. Pages 41 – 50.

Brave, Scott. 1996. Evolving deterministic finite automata using cellular encoding. In Koza, John R., Goldberg, David E., Fogel, David B., and Riolo, Rick L. (editors). 1996. *Genetic Programming 1996: Proceedings of the First Annual Conference, July 28-31, 1996, Stanford University.* Cambridge, MA: MIT Press. Pages 39–44.

Grimbleby, J. B. 1995. Automatic analogue network synthesis using genetic algorithms. *Proceedings of the First International Conference on Genetic Algorithms in Engineering Systems: Innovations and Applications.* London: Institution of Electrical Engineers. Pages 53–58.

Gruau, Frederic. 1992. *Cellular Encoding of Genetic Neural Networks.* Technical report 92-21. Laboratoire de l'Informatique du Parallélisme. Ecole Normale Supérieure de Lyon. May 1992.

Holland, John H. 1975. *Adaptation in Natural and Artificial Systems.* Ann Arbor, MI: University of Michigan Press.

Kinnear, Kenneth E. Jr. (editor). 1994. *Advances in Genetic Programming.* Cambridge, MA: The MIT Press.

Kitano, Hiroaki. 1990. Designing neural networks using genetic algorithms with graph generation system. *Complex Systems.* 4(1990) 461–476.

Koza, John R. 1992. *Genetic Programming: On the Programming of Computers by Means of Natural Selection.* Cambridge, MA: MIT Press.

Koza, John R. 1994a. *Genetic Programming II: Automatic Discovery of Reusable Programs.* Cambridge, MA: MIT Press.

Koza, John R. 1994b. *Genetic Programming II Videotape: The Next Generation.* Cambridge, MA: MIT Press.

Koza, John R. 1995. Evolving the architecture of a multi-part program in genetic programming using architecture-altering operations. In McDonnell, John R., Reynolds, Robert G., and Fogel, David B. (editors). 1995. *Evolutionary Programming IV: Proceedings of the Fourth Annual Conference on Evolutionary Programming.* Cambridge, MA: The MIT Press. Pages 695–717.

Koza, John R., Andre, David, Bennett III, Forrest H, and Keane, Martin A. 1996. Use of automatically defined functions and architecture-altering operations in automated circuit synthesis using genetic programming. In Koza, John R., Goldberg, David E., Fogel, David B., and Riolo, Rick L. (editors). 1996. *Genetic Programming 1996: Proceedings of the First Annual Conference.* Cambridge, MA: The MIT Press.

Koza, John R., Bennett III, Forrest H, Andre, David, and Keane, Martin A. 1996. Automated WYWIWYG design of both the topology and component values of analog electrical circuits using genetic programming. In Koza, John R., Goldberg, David E., Fogel, David B., and Riolo, Rick L. (editors). 1996. *Genetic Programming 1996: Proceedings of the First Annual Conference.* Cambridge, MA: The MIT Press.

Koza, John R., Bennett III, Forrest H, Andre, David, Keane, Martin A, and Dunlap, Frank. 1997. Automated synthesis of analog electrical circuits by means of genetic programming. *IEEE Transactions on Evolutionary Computation.* 1(2). Pages 109 – 128.

Koza, John R., Bennett III, Forrest H, Keane, Martin A., and Andre, David. 1997. Automatic programming of a time-optimal robot controller and an analog electrical circuit to implement the robot controller by means of genetic programming. *Proceedings of 1997 IEEE International Symposium on Computational Intelligence in Robotics and Automation.* Los Alamitos, CA; Computer Society Press. Pages 340 – 346.

Koza, John R., Bennett III, Forrest H, Lohn, Jason, Dunlap, Frank, Andre, David, and Keane, Martin A. 1997. Automated synthesis of computational circuits using genetic programming. *Proceedings of the 1997 IEEE Conference on Evolutionary Computation.* Piscataway, NJ: IEEE Press. 447–452.

Koza, John R., Bennett III, Forrest H, Lohn, Jason, Dunlap, Frank, Andre, David, and Keane, Martin A. 1997b. Use of architecture-altering operations to dynamically adapt a three-way analog source identification circuit to accommodate a new source. In Koza, John R., Deb, Kalyanmoy, Dorigo, Marco, Fogel, David B., Garzon, Max, Iba, Hitoshi, and Riolo, Rick L. (editors). 1997. *Genetic Programming 1997: Proceedings of the Second Annual Conference* San Francisco, CA: Morgan Kaufmann. 213 – 221.

Koza, John R., Deb, Kalyanmoy, Dorigo, Marco, Fogel, David B., Garzon, Max, Iba, Hitoshi, and Riolo, Rick L. (editors). 1997. *Genetic Programming 1997: Proceedings of the Second Annual Conference* San Francisco, CA: Morgan Kaufmann.

Koza, John R., Goldberg, David E., Fogel, David B., and Riolo, Rick L. (editors). 1996. *Genetic Programming 1996: Proceedings of the First Annual Conference.* Cambridge, MA: The MIT Press.

Koza, John R., and Rice, James P. 1992. *Genetic Programming: The Movie.* Cambridge, MA: MIT Press.

Kruiskamp Marinum Wilhelmus and Leenaerts, Domine. 1995. DARWIN: CMOS opamp synthesis by means of a genetic algorithm. *Proceedings of the 32nd Design Automation Conference.* New York, NY: Association for Computing Machinery. Pages 433–438.

Quarles, Thomas, Newton, A. R., Pederson, D. O., and Sangiovanni-Vincentelli, A. 1994. *SPICE 3 Version 3F5 User's Manual.* Department of Electrical Engineering and Computer Science, University of California, Berkeley, CA. March 1994.

Thompson, Adrian. 1996. Silicon evolution. In Koza, John R., Goldberg, David E., Fogel, David B., and Riolo, Rick L. (editors). 1996. *Genetic Programming 1996: Proceedings of the First Annual Conference.* Cambridge, MA: MIT Press.

Improving Engineering Design Models Using An Alternative Genetic Programming Approach

Andrew H. Watson.
P.E.D.C,
University of Plymouth,
Plymouth. PL4 8AA. UK.
awatson@plymouth.ac.uk

Ian C. Parmee
P.E.D.C,
University of Plymouth,
Plymouth. PL4 8AA. UK.
iparmee@plymouth.ac.uk

Abstract. This paper describes an alternative approach to Genetic Programming (GP) for engineering design model development. The algorithm is initially developed to solve Boolean induction and simple symbolic regression problems within a discrete search space. This technique, called "**DRAM-GP**" (i.e. **D**istributed, **R**apid, **A**ttenuated **M**emory **GP**), is based upon a steady state population utilising a novel constrained complexity crossover operator. Node complexity weightings are introduced to provide a basis for speciation. Separate species of solutions, classified by complexity can be established which act as discrete GP sub-populations communicating with each other via crossover. The technique is extended to incorporate *both* continuous and discrete search spaces (**HDRAM-GP**" i.e. Hybrid **DRAM-GP**). **HDRAM-GP** includes a real numbered Genetic Algorithm (GA) to aid search in the continuous space. Its application is demonstrated on engineering fluid dynamics systems.

1. Introduction

Preliminary engineering design software often includes unavoidable function approximation. This can be due to a requirement for keeping computational expense to a minimum or to lack of knowledge and the inclusion of empirically derived coefficients (i.e. friction loss, drag etc.). The objective of the research described here is to improve these approximate functions describing physical processes using GP and other complementary Adaptive Search (AS) techniques, to provide a better correlation to either empirical data or to results from more complex, definitive analysis techniques (i.e. Finite Element Analysis and Computational Fluid Dynamics). This approach could reduce the risk associated with the use of preliminary design software without increased CPU time thereby reducing design lead time by lessening the dependence upon computationally expensive detailed analysis.

The objective of the research described within this paper is to evolve solutions to simple discrete symbolic regression and also physical relationships in engineering systems described by both discrete *and* continuous variables using the GP paradigm. This work incorporates alternative GP approaches which address problems associated with standard GP. These problems have been reported by Iba et. al.[1] and are:-

1. Random sub-tree crossover disrupts beneficial sub-trees in tree structures.
2. No evaluation of tree descriptions. Trees can grow exponentially large or so small that they degrade search efficiency.

Traditional GP blindly combines sub-trees, by applying crossover operations. This can often disrupt beneficial sub-functions in tree structures. Thus, crossover operations seem ineffective as a means of constructing higher-order functions. Recombination operators (such as swapping sub-trees or nodes) often cause radical changes in the semantics of the trees. This *semantic disruption* [1] is due to the 'context-sensitive' representation of GP trees. As a result, useful sub-trees may not be able to contribute to higher fitness values of the whole tree, and the accumulation of useful sub-functions may be disturbed. To avoid this, Koza [2,3] proposes a strategy called Automatic Defining Functions (ADF's) for maintenance of useful sub-trees. However, research within the Plymouth Engineering Design Centre relating to the manipulation of fixed length design hierarchies described by both discrete and continuous variables has shown that speciation in terms of the discrete variables and the introduction of restricted crossover regimes can contribute significantly to the identification of high performance structures [11]. Although addressing a different domain, this research is relevant to the semantic disruption problems associated with standard GP representations and also to the issue of exploring both continuous and discrete search spaces. This previous work has provided a foundation for the development of the strategies described here.

The fitness definitions used in traditional GP do not include evaluations of the tree descriptions. Without the necessary control mechanisms, trees may grow exponentially large, increasing the evaluation procedures, or so small that they degrade search efficiency. Usually the maximum depth of trees is set in order to control tree sizes, but an appropriate depth is not always known beforehand. Kinnear [4] proposed using a size component in the fitness definition; i.e. the size of the tree is multiplied by a size factor, and the result is added to the raw fitness value. The use of a minimum description length (MDL) based fitness function for evaluating tree structures has been used together with a local hill-climber [1,5]. This fitness definition involves a trade-off between certain structural details of the tree and its fitting (or classification) of errors.

In order to produce an efficient guided crossover operator to search the symbolic search space a symbolic function classification is required which can then be used to minimise *semantic disruption*. A classification in terms of Node Complexity (NC), is introduced which includes information regarding the lengths of the individuals. Semantic disruption is therefore minimised whilst tree length is controlled. Earlier work has shown the effectiveness of steady state GP combined with NC controlled crossover. This earlier technique, called "**RAM-GP**" [10] (i.e. **R**apid, **A**ttenuated **M**emory **GP**), uses NC weightings as a basis for the crossover operator together with a high rate of mutation and steady state GP.

2. Principles of DRAM-GP

RAM-GP has been extended to incorporate sub-populations of solutions classified by the complexity of the root node of each individual. These sub-populations act as discrete GP sub-populations which communicate with each other via crossover. We have named this approach "**DRAM-GP**" [14]. The main concepts of DRAM-GP involve a steady state GP with constrained complexity crossover (CCC). Crossover is constrained by node complexity weighting values. The root node will give a complexity rating of the whole tree, and is thus used to speciate the population into smaller sub-populations. These points are discussed in detail below.

2.1 Steady State GP

In the classical GP model of evolution by generation [2] (the generation model), each reproductive phase involves the creation of a complete new population of individuals, by selecting parents from the old population and applying genetic operations. The new population then replaces the old in one atomic step. Steady state GP has been investigated by Kinnear [6]. The process involves evaluating an individual immediately for fitness, and then merging it into the population (or in this case a species), in place of the existing lowest fitness individual. There are no generations in steady state GP, a *generation equivalent* has passed when the number of new individuals that have been generated is equal to the population size. The population size being the *total* number of individuals (i.e. species population size x number of species).

2.2 Node Complexity (NC)

NC weighting is a measure of the complexity of a tree and all of its nodes. If for example we have a functional set and terminal set consisting of:-

$F = \{ +, -, *, \% \}$, and $T = \{ a, b, c, d, e, f \}$

(as in the two-box problem [3]), we can weight these (e.g. all terminals=1.0, plus=1.1, minus=1.1, multiply=1.2, divide=1.2) each NC value is then a function of the NC values of the nodes below it and the weighting of that node. An example of the NC weighting is shown in figure 1.

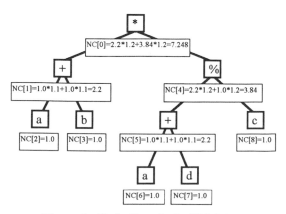

Figure 1 - Node Complexity Weighting

Each node has a specific weighting factor which is applied to the NC values below them. The NC value is then the sum of these adjusted lower node values. It can be seen that the complexity of the tree will *decrease* with tree depth, for example in figure 1 NC[0]=7.248 (the root node) and NC[7]=1.0 (a terminal). Crossover is then constrained by only crossing sub-trees with *similar* NC values. (All of the results within this paper are restricted to between ± 2.0 NC). This then provides a numerical complexity measure which controls crossover and minimises building block disruption by ensuring some similarity between crossed sub-trees. Tree lengths are also indirectly controlled.

2.3 Species Sub-Populations

The population is equally divided into sub-populations or 'species'. The run parameters that define the species groupings is the minimum and maximum NC values and the number of species used. The sub-populations are then divided equally between these two limits. If for example, the minimum NC value is set at 10 and the maximum NC value is set to 40 with 3 species and a total population size of 300, then each species will have 100 individuals with the following NC[0] values:-

Species 1 -   10.0 <NC[0]<20.0
Species 2 -   20.0 <NC[0]<30.0
Species 3 -   30.0 <NC[0]<40.0

Communication between sub-populations is then achieved through the action of crossover. As a new child individual is produced it is possible that its complexity changes, and it is placed in the correct species grouping. If a crossed individual's NC[0] value lies outside the species ranges then the individual is discarded.

2.4 Constrained Complexity Crossover (CCC)

CCC is initiated by randomly choosing parents P1 and P2 from the total population. A cross point CP1 is randomly chosen from P1 which then defines the root node of the sub-tree to be replaced. Crossover is then constrained by only crossing sub-trees with *similar* root NC values. The sub-tree with root node CP2 then replaces the sub tree with root node with each allele having a probability of being mutated. When mutating, functionals can only be mutated to other functionals, and terminals into any other terminals. Once crossed, only one child is produced which is then evaluated and placed into the correct root node grouped species, replacing the worst individual within that species. Communication between sub-populations is then achieved through the action of CCC.

Mutation occurs every IM crosses (usually set to IM=Population Size) and changes *only one* allele within each individual with a probability of mutation FMUTATE (set to 0.5 throughout the work presented here). The top 5 individuals of each species are elite and are *never* mutated, but are allowed to participate in crossover.

3. Boolean Induction with DRAM-GP

Boolean concept learning (or Boolean induction) is an important part of machine learning, and can be regarded as a type of pattern recognition, in which the input (independent) and output (dependent) variables are binary. The effectiveness of DRAM-GP is initially demonstrated through 4 experiments. It should be noted that all calculations within this paper are based on 100 runs for DRAM-GP. The results describe computational effort required to obtain one correct solution with a probability of 99%. Computational effort E, and other performance calculations are discussed in Koza [2,3].

3.1 Parity 3 (Koza [2,3])

To show the effectiveness of DRAM-GP as a Boolean concept learner, a simple experiment (Parity 3) in which the goal function is the even parity function f of 3 variables. f takes the values 1 if the 3 input variables x_1 , ... , x_3 have even parity. The NC weightings for the functionals, N_F were chosen based upon the number of outputs that are true for each functional. The AND functional has 1 of 4 true values, and is thus considered more complex than the OR function which has 3 of 4 true outputs. Table 1 shows the functional and terminal sets for this problem together with the NC weightings.

Functional set	F ={and ,or ,nand ,nor}
NC Functionals	N_F={1.2, 1.1, 1.1,1.2 }
Terminal set	T={d0,d1,d2}
NC Terminals	N_T={1,1,1}

Table 1 - Weightings For The Parity 3 Problem

Initially, the fitness was calculated after Koza [2], i.e. Fitness = Test points - hits. This leads to individuals with the same fitness values but vastly differing complexities. A solution with a fitness of 4.0 and a NC[0] value of 8.88 should be ranked above another individual with the same fitness but a higher complexity. It was for this reason that the fitness measure was adjusted to:- Fitness = (Test Points - Hits) + 0.001NC[0]. This then allows individuals of the same fitness but less complexity to be ranked above ones with higher complexity values. Table 2 shows a comparison of standard GP with **DRAM-GP**.

Method	Popsize (popsize x species)	Effort E
Koza[2](STD)	4,000 (4000x1)	80,000
Koza[3](STD)	16,000 (16,000x1)	96,000
Koza[3](ADF)	16,000 (16,000x1)	64,000
DRAM-GP	100 (20x5)	8,400
DRAM-GP	200 (20x10)	10,000
DRAM-GP	300 (20x15)	8,400
DRAM-GP	400 (20x20)	7,600

Table 2 - Parity3 Results

3.2 Parity 4 (Koza [2,3])

This problem is as parity 3 but using 4 input variables. Table 3 shows results using standard GP and **DRAM-GP**.

Method	Popsize (popsize x species)		Effort E
Koza[2](STD)	4000		1,276,000
Koza[3](STD)	16000		384,000
Koza[2](ADF)	4000		88,000
Koza[3](ADF)	16000		176,000
DRAM-GP	100	(20x5)	102,600
DRAM-GP	200	(40x5)	95,400
DRAM-GP	300	(20x15)	47,700
DRAM-GP	400	(40x10)	56,400
DRAM-GP	500	(50x10)	78,500

Table 3 - Parity4 Results

3.3 Parity 5 (Koza [2,3])

This problem has 5 input variables. Table 4 shows the results.

Method	Popsize (popsize x species)		Effort E
Koza[3](STD)	16,000		6,528,000
Koza[2](ADF)	4,000		152,000
Koza[3](ADF)	16,000		464,000
DRAM-GP	250	(25x10)	5,128,250
DRAM-GP	250	(50x5)	5,137,000
DRAM-GP	500	(100x5)	1,260,000
DRAM-GP	500	(50x10)	3,870,000

Table 4 - Parity5 Results

3.4 The 6-Multiplexer (Koza [2])

The input to the Boolean N-multiplexer function is the Boolean value (0 or 1) of the particular data bit that is singled out by the k address bits a_i and 2^k data bits d_i, where $N=k+2^k$. The value of the Boolean multiplexer function is the Boolean value (0 or 1) of the particular data bit that is singled out by the k address bits of the multiplexer. The experiments presented here have $k=2$, i.e. the 6-multiplexer. Table 5 shows the functional and terminal sets for this problem together with the NC weightings.

Functional set	F ={and ,or ,not ,if }
NC Functionals	N_F={1.2, 1.2, 1.1,1.3 }
Terminal set	T={a0,a1,d0,d1,d2,d3}
NC Terminals	N_T={1,1,1,1,1,1}

Table 5 - Weightings For The 6-Multiplexer Problem

Table 6 shows a comparison of GP and **DRAM-GP**.

Method	Popsize (popsize x species)	Effort E
Koza [2]	4000	160,000
Koza [3]	2000	18,000
DRAM-GP	20 (10x2)	15,600
DRAM-GP	40 (20x2)	13,320
DRAM-GP	50 (10x5)	14,250
DRAM-GP	100 (10x10)	16,500
DRAM-GP	100 (20x5)	14,100

Table 6 - 6-Multiplexer Results

4. Symbolic Regression with DRAM-GP

Symbolic regression (or function identification) involves finding a mathematical expression, in symbolic form, that provides a good, best, or perfect fit between a given finite sampling of values of the independent variables and the associated values of the dependent variables. That is, symbolic regression involves finding a model that fits a given sample of data. When the variables are real-valued, symbolic regression involves finding both the functional form and the numeric coefficients for the model [2]. This approach is also called nonparametric regression, the aim of which is to relax assumptions on the form of a regression function, and to let data search for a suitable function that adequately describes the available data. These approaches are powerful in exploring fine structural relationships and provide very useful diagnostic tools for parametric models [8,9].

The following two experiments show how **DRAM-GP** can be applied to symbolic regression within a discrete search domain.

4.1 Two-Box Problem (Koza [3])

The two-box problem concerns the identification of a relationship between six independent variables $(x_1 , \ldots , x_6)$, where this relationship relates to the difference y in the volumes of the first box whose length, width, and height are x_1 , x_2 , x_3 and the second box whose length, width, and height are x_4 , x_5 , x_6. Thus:-

$$y = (x_1 \ x_2 \ x_3) - (x_4 \ x_5 \ x_6).$$

The goal of this symbolic regression is to derive the above equation as a "complete form" when given a set of N observations. Table 7 shows the functional and terminal sets for this problem together with the NC weightings.

Functional set	F ={ + , - , * , % }
NC Functionals	N_F={1.1,1.1,1.2,1.2}
Terminal set	T={x1,x2,x3,x4,x5,x6}
NC Terminals	N_T={1,1,1,1,1,1}

Table 7 - Weightings For The Two-Box Problem

200

In this problem, where the raw fitness is a floating point number rather than an integer, there is no need to include the NC[0] weighting in the fitness calculation. The fitness measure is the mean squared-error (MSE) of all of the test points.

Table 8 shows a comparison of standard GP and **DRAM-GP**.

Method	Popsize (popsizex species)	Effort E
Koza [2] (STD)	4000	1,176,000
Koza [4] (ADF)	4000	2,220,000
DRAM-GP	20 (10x2)	163,800
DRAM-GP	50 (25x2)	76,500
DRAM-GP	100 (50x2)	213,000
DRAM-GP	150 (30x5)	112,500
DRAM-GP	200 (40x5)	156,000
DRAM-GP	250 (50x5)	169,500
DRAM-GP	300 (30x10)	130,500
DRAM-GP	400 (40x10)	136,000
DRAM-GP	500 (50x10)	237,000

Table 8 - Two-Box Problem Results

4.2 Complex Multiplication (Koza [2])

This problem attempts to find the unknown relationships between two independent variables, y_1 and y_2, and four dependent variables, x_1 ,x_2 ,x_3 , and x_4 given 50 six-tuples of data. Where the target function is vector multiplication, i.e.:-

$y_1 = x_1 x_3 - x_2 x_4$ and $y_2 = x_2 x_3 - x_1 x_4$.

Table 9 shows the functional and terminal sets for this problem together with the NC weightings.

Functional set	F = { +, -, *, %}
NC Functionals	N_F={1.1, 1.1, 1.2, 1.2 }
Terminal set	T = { x1 , x2 , x3 , x4 }
NC Terminals	N_T={ 1.0, 1.0, 1.0, 1.0}

Table 9 - Weightings For The Two-Box Problem

Table 10 shows a comparison of standard GP with **DRAM-GP**.

Method	Popsize (pop.x.species)	Effort E
Koza [2] (STD)	500	609,500
DRAM-GP	20 (10x2)	229,500
DRAM-GP	50 (10x5)	156,600
DRAM-GP	100 (20x5)	100,100
DRAM-GP	150 (30x5)	155,700
DRAM-GP	200 (40x5)	142,800
DRAM-GP	250 (50x5)	155,000

Table 10 - Complex Multiplication Problem Results

5. Engineering Test Problems

For mixed discrete and continuous search we require a search algorithm which will efficiently search through both spaces. In order to achieve this **DRAM-GP** is extended to use two alternating crossover operators. This technique is called "**HDRAM-GP**" (i.e. Hybrid **DRAM-GP**)[12]. The GP operator searches the discrete functional structure whilst the GA searches the continuous coefficient space. The GP crossover operator can select parents from any species and is thus called an inter-species crossover operator, while the GA crossover operator is limited to selecting parents from the same species and is thus intra-species. This crossover regime is possible due to the adopted computer representation of the individuals i.e. an array of characters for the functional description and an array of floating point numbers for the real number coefficients. GP crossover then operates within the character array to evolve the functional structures of the system, whilst GA crossover concurrently operates only within the real number array to evolve the coefficients of the structures. The GA crossover operator is restricted to individuals within the same species and *only* manipulates the real numbers stored in the floating point array within the individual structures. Two parents, P1 and P2 are randomly selected from the same species and a single crossover point, CP1, is selected. The real numbers of P1 and P2 are then swapped and the resulting child individuals evaluated. Every equivalent generation mutation occurs and changes only one allele within each individual with a probability of mutation of 0.5 The top 5 individuals are elite and are never mutated, but are allowed to participate in crossover.

5.1 Explicit Formula For Friction Factor In Turbulent Pipe Flow

For computation of pressure drop in turbulent pipe flow an expression is required for the friction factor f as a function of Reynolds number RE and the relative roughness K/D (where K is the equivalent sandroughness of the pipe and D the diameter of the pipe). The most accurate and accepted universal formula is Colebrook and White's. This formula is implicit, that is, f appears in two places in the transcendental equation, i.e. the equation is solved by iteration, or by finding f from a graph (Moody's chart), neither of which is convenient. Many formulae have been proposed for giving f directly for the entire range of K/D and RE. The best yet produced is probably that by S.E.Haaland [13]. It combines reasonable simplicity with acceptable accuracy (within 1.5% of Colebrook and White's formula). The best evolved formula using standard GP [9], is:-

$$\frac{1}{\sqrt{f}} = -3.8364 \log_{10}\left\{\frac{0.2097K}{D} + \frac{11.1001}{\text{Re}}\right\}$$

The average error being within 0.27% of Colebrook and White's formula, using a population size, P=1000 and generations, G=1000. This result was achieved by reducing the dimensionality of the problem to that of finding the sub-function y in the following equation:-

$$f^{-0.5} = a.\log_{10} y,$$

where a = constant and $y = f(Re, K/D)$.

The standard GP also incorporated a hill climber and required a total of 3 runs, seeding successive runs with a simplified equation of the best result from the preceding run. Without the function decomposition the method suffered from premature convergence. Using **HDRAM-GP** to evolve the sub-function y described above, and using P=40 with G=1000, the following equation was the best of 5 independent runs:-

$$\frac{1}{\sqrt{f}} = 3.7922 \log_{10}\left\{-15.0169 - \left[\frac{-5.4353}{\left(57.3353 / (717.8244 + \mathrm{Re})\right) + K / D}\right]\right\}$$

with an average error of 0.79%.

The second series of runs for this problem assumed no functional form for the resulting equation, i.e.

$$f^{-0.5} = y$$

where $y = f(Re, K/D)$.

This increases the complexity of the problem to a point where standard GP converges upon unsatisfactory solutions with 20% to 30% errors. The best result of 5 independent runs using **HDRAM-GP** (P=40, M=1000) is:-

$$\frac{1}{\sqrt{f}} = 3.5039 \log_{10}\left\{\frac{3.2917\mathrm{Re}}{18.7046 + K / D (0.41644 - \mathrm{Re})}\right\}.$$

The average error being within 1.82% of Colebrook and White's formula. All 5 runs converged to a solution of the correct form, i.e.

$$f^{-0.5} = a_1 \log_{10} y$$

where $y = f(Re, K/D)$.

This shows that **HDRAM-GP** is able to solve problems which present significant difficulty to standard GP representation.

5.2 Eddy Correlation's For Laminar Two-Dimensional Sudden Expansion Flows

The problem here is that of finding a general equation for the velocity vectors in laminar two-dimensional flow of an incompressible fluid past a sudden expansion. Data used for the fitness function is derived by using computational fluid dynamics (CFD). Figure 2 shows the expansion flow model. The CFD results produce a velocity vector and if this is divided into its x and y components the velocities can be represented as surfaces as shown in figures 3 and 4 (for Re=1000). The modelling of the velocity within a sudden expansion can be achieved by the fitting of these two surfaces.

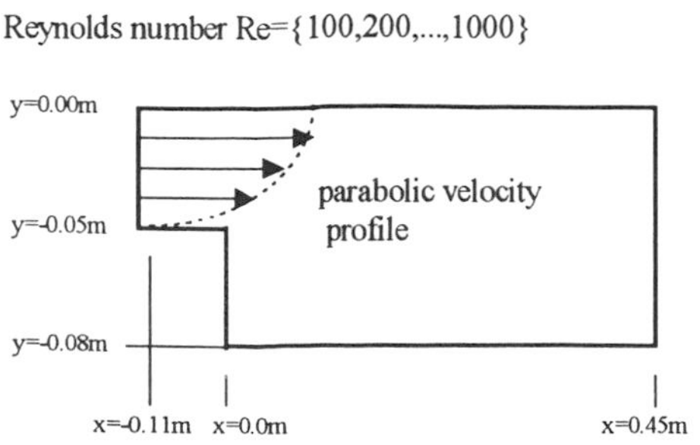

Reynolds number Re={100,200,....,1000}

Figure 2 - Model For Sudden Expansion

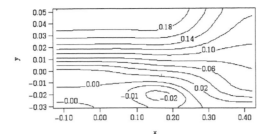

Figure 3 - X-Velocity Test Surface

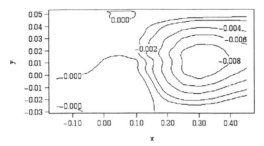

Figure 4 - Y-Velocity Test Surface

Previous work on sudden expansion flow [9] showed that although the standard GP paradigm can model this system to some extent, it requires multiple runs to achieve a satisfactory solution. The results were obtained by reducing the dimensions of the problem for an initial run i.e. only using the data at Re=1000, and seeding a second run with the best result of the first run, but using the whole range of test data for Re= 100,200,...,1000.

Using **HDRAM-GP**, equations are evolved which describe the X and Y velocities using only one run of the algorithm and all the data i.e Re=100,200,...,1000. The resulting equation can produce velocity vectors for Re in the range 0→1000. Figures 5 and 6 show the evolved X and Y velocity at Re=1000 respectively .

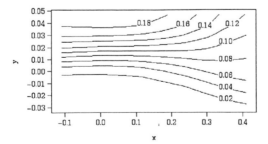

Figure 5 - Evolved X-velocity

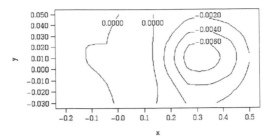

Figure 6 - Evolved Y-velocity

5.3 Thermal Paint Jet Turbine Blade Data

The final problem under investigation is the modelling of the surface temperature of a turbine blade under set operating conditions. Figure 7 shows a surface of a turbine blade used to produce the test data. Using standard GP, acceptable results have been produced for 1-dimensional curves (sections through the surface) and the best result for the whole surface has an average error of ± 17.5 °C, with popsize=1000, and generations=10,000. Using **HDRAM-GP**, with popsize=400 (40 species of 10 individuals), and generations=1000, the average error is ± 9.8 °C. The resulting surface is shown in figure 8.

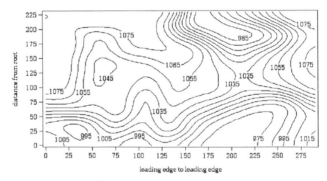

Figure 7 - Turbine Surface (°C)

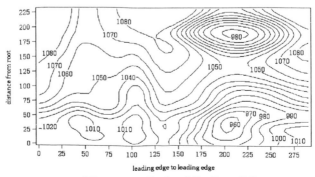

Figure 8 - Evolved Surface (°C)

6. Conclusions

Previous work [14] has shown that **DRAM-GP** solves discrete problems such as Boolean induction and simple symbolic regression. The results here show that **HDRAM-GP** can be applied to model approximate equations to simple engineering systems. The only inputs required to solve the problems is the number of species and the maximum and minimum NC species groupings. We therefore have a robust method for the approximate modelling of simple engineering systems which utilises concurrent search of both the discrete and continuous variable spaces. The crossover mechanism is responsible for the transfer of information from species to species, and fit solutions are rapidly transferred to other species, usually of a lower NC grouping, thus producing less complex solutions. This then provides a numerical complexity measure which controls crossover, and minimises disruption by ensuring some similarity between crossed sub-trees. The technique also controls tree growth.

The main advantages of **HDRAM-GP** over standard GP are summarised as follows:-

- A reduced overall population size.
- A reduction in associated memory required to run the evolutionary program.
- The ability to search coefficients within functional structures and the functional structures.
- A length control mechanism.
- Rapid evolution of results.

All runs were performed on a 6-processor Sun-Ultra Enterprise 4000 using Gnu C++, and a DX4-100 PC using Borland Turbo C++ 3.0, using 1st author code.

Acknowledgements
The research has been carried out within the Plymouth Engineering Design Centre, at the University of Plymouth, with financial support from both the Engineering & Physical Science Research Council and Rolls-Royce Plc. We wish to thank both parties for their continuing support.

References
[1] Iba, H., DeGaris, H. & Sato, T. A. 1996; A Numerical Approach to Genetic Programming for System Identification. Evolutionary Computation 3(4): p417-452.
[2] Koza , J. 1992; Genetic Programming; MIT Press.
[3] Koza,J. 1994; Genetic Programming II, MIT Press.
[4] Kinnear Jr. K. E. 1993,Generality and difficulty in Genetic Programming: Evolving a Sort. Proc. of 5th International Joint Conference on Genetic Algorithms.

[5] Iba, H.; Sato, T. & De Garis, H. 1993; System Identification Approach to Genetic Programming. Proceedings of the 5th international conference on genetic algorithms: p279-286. San Mateo, CA: Morgan Kaufmann.

[6] Kinnear Jr. K. E. 1993, Evolving a Sort: Lessons in Genetic Programming. Proc. of 1993 International Conference on Neural Networks. IEEE Press.

[7] Kinnear Jr. K. E, Advances in Genetic Programming, MIT Press, 1994.

[8] Watson, A. H.; Parmee, I. C. 1996, Systems Identification Using Genetic Programming. Proceedings of ACEDC: p248-255. University of Plymouth.

[9] Watson, A. H.; Parmee, I. C. 1996, Identification Of Fluid Systems Using Genetic Programming. Proceedings EUFIT '96. Vol I, p395-399. ELITE Foundation.

[10] Watson, A. H.; Parmee, I. C. 1997, Steady State Genetic Programming with Constrained Complexity Crossover. Proceedings of Genetic Programming '97 (GP97): in publication. MIT Press.

[11] Parmee, I.C. 1996, The Development Of A Dual-Agent Strategy For Efficient Search Across Whole System Engineering Design Hierarchies. Proceedings of Parallel Problem Solving from Nature. (PPSN IV), Lecture notes in Computer Science No. 1141 : p523-532. Springer-Verlag, Berlin.

[12] Watson, A. H.; Parmee, I. C. 1997, An Improved Genetic Programming Strategy For Preliminary Design Model Development. Proceedings EUFIT '97. Vol I, p682-686. ELITE Foundation.

[13] Haaland, S.E.; Simple And Explicit Formulas For The Friction Factor In Turbulent Pipe Flow. Journal of Fluids Engineering 105. 1983.

[14] Watson, A.H.; Parmee, I. C. Steady State Genetic Programming with Constrained Complexity Crossover Using Species Sub-Populations. Procs. 7th International Conference on Genetic Algorithms. (ICGA 97). Morgan Kaufmann. 1997(b).

From Mondrian to Frank Lloyd Wright: Transforming Evolving Representations

Thorsten Schnier
John S. Gero

Key Centre of Design Computing
Department of Architectural and Design Science
Faculty of Architecture
University of Sydney, NSW, 2006, Australia
fax: +61-2-9351-3031
email : {thorsten,john}@arch.usyd.edu.au

Abstract. If a computer is to create designs with the goal of following a certain style it has to have information about this style. Unfortunately, the most often used method of formal representations of style, shape grammars, does not lend itself to automated implementation. However, It has been shown how an evolutionary system with evolving representation can provide an alternative approach that allows a system to learn style knowledge automatically and without the need for an explicit representation. This paper shows how the applicability of evolved representation can be extended by the introduction of transformations of the representation. One such transformation allows mixing of style knowledge, similar to the cross-breeding of animals of different races, with the added possibility of controlling exactly what features are used from which source. This can be achieved through different ways of mixing representations learned from different examples and then using the new, combined representation to create new designs. In a similar manner, information learned in one application domain can be used in a different domain. To achieve this, either the representation or the genotype-phenotype transformation has to be adapted. The same operations also allow mixing of knowledge from different domains. As an example, we show how style information learned from a set of Mondrian paintings can be combined with style information from a Frank Lloyd Wright window design, to create new window designs. Also, we show how the combined style information can then be used to create three-dimensional objects, showing style features similar to the newly designed windows.

1. Introduction

Any design system that is intended to create designs that follow a certain style requires knowledge about the style, given to the system represented either explicitly or implicitly in design data or coded into the design system. Collecting this knowledge is usually done by hand, for example by creating a shape grammar for a representative set of designs. Often, this also involves research about the work and methodology of the designer. In an earlier paper [1], we have proposed an alternative, machine learning approach. It uses the fact that it is possible to categorize style from its visible features [2]. The approach creates an implicit representation of style features, requiring only a set of sample designs. The acquired knowledge can be used directly to create new designs that show style similarities to the example designs, but at the same time are adapted towards different design conditions. In this paper, we show how manipulating the learned style knowledge can transform it, allowing the creation of a much wider, possibly more 'interesting' set of designs.

2. Learning Style Feature using Genetic Engineering

The approach used to learn style knowledge is based on evolutionary systems. Evolutionary systems are population-based search algorithms. The population consists of individuals, represented by their genetic code, the 'genotype'. A transformation exists that transforms the genotypes into 'phenotypes', and a measure for the performance of individual phenotypes, the 'fitness' can be calculated. New individuals are created using genetic operations from genetic material from one (mutation) or two (cross-over) genotypes with high fitness; genotypes with low fitness are removed.

To learn style features, the system is programmed to try to create copies of the example or examples given to it; the fitness function is a measure of the distance between the current phenotype and the example design. At the same time, particularly successful combinations of genes in the genotypes are identified and encapsulated into 'evolved' genes. As a result of this 'genetic engineering', the representation evolves, and the search space is transformed in a way that the search is more and more biased towards designs similar to the examples.

To create new designs a second evolutionary system is run with a fitness function evaluating the phenotypes with respect to the design criteria. This evolutionary system uses the evolved representation that has been created in the first step, thus incorporating the knowledge acquired in this step into the new designs. The evolved genes encapsulate sets of basic genes, protecting them from the genetic operations. This is relatively easy if the evolved genes contain only sequences of directly successive genes. If however the evolved genes are allowed to consist of non-successive basic genes (complex evolved genes), the genetic operations can lead to situations where conflicts between different evolved genes arise. A solution, using diploid genetic code with dominant and recessive genes, has been described in [3].

3. Transforming Evolved Representations

Using the evolved representation, it is possible to create new individuals that show similarities with the examples. However, while interesting, the results would usually not be called 'creative'. To make the new designs more 'interesting' and 'surprising' two operations are introduced: combining elements from different sources, and transforming elements into a different domain. Thanks to the flexibility of evolutionary systems, both operations are possible with the evolved representation. It is interesting to contrast this approach with that of Knight [4]. While Knight uses shape grammars as an analytical tool to describe the transformation of styles, we use transformations of style knowledge, represented in an evolved representation, as a generative tool.

3.1. Combining Genetic Material

In nature it is sometimes possible to combine the genetic material from two individuals from two different 'groups' of animals; the resulting offspring includes features from both groups. The most common example is probably cross-breeding between different races, for example in dogs; the results are generally referred to as 'hybrids'. In general, this cross-breeding is not possible, in fact 'species' are defined by the fact that they cannot interbreed. For a successful combination it seems that three conditions have to be met.

- The genetic material of both 'parents' has to be such that the same genotype-phenotype transformation can be used to transform it into a living individual. For life on earth, this is rarely a problem, since the vast majority of life forms use the same universal RNA/DNA-based genetic material.
- The environment in which the 'transformation engine' works has to be compatible. For example, mixing dog breeds of different sizes is generally only successful if the female dog belong to the larger breed, otherwise its womb might not be able to support the developing puppy.
- The genetic materials have to be compatible. In other words, the transformation engine has to be able to transform a genotype consisting of material from both sources into a functioning individual. This, in nature, is the most important obstacle in interbreeding.

While biological systems use a common representation and achieve a huge variety of organisms by a highly interactive multi-level development process, most evolutionary systems use a specialized representation, with a simple, usually linear genotype-phenotype transformation. The three conditions therefore have very different importance for evolutionary systems.

- Contrary to biological systems, evolutionary systems use many different genotype-phenotype transformations, often designed for specific applications. As a result, this condition prevents 'interbreeding' in most cases.
- In the vast majority of evolutionary systems implementations, the genotype-phenotype transformation is very simple, without any interaction with the

environment. This point is therefore usually unimportant for evolutionary systems.

- To make the evolutionary search as efficient as possible, the genotype-phenotype transformation is usually designed so that most or all of the possible genotypes can be transformed into phenotypes. If all the genetic material used to create a new individual comes from genotypes that use the same genotype-phenotype transformation, the offspring is equally likely to be a valid individual. However, this does not guarantee that the offspring will have a high fitness.

In evolutionary systems, the most important condition is therefore that the sources of the genetic material are systems that use the same genotype-phenotype transformation.

3.1.1 What to combine?

If, as defined in the previous section, the sources for the genetic material use the same genotype-phenotype translation, they also use the same basic representation. As a result, an initial, random population will look similar in any of the sources. Features specific to an application are only present in the form of certain gene configurations in individuals in later stages of the evolutionary process. The combination of genetic material from different sources is therefore only possible by combining individuals, for example with a cross-over operation. The resulting individual will show features of both individuals, and therefore both sources; however in following generations the genetic operations can destroy any such features.

Evolving representations present an alternative: features of an application are integrated into the representation and, while all applications have the same basic representation, they can have very different evolved representations. As a result, random individuals created from different evolved representations will look different and if the evolved representations are combined, the random initial individuals will show features from all the sources. During the evolutionary process the evolved features are protected: while it is still possible that the evolutionary process leads to individuals that use only evolved genes, and therefore features, from one source, the genetic operations cannot disrupt the evolved genes. This is similar to the case in nature: the 'basic coding' are the base pairs on the DNA (or any variant of RNA), but the units of inheritance are long sequences of base pairs, the genes.

3.2. Extending Genetic Material

Since in evolutionary systems genes and the genotype-phenotype transformation are just data and programs, they can be adapted as required. This allows the use of genetic material in different domains from where it originally was produced. To allow the genotype-phenotype transformation to use the 'foreign' genetic material, we can either modify the genetic material so that it fits into the new domain, or we can modify the transformation so that it is able to directly use the material from a different domain. The same procedures allow us to combine genetic material from two different domains.

4. Example

As an example, we show how style features learned from Mondrian paintings can be combined with style features from Frank Lloyd Wright window designs. For both the paintings and the window design, we use the same basic representation, combining the evolved genes therefore corresponds to cross-breeding between two breeds. To show how a representation can be transformed for a different domain, we use the evolved genes created in the two two-dimensional example applications to create three-dimensional objects.

4.1. Basic Representation

To represent Mondrian paintings a tree-coding is used where every node of a tree corresponds to a division of a rectangle into two smaller rectangles. The position and direction of the division, the thickness of the dividing line, and the colour of one of the two resulting rectangles are encoded in four variables at every node. Every node can also have two subtrees that describe further subdivisions of the rectangles, Figure 4.1 shows an example[1].

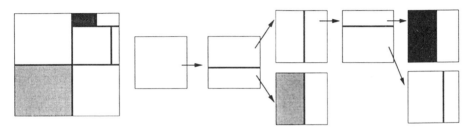

Figure 4.1: Representation of a Mondrian painting as tree-structured series of rectangle divisions.

This representation allows the creation of a large set of rectangle-based two-dimensional designs. Some additional designs, including for example the 'pinwheel' shapes used in paintings by Vantongerloo [5], can be represented as well if 'invisible' lines are allowed, as shown in Figure 4.2. The invisible line, which can be located anywhere in the painting as long as it intersects the middle rectangle, splits the design into two halves that can be represented.

---

[1] For colour versions of this and other figures in this paper, please refer to http://www.arch.usyd.edu.au/~thorsten/publications/acdm98/acdm98.html or contact the authors.

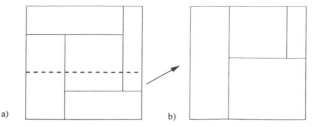

Figure 4.2: (a) Representing a pin-wheel figure by dividing it using an 'invisible' line, (b) the lower half, which can be represented in the basic representation without problems.

4.2 Learning Representations

Figure 4.3 shows the examples used to generate an evolved representation based on Mondrian paintings. The paintings have eight (Figures 4.3(a) and (c) and seven (Figure (b)) areas, therefore a tree with eight nodes is required to completely represent a single painting. At each node, the evolutionary system can choose between 4 different positions (top, bottom, left, right), 15 fractions, 4 line-widths and 12 colours.

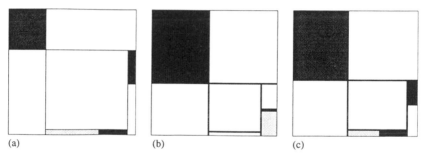

(a)   (b)   (c)

Figure 4.3: Mondrian paintings used to create evolved representation.

For each phenotype a Pareto fitness vector with fifteen elements is calculated. For each of the three examples five fitness values describe how 'close' the phenotype is to the example in terms of positions of the divisions, correctness of colours and line-widths, completeness and absence of additional divisions.

The gene extraction method is very similar to the one described in detail in [1]. In the system used here, an additional function has been introduced to improve the quality of the evolved genes. During the evolutionary process individuals with high fitness compared to the current population will still have some wrong colours, line-widths, etc. As a result, it is possible that genes are generated that incorporate these incorrect features. To prevent this, a function has been added that checks if the phenotype is a true subset of the example, ie. if it is possible to convert the phenotype into the example by adding further divisions. Gene combinations that occur in at least one phenotype that is a true subset of at least one example are guaranteed not to have any incorrect features, only those are therefore considered as new evolved genes.

The run produced 110 evolved genes, the first and the last seven genes created are shown in Figure 4.4.

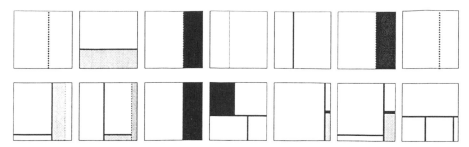

Figure 4.4: Evolved genes created from the examples in Figure 3. To represent genes with incomplete nodes the following convention was used: no colour: light pink; no line-width : line-width 2; no direction: vertical, and line stippled; no fraction: 1/3, and line coloured red.

As the second example, a window design, created by Frank Lloyd Wright for Hollyhock House [6] was used. Three segments from the centre of the window were coded. Figure 4.5(a) shows the drawing used to generate the evolved representation. In the original, the light blue rectangles have a wider frame, this had to be changed because it could not be represented in the basic representation. To compare the example with phenotypes, the outer rectangle of the window is transformed into a square, see Figure 4.5(b). Due to the higher complexity of this example, the system learned more evolved genes, 159. The last 11 evolved genes created are shown in Figure 4.5(c).

4.3 From Mondrian to Frank Lloyd Wright

In order to make use of evolved representations a set of initial individuals is created using the representation and then an evolutionary system is run with a fitness describing the desired new design. Since the evolved representations from the paintings and the window design use the same basic representations, they can simply be combined by creating a random initial population using evolved genes from both sets. However, it is also possible to add additional control over the way the two representations are used. For example, it is possible to remove some evolved genes from the representations before they are combined. Another possibility is to remove parts from the genes before they are combined. To maximize the influence of the evolved representation the initial individuals are generated so that they do not contain any basic genes. To achieve this the system makes use of a diploid representation, described in more detail in [3]

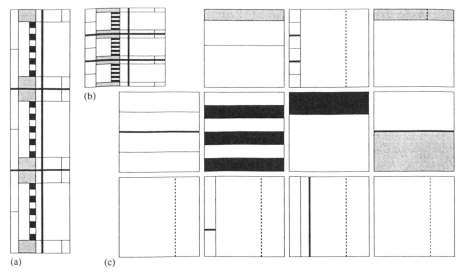

Figure 4.5: Evolved representation from Frank Lloyd window: (a) part of the design used as example, (b) the design as seen by the evolutionary system, (c) the last 11 evolved genes.

The fitness function used in this example specifies that the new designs have to have a vertical line in the middle of the frame, 22 panes, and that the sides of all panes are longer than 7% of the size of the outer square. The fitness function is intentionally chosen in a way that it neutral to the use of the style features, it neither prevents nor promotes them. In fact, the influence of the evolved representations can easily be seen in the initial, random individuals as well. The fitness function is not very difficult, the system usually finds perfect phenotypes within a few minutes.

For comparison, Figure 4.6(a) show new designs using only the basic representation and Figures 4.6(b) and 4.6(c) show the results using the representations created from only the Mondrian paintings and from only the Frank Lloyd Wright window respectively. Crossover and mutation were able to adapt the designs quickly to the new conditions, while the style information has mostly been preserved through the evolution. Mutation was implemented so that in rare occasions, it would replace an evolved gene with basic genes, this can be noticed in a few places that contain features that were not part of any of the examples.

Figures 4.6(d) and 4.6(e) shows the results of combining the evolved representations created from the paintings and the window in two different ways. To create the results in Figures 4.6(d) the initial population was created choosing random evolved genes from both sources. Features in the designs can easily be identified corresponding to either one of the paintings or to the window. The initial individuals and the final designs contain more genes from the windows than from the paintings, this reflects the fact that more evolved genes from the window design were available.

To create the designs in Figure 4.6(e) the evolved representations were modified before they were used to create initial individuals: in all evolved genes created from the Mondrian paintings, information about position and fraction of the rectangle division was removed. Similarly, in all evolved genes from the window design, all information about colour and line thickness was removed. The resulting genes were then used to create the initial individuals. The results show, as expected, topological features inherited from the Frank Lloyd Wright window design, and colouring and line thicknesses from the Mondrian paintings. The imbalance in the number of evolved genes does not matter in this case, since the different sets of evolved genes are responsible for different aspects of the new designs.

Figure 4.7 shows how the fourth design in Figure 4.6(e) can be assembled into a complete window, similar to the original Frank Lloyd Wright design.

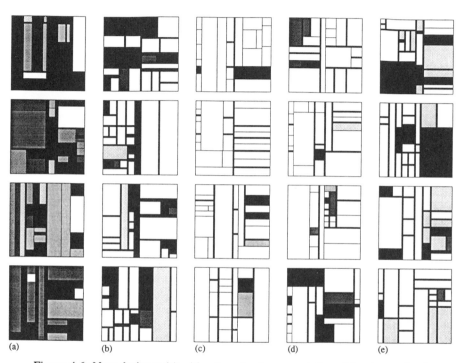

Figure 4.6: New designs: (a) without evolved representation, (b) using evolved representation from Mondrian paintings, (c) using evolved representation from Frank Lloyd Wright window, (d) using full genes from both representations, (e) using topology from window design and colour and line-width information from paintings.

Figure 4.7: One of the designs in Figure 4.6(e), assembled into a full window (shown rotated).

4.4 Creating 3-dimensional Objects from 2-dimensional Examples

As mentioned in Section 3.2, it is possible to use evolved representations in domains different from where they were initially created, if either the genotype-phenotype transformation or the evolved representations are adapted for this purpose. The new domain used in this example is the creation of coloured cubes. To create the three-dimensional objects a basic representation is used that is similar to the one used for the paintings. However, instead of rectangles divided by lines, the nodes specify planes intersecting cubes. Each node has an additional variable specifying whether the intersecting plane is perpendicular to the x-y plane, to the y-z plane or to the z-x plane. In other words, to divide a cube, its projection into one of those three planes is used and the resulting two-dimensional shape then cut as in the Mondrian painting.

Due to the similar nature of the representations, it is easy to adapt the evolved genes created from the paintings and the window. Obviously, none of the evolved genes provides a value for the cutting plane. One possibility is to assemble initial individuals without this value and then provide a random value for this at every node. However, this makes the topology features learned from the window less recognizable in the phenotype. The reason is that genes specifying topological features span a number of connected nodes and are only recognizable in the resulting phenotype if they are all perpendicular to the same plane. For example, a series of parallel divisions only remains parallel if all divisions are perpendicular to the same plane, in other words, have the same value for the added variable. Because of the random assignment of this value, this is not the case.

Better results can be produced by adapting the evolved genes before use for the three-dimensional application. For every gene, a value for the intersection plane is randomly chosen and added to every node of that gene. For the example, all topology features were removed from the evolved genes produced form the example paintings and all colour and line-width information removed from the evolved genes created from the window. Then, information about the intersection plane was added to all window-genes in the manner described above. The resulting genes were then used to create the initial population. The fitness function is similar to the one used to create new window designs: 70 sub-cubes, a plane intersecting the resulting cube in the middle, and no edges shorter than 7% of the edge length of the outer cube.

The result is shown in Figure 4.8(a). As a comparison, Figure 4.8(b) shows a cube created with the same fitness function and genes, but assigning the planes randomly to the individual nodes. While similarities in colouring and line-width to the Mondrian paintings exist in both cubes, the cube in Figure 8(a) shows more topological features relating to the Frank Lloyd window example.

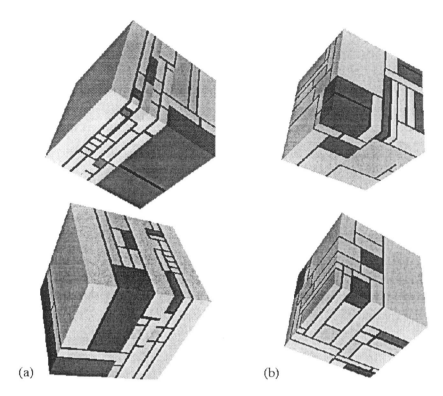

(a)   (b)

Figure 4.8: Three-dimensional objects created from 2-d representation, views show opposing corners: (a) plane information added to evolved genes, (b) plane information added to nodes.

5 Conclusions

This work shows that it is possible to use evolutionary systems to produce designs that are 'hybrids', incorporating different styles, by using a mechanism similar to crossbreeding between different races in nature. However, the designer using this computational process has far more control over the mixing process, allowing the inclusion and exclusion of specific features from either of the sources into the new designs. Compared with crossbreeding in nature, it also adds the ability to combine features from different domains, somewhat the equivalent of crossbreeding between different species.

Without evolved representations it still is possible to create single hybrid individuals, for example by using a cross-over between two individuals from different, highly adapted populations. However, the mixing is restricted by the genetic operation used to combine the parents, and the individual features are not protected during further genetic operations. Using evolved representations, it is possible to create high numbers of random initial individuals that show the different style features in a probabilistic distribution. These features are protected through the course of the evolution.

Another way to conceive of this work is to relate it to analogy where the designer draws ideas from a source design and introduces them into the target design [7]. Of the evolved representations one may be considered the target and the other the source. Whereas in analogically-based design there is considerable difficulty in reformulating the ideas from the source design so that they can be introduced into the target design, there is no such problem in this approach. This work may be considered in terms of creative design processes, ie computational processes which are potentially capable of producing "creative" designs. One computational model of creative design involves the concept of processes which are capable of changing the state space of possible designs [8]. The combination of two evolved representations fits that model well and is clearly capable of producing novel designs which lie outside the space of designs which could possibly be produced using only either of the evolved representations, implying that the new state space is not simply the union of the previous two state spaces.

To extend this work it would be interesting to create and mix evolved representations from different instances in the same domain used for the examples and from other, different domains. Especially if more complex genotype-phenotype representations are used, together with complex, possibly multi-level evolved genes, the results promise to be very interesting. Also, while in this implementation the combination of genes from different sources and the transformation into other domains is initiated and controlled by the designer, it would be interesting to develop a system with a number of sets of evolved genes from different domains, that can suggest suitable transformations to the designer.

6 Acknowledgments

This work is supported by a grant from the Australian Research Council and by a University of Sydney Postgraduate Research Award. Computing resources have been provided by the Key Centre of Design Computing.

References

1. Chan,C.-S. (1995) A cognitive theory of style, *Environment and Planning B: Planning and Design* **22**: 461-47.

2. Coyne, R. D., Rosenman, M. A., Radford, A. D., Balachandran, M. & Gero, J. S. (1990). *Knowledge-Based Design Systems,* Addison-Wesley, Reading, Massachusetts.

3. Gero, J. S. (1994). Computational models of creative design processes, *in* T. Dartnall (ed.), *AI and Creativity*, Kluwer, Dordrecht, pp. 269-281.

4. Hanks, D. A. (1989). *Frank Lloyd Wright: Preserving an Architectural Heritage*, Studio Vista, London.

5. Knight, T. W. (1989). Transformations of De Stijl art: the paintings of Georges Vantongerloo and Fritz Glarner, *Environment and Planning B* **16**: 51-98.

6. Knight, T. W. (1994). *Transformations in Design: a Formal Approach to Stylistic Change and Innovation in the Visual Arts*, Cambridge University Press, Cambridge.

7. Schnier, T. & Gero, J. S. (1996). Learning genetic representations as alternative to hand-coded shape grammars, *in* J. S. Gero & F. Sudweeks (eds), *Artificial Intelligence in Design '96*, Kluwer, Dordrecht, pp. 39-57.

8. Schnier, T. & Gero, J. S. (1997). Dominant and recessive genes in evolutionary systems applied to spatial reasoning, *Tenth Australian Joint Conference on Artificial Intelligence* (to be published).

A Comparison of Evolutionary-Based Strategies for Mixed Discrete Multilevel Design Problems

Kai Chen and Ian C. Parmee

Plymouth Engineering Design Centre (PEDC)
Plymouth University
Drake Circus, Plymouth PL4 8AA, UK
Tel: +44(0)1752 233508
Fax: +44(0)1752 233529
E-mail: iparmee@plym.ac.uk

Abstract

This paper presents a comparison of evolutionary strategies which employ individual mutation schemes for different types of decision variables in optimal design and control problems. The work utilises the GAANT[9] algorithm as an improvement on previous work involving a dual-agent GA integrated with a nuclear power station whole plant design problem. The objective of the algorithm is to maintain diversity across both discrete and continuous variables. The algorithm is important during the preliminary design stages of industrial design problems with a limited number of discrete paths and heavy constraint. Particularly a nuclear power station re-design problem has been studied in depth.

1 Introduction

This paper is mainly concerned with the search of optima within a mixed-discrete space with a limited number of discrete searching paths. The applications are for the preliminary stages of whole system engineering design. Typically the system of interest is a nuclear power station re-design feasibility study problem. Previous work studied a dual-agent Genetic Algorithm (GA) strategy which employs individual mutation schemes for different types of decision variables in optimal design and control problems[2]. The dual-agent GA was integrated with a nuclear power station whole plant design problem. In this case an actual thermal system plant design case has been studied successfully[11, 2, 3]. The obvious advantage of the dual-agent GA is that the diversity across the discrete search space can be maintained. However, the high probability of mutation in discrete space is highly unstable and potentially good configurations can disappear easily at the early stages due to selection and high-rate mutation.

The dual-agent scheme has led to the utilisation of a dual-agent strategy involving a GA and elements of an ant colony search paradigm (GAANT) [9]. Elements of both the GAANT algorithm and earlier work[8] involving dual mutation strategies operating within a structured GA framework [5] have been utilised to overcome problems related to mixed-integer representation and variable dependence. The paper utilises the GAANT algorithm in which discrete design options are speciated in each generation and the related fitness contributes to the evolution process. Ant colony operators such as evaporation, proportional distribution, conservation etc. are applied periodically. In every cycle the discrete species are allowed plenty of opportunity to improve by a GA optimisation of their related continuous variables. The dual-agent GA and the GAANT algorithm are applied to a variation of the generalised Rosenbrock's function and a real-life nuclear power station thermal system re-design problem. The comparison shows that the GAANT algorithm has better diversity in both discrete and continuous spaces.

2 The Optimisation Problem

The general formulation of an optimal design and control problem of interest is a mixed-integer non-linear programming as follows:

$$\max(\min) f(\mathbf{X}_1, \mathbf{X}_2, \mathbf{X}_3) \tag{1}$$

$$\text{subject to } \mathbf{g}(\mathbf{X}_1, \mathbf{X}_2, \mathbf{X}_3) \leq \mathbf{0} \tag{2}$$

$$\mathbf{X}_{i,l} \leq \mathbf{X}_i \leq \mathbf{X}_{i,u}, \quad i \in \{1, 2, 3\} \tag{3}$$

where
$\mathbf{X}_1 \in R^{n_1}$ Operational Control Variables (OCV), continuous;
$\mathbf{X}_2 \in N^{n_2}$ Configuration Decision Variables (CDV), discrete;
$\mathbf{X}_3 \in R^{n_3}$ Configuration Related Decision Variables (CRDV), continuous.

The OCVs are the continuous system control inputs for a desired time period. For a typical Advanced Gas-cooled Reactor (AGR) nuclear power station, an OCV could include boiler feed water flow rate, boiler mean gas ($CO2$) flow and high pressure (HP) and intermediate pressure (IP) turbine control valve settings, etc. A property of OCVs which occurs in nuclear thermal cycle design is that the feasible set and the optimal values of OCVs do not change significantly when the configuration of the design is changed.

CDVs are the discrete configuration decision variables. They can be the design variables and/or control variables which decide the system structure and design configuration. In a nuclear power station, the CDVs can be the number of new turbines and feed heaters, and the connection methods of the system components. These decision variables determine the system configuration (e.g. the number, connection and layout of feed heaters in power stations). CRDVs are the continuous decision variables which are directly related to the CDVs. That is, they are related to certain CDVs and are only active when certain configurations are effective. This gives rise to problems related to variable dependence. Examples of CRDVs are the parameters governing feed heater performance aspects in a nuclear power station feed heater design problem. The functions $f(\cdot)$ and $\mathbf{g}(\cdot)$ are not normally available explicitly for a practical design problem. A complex numerical simulation model of the plant is required to evaluate the values of the functions. For certain configuration some variables are not active at all. These contribute to the fact that a straightforward general-purpose numerical mixed-integer programming solver is not available.

3 A Dual-Agent GA Solver

The previous work has investigated the feasibility of using a hybrid dual-agent GA structure for solving mixed-discrete design and operation optimisation problems [2]. The strategy has evolved during the process of the re-design project of a nuclear power station thermal system. It employed different mutation and crossover operations for the discrete and continuous decision variables. The key features of the dual-agent GA algorithm are listed below:

1. **Dual Mutation:** To improve diversity of search across the design space, a dual mutation scheme is introduced. For integer variables a high mutation rate is assigned (e.g. >0.1) to maintain search diversity and a low mutation rate (e.g. 0.001) is assigned to the rest of the chromosome (i.e. OCV and CRDV sub-strings). Care is taken to maintain at least one individual for each configuration at the first stages.

2. **Crossover:** A special one-point crossover operation is adopted. Only parents of like configurations and active CRDVs are selected [9].

3. **Selection:** A two-stage selection scheme is used. In the first stage (e.g. for the first 5 or 10 generations), all configurations are conserved. In the next stage, the worst configurations are rejected and a new configuration conservation is maintained. The first stage increases the possibility of searching widely across the whole search space. The subsequent stages reduce the number of search paths and increase the search efficiency.

4. **Periodical Hillclimbing:** A nonlinear optimiser is employed periodically (e.g. at every 5th generation)to the continuous variables for the best individuals. The purpose to fine-tune the continuous variables for the configurations with good performance.

4 A GA with Ant Colony and Speciation

The dual-agent GA described in the previous section has the advantage of being able to classify the configurations inside a population of individuals and to treat configuration (discrete) and continuous variables differently in selection, crossover and mutation operations. But a high mutation rate applied to the configuration segment of the chromosome implies that the search in the configuration space is more or less like a random walk and hence highly unstable, although it can maintain certain degree of search diversity in the configuration search space.

In this section, the GAANT algorithm[9] is described. GAANT maintains both diversity and relative stability in both the configuration (discrete) search space and the continuous search space describing each configuration. The GAANT strategy has been utilised in this paper with variations relating to ant-colony operations on species rather than on **each** chromosome, periodical hill-climbing in the continuous search space, elitist operation inside species and the maintenance of diversity in continuous space by applying a special crossover. In particular a species is a design path which is composed of one or several close configurations. Each configuration is corresponding to a distinct combination of the discrete decision variables. The closeness of the configurations are defined by the user of the algorithm based on related engineering knowledge.

4.1 Ant Colony Analogy

Aspects of the Ant Colony metaphor have been utilised[4, 1] in terms of the following operators:

1. **Fitness proportional distribution:** Similar to fitness proportional reproduction in simple genetic algorithms, the number of individuals distributed to each species is proportional to the relative fitness of the whole species.

2. **Evaporation:** If the relative fitness of a species is consistently poor, the species is 'evaporated'. That is, the species is eliminated and the resource (number of individuals) is re-distributed to other species by proportional distribution or by the application of species perturbation.

3. **Conservation:** The species with the best relative performance are maintained in future generations whereas the discrete parameters of medium performance species are mutated slightly.

4. **Species elitist:** When the relative performance of a species is better than a certain threshold, the best individual in the species is maintained.

The ant colony operations are applied periodically, typically every 10 generations. During each such *cycle*, there is no mutation operation applied to the discrete part of the chromosome and no species is allowed to become extinct. For the crossover operation, only individuals of the same species are chosen as parents to avoid the creation of non-feasible parameter sets. These operation allow each species to evolve for a period of time and avoid the early extinction of potentially good species. The local mutation operation involving medium fitness species allows the possibility that evaporated species still have a chance to be re-introduced to the population.

4.2 A GAANT algorithm

The flow-chart of the original GAANT algorithm is available in [9]. The initial generation is generated such that the individuals are evenly distributed across the species. A GA is then applied to the population for a cycle of generations. Inside a cycle, mutation is applied to continuous variables only. Crossover only happens between parents of the same species. To maintain the diversity of search inside a species, parents for crossover are selected so that they are most likely to be distant in terms of performance. At the end of each generation, speciation (classification) is done and statistical information is calculated every generation. Three important indices are:

1. *fit_spec* — The average fitness of the species in the current generation (transformed to maximising with the best possible fitness being 1.0);

2. *fit_all* – The average fitness of the current generation;

3. *rfit* — The relative fitness of the species, i.e. $rfit = fit\_spec/fit\_all$.

Two threshold values of the relative fitness, *rfit*, are set by the user, RF_1 and RF_2 (with $RF_1 < RF_2$). At the end of each generation, species elitist operation is applied. That is, if *rfit* is greater than RF_1 then the best individual of the species is preserved. At the end of each cycle, different ant colony operations are applied to each species according to its relative fitness, *rfit*. If *rfit* is less than RF_1, then the species is evaporated, i.e. the individuals of the species are assigned to other species. If *rfit* is greater than RF_2 then the species is maintained. If *rfit* is greater than RF_1 but less than RF_2, a small value of mutation is applied to the species such that the majority of the species is preserved but there is a small chance to mutate to other species.

In every generation, simple mutation with a low mutation rate is applied to continuous variables. Crossover in continuous variables is applied in the same species only and diversity in continuous variables is maintained by selecting the two parent for crossover which are most distant in terms of performance. Elitism is applied to each species, i.e. the best individual of each species is preserved if the relative performance of the species is over certain threshold.

5 Test Results

5.1 A simple test function

A simple mixed-discrete function is used for testing the performance of the algorithms. The function is a variation of the generalised Rosenbrock's function[6]. The optimisation problem is formulated as follows:

$$\min f_{2a}(\mathbf{x}) = \sum_{i=1}^{5}(100 \cdot (x_{i+1} - x_i^2)^2 + (x_i - 1)^2) \tag{4}$$

$$x_1 \in \{-1, 0, 1\}, x_2 \in \{1, 2, 3\}, x_3 \in \{0, 1, 2\};$$
$$x_4 \in [-5.12, 5.12], x_5 \in [-5.12, 5.12], x_6 \in [-5.12, 5.12]$$

Notice that the dimension of the decision variable vector is 6. The first three decision variables are now discrete variables with different upper and lower bounds. The number of species is chosen as 27 because this is actually the total number of different combinations of the discrete variables. The optimal solution is at $\mathbf{x}^* = (1, 1, 1, 1.0, 1.0, 1.0)$, the best fitness is $f_{2a}^* = f_{2a}(\mathbf{x}^*) = 0.0$. A test composed of 100 random experiments has been conducted for a simple GA, the dual-agent GA and the GAANT algorithm presented in this paper, respectively. Each experiments has 100000 evaluations (function calls). The test results are summarised in Table 1. It is shown that the GAANT can give the best solution (9.45E-7), while simple GA (2.98E-5) and dual-agent GA (1.24E-6) give slightly worse solutions. Of the 100 tests for each algorithm, GAANT produces 32 successful runs (i.e. $\leq$ 1.0E-4), while the dual GA has 15 successful runs and the simple GA has only 2. This shows the consistency and robustness of GAANT. Notice that for the dual-agent GA, there is a considerable number (20) of runs reaching the region near the optimum (fitness $\in$ [1.0E-4, 1.0E-2], but which fail to push further due to pre-maturity although making more than twice as many calls to the best path as the GAANT. The small average number of calls to the best path and its standard deviation show the good diversity and robust performance of the GAANT.

Table 1: Test results of problem f_{2a}

	Simple GA	Dual Mutation GA	GAANT
Best fitness	2.98e-5	1.24e-6	9.54e-7
No. of successful runs (fitness < 1.0e-4)	2	15	32
No. of solutions with 1.0e-4 $\leq$ fitness <1.0e-2	8	20	9
Average No. of calls along the best path	5601.6	33751	15071
SD* of calls to best species	11%	12%	6%
No. of missed species	0	0	0

SD — Standard deviation, as a percentage of the mean

5.2 A Design Application in a Nuclear Station

The dual-agent GA and the GAANT algorithm have been applied to an actual design study for an AGR station. The purpose is to introduce up to two tubular feed heaters to the dashed block in the feed water circuit. These feed heaters draw steam from the low pressure (LP) turbines and heat the feed water flowing through them. Functionally they can reduce water in the last stages of the LP turbines to avoid turbine blade corrosion and increase the overall plant efficiency. The details of the design problem have been described in [2]. The objective is to find the best design configuration and design parameters together with the best OCVs, so that the whole plant electricity output is maximised, subject to 28 plant safety and integrity constraints. The problem has 18 decision variables. There are four discrete variables and 14 continuous variables. There are 48 different configurations (species). Fig. 1 shows part of

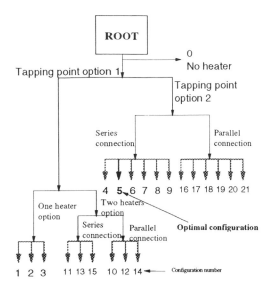

Figure 1: Configuration decision tree of the feed heater design problem

the decision tree. A distributed structure of evaluation is employed to reduced the time consumed by the expensive evaluation program. The PVM library[7] is used so that the evaluation of a population is done simultaneously across a group of networked machines of different platforms. Typically this results in 85% saving in overall computation time. A number of tests have been run for each algorithm. Each tests has 100000 evaluations and it took on average 7 hours in a SUN network led by an Enterprise 4000 with 6 processors of 140MHz. (Due to the limited time available only several runs have been conducted for each algorithm, the final paper will include results from more tests.) The average results are shown in Fig. 2 and Fig. 3. Both algorithms perform well in finding the best solution. Due the property of this particular problem, a considerable number of species have similar performance and due to this reason the GAANT algorithm can maintain all the 'good' candidate configurations till the last generation. Each good species has been given opportunities to evolve. As a result, in the last generation, all the good configurations with their corresponding continuous parameters are available. There are more configurations found with performance higher than 0.98 by GAANT. The dual-agent GA can maintain good diversity at the starting stage of the GA, but due to the continuously applied high mutation rate, good configurations are not always maintained and only one 'good' configuration survived and dominated in the last generation. As a result, the other configurations are either lost or have non-optimal continuous variables in the last generation. For GAANT much fewer generations are needed to obtain the best solution and a good diversity is always maintained.

6 Conclusions

A general formulation of a mixed-integer type optimal design and control problems with dependencies between decision variable sets has been presented. Previous work used a dual-agent GA combined with nonlinear local optimiser which can increase the efficiency of search and increase the probability of finding the global optimum. The disadvantage of the dual-agent techniques is that the mutation for discrete variables is highly disruptive and a relatively stable evolution of the configurations is not maintained. The GAANT strategy used in this paper can maintain the diversity of dual-agent algorithm and stable evolution of the configurations by a classification mechanism. Two test cases have been studied and both have shown the good diversity and robustness of GAANT.

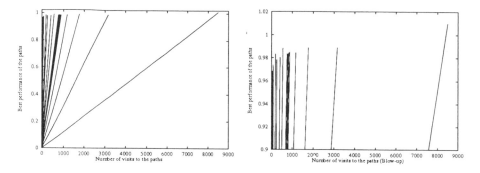

Figure 2: Performance of dual-agent algorithm

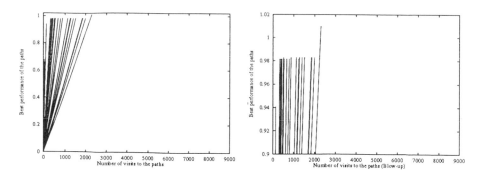

Figure 3: Performance of GAANT algorithm

Previous application of the GAANT approach has shown similar benefits for preliminary hydro-power system design [10, 9]. This indicates a generic capability of GAANT for engineering design problems. Further research will be conducted to improve the performance of the GAANT optimiser by studying the influence of different parameters and methods of classification on the performance of the GAANT algorithms.

7 Acknowledgements

This work was funded jointly by the UK Engineering and Physical Science Research Council (EPSRC) and Nuclear Electric Ltd. The authors wish to thank these two organisations for their continuing commitment to research at the Centre.

References

[1] G. Bilchev and I. C. Parmee. The ant colony metaphor for searching continuous design spaces. In Terence C. Fogarty, editor, *Proceedings of AISB Workshop on Evolutionary Computing, Lecture notes in computer science 993*, pages 25–39. Springer-Verlag, ISBN3540604693, 1995.

[2] K. Chen, I. C. Parmee, and C. R. Gane. Dual mutation strategies for mixed-integer optimisation in power station design. In *Proceedings of the 1997 IEEE International Conference on Evolutionary Computations (ICEC'97)*, pages 385–390. Indianapolis, USA, 13-16 April, 1997.

[3] K. Chen, I. C. Parmee, and C. R. Gane. A genetic algorithm for mixed-integer optimisation in power and water system design and control. In D. Ruan, editor, *Intelligent hybrid systems: Fuzzy logic, neural networks, and genetic algorithms* , pages 311–351. Kluwer Academic Publishers, Boston, September, 1997.

[4] A. Coloni, Dorigo M., and Maniezzo V. An investigation of some properties of the ant algorithm. In *Proceedings of PPSN'92*, pages 509–520. Elsevier Publishing, 1992.

[5] D. Dasguta and D. McGregor. A structured genetic algorithm. Technical Report Research Report IKBS-2-91, University of Strathclyde, Gloasgow, UK, 1991.

[6] K. De Jong. *An analysis of the behaviour of a class of genetic adaptive systems.* PhD thesis, University of Michigan, Diss. Abstr. Int. 36(10), 5140B, University Microfilms No. 76-9381, 1975.

[7] A. Geist and et al. *PVM: Parallel Virtual Machine.* The MIT Press, Cambridge, Massachusetts, 1994.

[8] I. C. Parmee. Diverse evolutionary search for preliminary whole system design. In *Proceedings of 4th International Conference on AI in Civil and Structural Engineering.* Cambridge University, Civil-Comp Press, August, 1995.

[9] I. C. Parmee. The development of a dual-agent strategy for efficient search across whole system engineering hierarchies. In *Proceedings of the 4th International Conference on Parallel Solving from Nature.* Berlin, 22-27 September, 1996.

[10] I. C. Parmee. Towards an optimal engineering design process using appropriate adaptive search techniques. *Journal of Engineering Design*, 7(4):341–362, December 1996.

[11] I. C. Parmee, C. R. Gane, M. Donne, and K. Chen. Genetic strategies for the design and optimal operation of thermal systems. In *Proceedings of the 4th European Congress on Intelligent Techniques and Soft Computing.* Aachen, Germany, 2-5 September, 1996.

Chapter 5

Aerospace Applications

Design Optimisation of a Simple 2-D Aerofoil Using Stochastic Search Methods.
W.A. Wright, C.M.E. Holden

Adaptive Strategy Based on the Genetic Algorithm and a Sequence of Discrete Models in Aircraft Structural Optimization.
V.A. Zarubin.

Multi-objective Optimisation and Preliminary Airframe Design.
D.Cvetkovic, I.Parmee, E.Webb

Evolving Robust Strategies for Autonomous Flight: A Challenge to Optimal Control Theory.
P.W. Blythe

Design Optimisation of a Simple 2-D Aerofoil Using Stochastic Search Methods

W.A. Wright
Sowerby Research Centre
British Aerospace Plc, Bristol
Andy.Wright@src.bae.co.uk
C.M.E. Holden
Sowerby Research Centre
British Aerospace Plc, Bristol
Carren.Holden@src.bae.co.uk

Abstract. This paper describes the results of a study which considers the aerodynamic optimisation of 2-D aerofoil given a number of realistic but simple constraints. It is shown that this simple problem, with a realistic objective function, has a number of local optima. It is demonstrated that stochastic optimisation methods provide a distinct advantage over the more conventional methods and it is further shown that the most optimal results are obtained when these methods are combined with the more conventional methods

1. Introduction

Although there are a large number of investigations into the use of stochastic optimisation methods (e.g. Genetic Algorithms [4] and Simulated Annealing [7]) in design (for example see [5] and [2]) few of these look at the performance on *real* aerospace problems. So far only limited studies have been reported in the open literature [1] and most of these concentrate on the use of more conventional optimisation methods.

To test further the performance of these more advance optimisation methods a problem is considered here which looks at the optimisation of the drag performance of this 2-D aerofoil. Although relatively simple the 2-D aerofoil provides a flexible and *real* shape optimisation problem which not only requires the uses of a real objective, or cost, function but also a number of constraints (on both the general shape and structure of the aerofoil together with its aerodynamic characteristics) to ensure that a realistic and plausible aerodynamic shape is produced. This paper reports the findings of an investigation into the relative performance of the Genetic Algorithm and Simulated Annealing methods on this problem.

2. Problem Description

2.1. B-spline representation

The complexity of a shape optimisation problem is often governed by the representation used for the shape of the object to optimised. Where possible it is desirable to incorporate any symmetry or constraints inherent in the shape of the object. In the aerospace industry standards (STEP) have been laid down which outline appropriate representations for aerodynamic geometries. However, what constitutes (within this standard) the most appropriate representation is still an open question.

Here the aerodynamic 2-D cross-sectional shape of a NACA0012 aerofoil was defined by a set of B-spline poles (see figure 1). These poles were fixed in x and varied in y to allow the shape of the aerofoil to be adjusted. The B-spline representation provides a number of advantages. In particular the method allows a smooth but flexible shape to be generated from a relatively small set of similar parameters. The similar nature of all the parameters also allows for the easy encoding of the optimisation problem since no one parameter requires a different treatment. However, the flexibility of the representation, as will be shown, does lead to difficulties.

To optimise this representation an inviscid Euler method was used to compare the pressure distribution induced by the flow round the aerofoil for various B-spline representations with that of the true geometry. From this it was determined that a B-spline representation with 24 8th order poles provided a geometry with sufficient accuracy for the purpose of this study. Ideally the optimisation with other representations should be undertaken. This will be the focus of future studies on this problem. For the GA optimisation each pole position was encoded using 12 bits.

2.2. Objective Function

For this study it was important to obtain a realistic objective function (cost function) against which to undertake the optimisation. In general most aircraft optimisation problems use complex objective functions such as *Direct Operating Cost*. This measures at the "in-service" cost of the resulting aircraft to the purchaser. Not surprisingly a significant element of this complex function depends on the fuel usage and this is related to the ratio of wing's *drag* and *lift*. Here the drag coefficient, C_D, of the aerofoil for a given coefficient of lift, C_L is used as the objective function. This provides a convenient, *non-dimensional*, measure of the aerofoil's performance.

For the purpose of this study two objective functions were constructed.

- A single point objective function. The coefficient of drag at a coefficient of lift of 0.5 (i.e $(C_D)_{C_L=0.5}$). This is coefficient of drag typical of the aerofoil in cruise flight.

- A multi-point objective function. The sum of the coefficients of drag at two specified coefficients of lift of 0.2 and 0.6 (i.e. $(C_D)_{C_L=0.2} + (C_D)_{C_L=0.6}$).

In addition to these objective functions a number of simple, but real, constraints were also chosen. These were that:

- the cross sectional area of the aerofoil (A) between the leading and trailing edge spars (i.e. wing box area) non-dimensionalised with respect to chord (the thickness of the wing) was greater than 0.052. This restriction ensures that the, wing is not too heavy and that there is sufficient capacity for fuel within the wing-box.

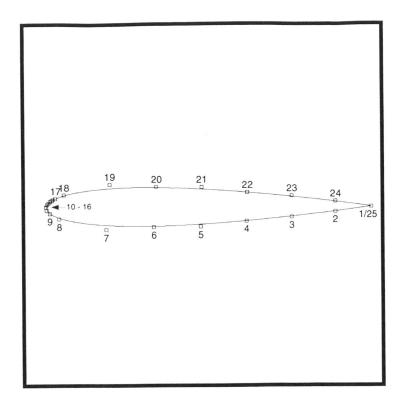

Figure 1: B-Spline representation of the NACA 0012 Aerofoil. The B-spline poles are shown here numbered from 1 to 25.

- the value of the zero lift pitching moment (C_m) was restricted such that $-0.1 < C_m < 0.02$. A larger pitching moment would require a larger tail with a resultant drag and weight penalty.

- The thickness at 15% chord and 65% chord (corresponding notionally to the positions of the front and rear spar respectively) were also limited such that they did not go below 9.66% chord and 7.34% chord respectively. This again limits the thickness of the wing keeping the weight to a manageable level. Note the weight of a wing is inversely proportional to its thickness.

The objective functions were calculated using the panel method of Eppler and Somers [3]. This technique is an inviscid potential method which coupled to an engineering boundary layer allows for the viscous effects in the flow around the aerofoil. The method assumes an incompressible flow (i.e. zero Mach number $M = 0.0$). For the purpose of this study a Reynolds number, based on aerofoil chord, of $Re = 6.0 \times 10^6$ (where chord= 1.0) was used. It should be noted that in cruise the typical modern jet passenger aircraft will fly at compressible speeds (i.e. greater than $M = 0.3$) any "wave drag" contributions to the total aerofoil drag have, therefore, been ignored here. Codes which are able to cope with the shocks produced by the compressible flow together with appropriate viscous effects are more expensive and so were not used here. However, this does not diminish the aerodynamic plausibility of the study since the effect on profile drag, which is not a function of compressible flow, are still of great interest and can be examined with this model. Furthermore, this work is directly applicable to the design of high lift devices and UAVs (unmanned airborne vehicles). Both of which operate in a regime where the flow is incompressible. For both the simulated annealing and genetic algorithm methods the constraints were combined with the objective function using the modified Fiacco and McCormick function [6].

3. Results

To undertake this study a number of optimisation methods were utilised from the OPTIONS suite of programmes [6] produced by Dynamics Consulting Ltd. Two stochastic methods were used.

- Simulated Annealing (SA). This method used a temperature dependent Metropolis algorithm [8] with a geometric temperature schedule.

- Genetic Algorithm (GA). This was an elitest method which uses both niche forming and clustering.

As an alternative two convention optimisation methods were also considered. These were:

- Powell's Direct Search Method [10], which uses a direct search in conjugate directions with quadratic convergence.

- Nelder and Meade's Simplex Method [9].

To obtain an understanding of the complexibility of this simple problem the objective function as a function of the y position of poles 2 and 24 for four fixed values of pole 1 (the pole at end of the trailing edge) was plotted (see figure 2). From these plots it can be seen that even for this problem the objective function behaves in quite a complex manner. For most fixed values of pole 1 the objective function clearly has more than one minima when plotted as a function of the other variables (poles 2 and 24).

Furthermore, as pole 1 is varied the shape and position of these minima evolve. Because of the presents of these local minima the non-stochastic methods did not perform well on the full aerofoil. In fact it was found that for both objective functions the NACA0012 aerofoil sits in a local minima. The simplex and gradient based methods were ineffective returning the aerofoil to the original design when the shape was perturbed by moving the poles by 3% of their original position.

Initially other problems were encountered. On using the stochastic optimisation methods non-physical designs were obtained. Most aerodynamic methods only provide an accurate approximation of the drag within certain strict assumptions. The Eppler method is unable to cope with separation of the boundary layer caused by "bumps" on the leading edge of the aerofoil. Such "bump" geometries were often produced by the stochastic algorithms. For such geometries the Eppler method breaks down and returns an *unrepresentative* low estimate of drag. The generation of such shapes, therefore, had to be eliminated to prevent the stochastic methods from finding a minimum which was not physical. This was achieved by applying an additional constraint on the shape of the leading edge. The constraint simply restricted the shape of the leading edge of the aerofoil ensuring that the y coordinate of the upper surface of the leading edge increased towards the front spar and conversely the lower surface y coordinate decreased towards the front spar thus preventing "bumps" from occurring.

3.1. Single Point Optimisation

For the single point objective function both the simulated annealing and genetic algorithm methods were used. In both cases the methods were able to give physically realistic results. Of the two the simulated annealing method (with a relatively quick annealing schedule, see table 1) produced the largest reduction in the objective function. The method reduced $(C_D)_{C_L=0.5}$ by 24% after 2200 object function evaluations compared to the genetic algorithm (with a population of 100) which achieved a reduction of 13% after 8000 evaluations. The resulting GA and SA aerofoils (see figure 3) have some unevenness around the leading edge. This was smoothed by further local optimisation using the PDS method reducing the objective function to 30% of the original NACA0012 value.

3.2. Multi-point Optimisation

Here only the GA was used. The population size was increased to 300 to allow for the extra complexity of the optimisation task. Two scenarios were computed optimum of the sum $(C_D)_{C_L=0.2} + (C_D)_{C_L=0.6}$ (a two point design) and, for comparison, the optimum of $(C_D)_{C_L=0.35}$ (a single point design). Figure 4 shows the final aerodynamic designs resulting from the optimisation using the genetic algorithm. These shapes are characterised by the aerofoil maximum thickness having moved aft to maintain large tracts of laminar flow. Figure 5 shows the impact of large tracts of laminar flow on the aerodynamic characteristics of the aerofoil. The single point design result has a narrow "bucket" giving a low drag at just one point in the graph. The two point design has a wider "bucket" giving a better result across a range of C_L values. Below a C_L of about 0.1 and above a C_L of 0.8, the C_D is better for the NACA 0012 aerofoil.

238

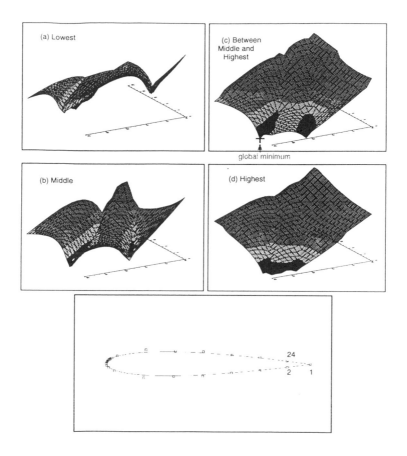

Figure 2: Four isometric plots of the objective function plotted as a function of the y position of pole 2 (x axis) and pole 24 (y axis) for pole 1 fixed at a value of (a) -0.016079, (b) -0.23743×10^{-4}, (c) 0.0080039, (d) 0.016032.

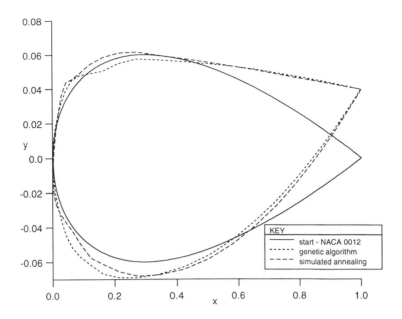

Figure 3: Shape of the complete aerofoil produced by the Genetic Algorithm compared with that produced by the Simulated Annealing optimisation. The uneven shape is smoothed through further optimisation using the PDS method. Note that the aspect ratio of the aerofoil has been altered artificially to excentuate the differences between the shapes.

Temp No	Temp	No of Iterations
1	50.	200
2	10.	200
3	2.	200
4	0.4	200
5	0.08	200
6	0.016	200
7	0.0032	200
8	0.00064	200
9	0.000128	200
10	0.0000256	200
11	0.0000000001	200

Table 1: Table describing the first annealing schedule used for by the Simulated Annealing algorithm for the optimisation of the complete aerofoil.

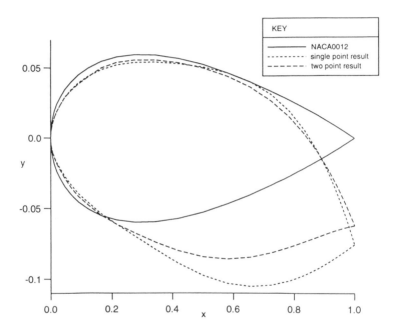

Figure 4: Initial and final aerofoil shapes produced after optimisation of the complete aerofoil by the Genetic Algorithm, for the single and two point optimisations. Again not the artificially increased aspect ratio.

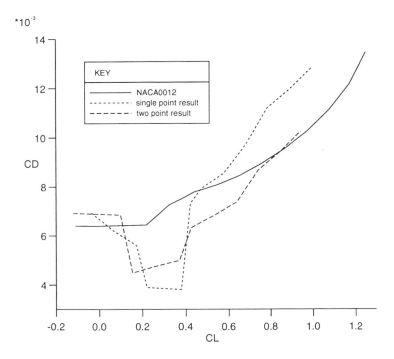

Figure 5: Graph of the drag coefficient versus lift coefficient for the NACA 0012 aerofoil and the single and two point designs from the Genetic Algorithm.

242

4. Conclusion

This investigation has looked at the optimisation of a simple but realistic problem, the shape optimisation of a 2-D aerofoil. Because of the presents of local optima it was found necessary to use stochastic optimisation method to search the design space. For the first problem two methods were used; Simulated Annealing and a Genetic Algorithm. Here it was found that the Simulated Annealing gave a, marginally, better solution with fewer objective function evaluations.

To allow physical drag calculations to be produced it was necessary constrain the shape of the trial aerofoils by restricting the geometry of the leading edge. The need for such constraints highlights a potential difficulty when using these methods. If the space of designs is not appropriately constrained then the optimisation method may produce shapes which fall outside the working assumptions of the aerodynamic code used to calculate its viability. This was seen in these experiments where the aerodynamic code produced very low drag values for unphysical aerofoil shapes.

In the second experiment both the single and two point optimisation results gave physically believable answers. Here the maximum thickness of the aerofoil moved to the rear in order to maintain large tracts of laminar flow. Indeed the drag polars contained the sort of laminar flow "buckets" that may be expected from a laminar flow aerofoil.

It is considered that results obtained using these methods could provide the designer with a useful starting point in those situations where the velocity profile of a similar aerofoil is not available. In reality of course an actual design study would be more complex than the test case investigated here. This example uses a relatively simple objective function with only a limited number of constraints. Clearly further investigation on more complex (possibly multi-disciplinary) problems needs to be undertaken to determine what the true potential of these methods are.

Acknowledgements

The authors would like to thank Professor Keane of the University of Southampton for both his help in using the OPTIONS suite of programs and advice on how best to approach this project. CH would also like to acknowledge Clyde Warsop of BAe Sowerby Research Centre without whose help and understanding of wing design this work would not have been possible.

5. References

[1] AGARD. *Optimum design methods for aerodynamics*, volume AGARD-R-803, 1994.

[2] D. Dasgupta and Z. Michalewicz. *Evolutionary Algorithms in Engineering Applications*. Springer-Verlag, 1997.

[3] R. Eppler and D.M. Somers. A computer program for the design and analysis of low speed airfoils. Technical Report NASA-TM-80210, NASA Langley Research Center, 1980.

[4] J.H. Holland. *Adaption in natural artificial systems*. University of Michigan Press, 1975.

[5] A.J. Keane. Experiences with optimizers in strutural design. In *Conference on Adaptive Computing in Engineering Design and Control*, pages 14–27, 1994.

[6] A.J. Keane. *The OPTIONS Design Exploration System: Reference Manual and User Guide*, 1995.

[7] C.D. Kirkpatrick, C.D. Gellatt Jr, and M.P. Vecchi. Optimisation by simulated annealing. *Science*, 220:671–680, 1983.

[8] N. Metropolis, A. Rosenbluth, Rosenbluth M., A. Teller, and E. Teller. Equation of state calculations by fast computing machines. *J. Chem. Phys.*, 6:1087, 1953.

[9] J.A. Nelder and R. Meade. A simplex method for function minimisation. *Computer Journal*, 7:308–313, 1965.

[10] M.J.D. Powell. An efficient method for finding the minimum of a function of several variables without calculating derivatives. *Computer Journal*, 7(4):303–307, 1964.

Adaptive Strategy Based on the Genetic Algorithm and a Sequence of Discrete Models in Aircraft Structural Optimization

V.A. Zarubin

Samara State Aerospace University

zarubin@mb.ssau.samara.ru

Abstract. Airframe structural optimization problem is formulated as a task of aircraft relative expenses minimization subject to minimum of structural mass and satisfaction of stress, displacements and many other constraints imposed on a structure. Vector of design variables (DVs) is presented as a set of material distribution (m-type), shape and skeleton (r-type) and conceptual (s-type) subvectors. Method, algorithm and software are created for posed problem solution. GA is used in an external loop of the algorithm for search of the best set of s-type of DVs which are treated as structural genes. In inner loops the genotype string of each individual is decoded into phenotype which is developed by the sequence of numerical models. Skeleton (r-type) and muscular (m-type) parameters are optimized by ordinary methods of structural mass minimization. Conceptual design of short-range passenger aircraft is presented as a numerical test. The test demonstrates the workability of method, algorithm and software created in this research.

1. Introduction

The aim of this paper is to present the multidisciplinary approach to aviation structure optimization based on complex structural and efficiency analysis and combination of the Genetic Algorithms with ordinary optimization methods [1,2]. The GA is used for forming the idea (conceptual design) which is then developed into detailed design with the help of the totality of computer models and Sequential Linearization in the Optimality Criteria Methods (SLOCM) [3] optimization.

The tasks which have to be solved to reach this aim are.
1. Analysis of the optimality criteria and optimal problem formulation.
2. Methods, algorithms and software creation for optimal problem solution.
3. Numerical testing of created software.

2. Optimal problem formulation

2.1. Design variables

Parameters which are varied during different design stages can be presented as a vector $\{x\}$ which consists of three subvectors: $\{m\}$, $\{r\}$ and $\{s\}$.

$\{m\}$ is a vector of structural material distribution parameters, for example, areas or thickness of structural elements cross sections. Usually these DVs are defined in FE-model as geometrical parameters of finite elements and they don't influence FE-mesh.

$\{r\}$ is a vector of shape and skeleton parameters. In FE-model these parameters are presented as height, radius, angle, position, etc.. These parameters define FE-mesh geometry, but don't influence topology. For example, the position of spar in wing chord percentage or stringers orientation in wing panel don't influence finite element topology description but they change the node coordinates if varied.

$\{s\}$ is a vector of main parameters which describes aircraft appearance, structural type, material, topology, etc.. The variation of these parameters may lead to a new FE-mesh with different type and number of elements and node coordinates. For example, in [1] the helicopter tail-boom structure is presented by two load bearing scheme types: truss and semimonocock, which led to principally different FE-models for each type of structural scheme.

2.2. Constraints

During structural design we have to satisfy many demands and constraints imposed upon the structure. Models of FE-analysis allow us to calculate values of many of constraint functions like stresses, displacements, frequencies, critical loads, critical velocities, etc. Preventing of violation of these constraints results in structural mass increase. Therefore minimum of structural mass is a very important demand for successful structure. Usually variations of m- and r-type parameters are used for structural mass minimization. The parameters of s-type have to be defined on the basis of more general criteria, like aircraft efficiencies.

2.3. Goal function

Choice of the criteria for aircraft structure evaluation during the design stage is a very important question. These criteria influence not only the aircraft performances but its future life and fate.

If main parameters $\{s\}$ are defined it is a common practice to use structural mass as a goal function of $\{m\}$ and $\{r\}$ DVs, but for $\{s\}$ parameters this function doesn't work properly. For example, wing aspect ratio λ has a positive sensitivity coefficient respect to the structural mass, what means that λ increase leads to structural mass increase. But there is a desire to make λ bigger with the purpose to reduce the induced drag. Structural mass in the role of goal function doesn't allow to do this and λ will be placed to its minimum value in such optimization.

There are many other main parameters from s-type group which in the case of structural mass minimization take its boundary values. For example, another three paramenters: wing profile thickness will try to attain its maximum, angle of sweepback - minimum, fuselage length aspect ratio - minimum too. But these three parameters influence drag, cruise velocity, fuel and transportation efficiency with opposite effect to structural mass. It may happen that the reduction of structural mass will lead to reduction of aircraft efficiency.

Take-off mass seems to be more general criterium which includes links between these conceptual DVs and fuel efficiency. But from the numerical tests [4] it follows that structural mass is more sensible to changes in comparison with the mass of fuel. Therefore after all these improvements of aerodynamics the take-off mass becomes bigger due to structural mass increase. Some reduction of fuel mass is not considerable.

It is obvious that we must take more general criteria of structural efficiency than structural or even take-off mass.

Two criteria of transportation efficiency were taken into consideration [5,6]. Numerical tests [4] of these criteria revealed that they influence upon some parameters as it could be predicted, for example, they resist the reduction of wing aspect ratio. But for many other parameters, for example angle of wing sweepback or material type, they don't work.

It became obvious, that for s-type of DVs determination we must use more general criterion or set of criteria. Therefore we chose a_{rel}, a relative expenses [6], as more advanced one. Structural mass is presented in a_{rel}. The expenses becomes smaller with structural mass decrease. This goal function takes into consideration most of the parameters which are interested for us at early design stages. But of course this criterion should be developed in the future works.

2.4. Optimal problem formulation

Optimal problem is formulated as following.

Minimize $a_{rel}(x)$ subject to $G_j(x) \leq 0$ and $x^l_i \leq x_i \leq x^u_i$, $\qquad\qquad$ (1)

where vector $\{x\} = \{s, r, m\}$; $i = 1,...,n$; n is number of DVs; $G_j(x)$ is the value of constraint function; x_i^l and x_i^u are lower and upper values of DVs, $j = 1,...,l$; l is number of constraints.

Expressions (1) mean that we want to find such parameters of aircraft appearance, shape, skeleton and material distribution which minimize relative expenses and satisfy the constraints imposed upon the aircraft structure.

3. Method of problem solution

Method is based on the aircraft models adaptation during design process. In accordance with [7] the mechanisms of adaptation provide a progressive modification of some parameters which in GA approach are considered as a set of genes combined in a chromosome. In structural design this approach suggests that all DVs are encoded as genes and combined in a string or chromosome. In our research only principal DVs are treated like genes, other DVs are the parameters which describe not the set of chromosomes, but the external sizes of the object. It is very natural to propose that s-type of the DVs are the principal parameters, which have to be encoded and treated as genes. And other DVs are the parameters of structural "skeleton" (r-type) and "muscular" systems (m-type) which have to be treated by correction and training accordingly [1].

Another distinctive peculiarity of our approach is the presence of special procedures of structural development, interpreted in [2] as training or life simulation and correction of body (not genes) parameters in the mechanisms of adaptation.

4. Algorithm of problem solution

Following steps and actions lead to problem solution.
0. Preliminary research. Formulation of main demands. Initial data forming.
1. Optimal problem formulation.
2. The initial population of several possible variants of load bearing structure is created. s-type parameters are encoded in a chromosome (string) for each individual.
3. The life of each individual is simulated in accordance with the constraints and demands imposed.

    3.1. Genotype is decoded into phenotype.

    3.2. Structural mass is calculated in accordance with the simplest empirical model of aircraft existence. Prototype parameters and statistics data are used.

    3.3. Calculation of the take-off mass of the first approximation.

    3.4. Load definition.

    3.5. Creation of the FE-model of the first level.

    3.6. Structural mass of FSD calculation.

    3.7. Calculation of the take-off mass of the second approximation.

    3.8. Conceptual design.

    3.9. Creation of the FE-model of the second level.

    3.10. Creation of the aerodynamic, inertia and other computer models of totality of discrete models of the second level.

    3.11. Minimum structural mass problem formulation.

    3.12. Structural optimization by SLOCM.

    3.13. Economical model tuning.

    3.14. Goal function calculation.

4. The exit conditions are checked. If the number of iterations is exhausted or the process converged, for example the difference between best and worst individuals is smaller than prescribed one, then go to 6, else go to 5.

5. The GA's selection, crossover and mutation procedures from [8] are used. The new generation is created. Go to 3.

6. Stop.

5. Description of the software

Goldberg's programs from [8] are taken as a basis for the GA realization. The POLINA FE-package [9] is used for structural simulation and structural mass evaluation. The package consists of following blocks and stand-alone programs.

1. Main block provides initial data (like size of population, number of iterations, chromosome structure, details of interactions with FE-package) input, program tuning and control.

2. Block of encoding-decoding transformations of genotype into phenotype and back.

3. Block of structural mass of the first approximation level evaluation.

4. Block of take-off mass of the first approximation level evaluation.

5. Generator of FE-model of the first level.

6. Generator of loads for FE-model of the first level.

7. Block of structural mass of the second approximation level evaluation.

8. Block of take-off mass of the second approximation level evaluation.

9. Block of goal function value calculation.

10. Block of GA selection.

11. Block of GA crossover and mutation.

POLINA FE-element package [9] is used for structural behavior simulation and structural mass of FSD estimation.

6. Numerical example

The short range passenger aircraft was taken as a subject for numerical demonstration.

Preliminary research [4] revealed that the 100 seats plane with the range of 1500-2000 km will be in demand for next 20 years.

The main requirements to the aircraft under designing formulated in the research [4] prescribe cabin layout, crew staff of 2 pilots and 3 stewards, 800 km/h of cruise speed, 10000 m of cruise flight height, 1800 m of runway length, 220 km/h of landing speed, 2000 km of range with 12000 kg of payload. Other demands define the levels of reliability, safe damages, safety, comfort.

Optimal problem was formulated as relative expenses minimization subject to minimum structural mass and behavior constraints.

Vector $\{s\}$ consists of the following parameters: λ is wing aspect ratio; λ_f is fuselage aspect ratio; c_0 is relative aerofoil thickness; η is wing's narrowing; χ is backsweep angle; H_{cr} is cruise height; V_{cr} is cruise speed; mat is type of material. Genotype-phenotype relations are presented in Tables 1 and 2.

Table 1. Genotype-phenotype relations, 2 bits strings

DV:	c_0	η	λ_f	mat
String:	--	--	--	--
00	0.10	2	6.5	D16
01	0.11	2.5	7.5	B95
10	0.12	3	8.5	AlLi1
11	0.13	3.5	10	AlLi2

Table 2. Genotype-phenotype relations, 3 bits strings

DV:	λ	H_{cr}	V_{cr}	χ
String:	---	---	---	---
000	7	3	500	0
001	8	6	600	5
010	9	7	650	10
011	10	7.5	700	15
100	10.5	8	750	20
101	11	8.5	800	25
110	11.5	9	850	30
111	12	10	900	35

The initial population of 50 individuals was created by random way. The best individual in this generation has the parameters shown in Table 3. The parameters of the worst one are placed in Table 4.

Table 3. The best individual of the initial population

λ	λ_f	V_{cr}	χ	c_0	η	H_{cr}	mat	fitness
10	8.5	700	35	0.11	2.0	10	D16	11.12535

Table 4. The worst individual of the initial population

λ	λ_f	V_{cr}	χ	c_0	η	H_{cr}	mat	fitness
10.5	6.5	900	35	0.13	3.0	8	AlLi1	29.65755

Then algorithm of the fourth section is implemented. In this research the algorithm is reduced and steps 3.8-3.12 are excluded, what means that structural optimization is done by FSD approach only without SLOCM optimization implementation. Each genotype is decoded into phenotype and simple empirical models [5,6] are used for aircraft appearance forming and take-off mass evaluation. Finite element model and loads are determined for each individual. Several examples of FE-models are presented at figure1. Then each individual is optimized by FSD approach, its structural and take-off masses of second approximation are calculated and goal function (relative expenses) is evaluated too. The diagram of fitness values are presented at figure2.

Figure1. Examples of FE-models of six individuals from initial population

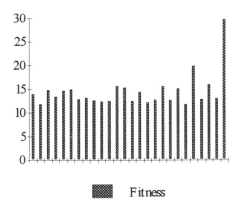

Fitness

Figure2. Fitness diagram of the initial population

Thirty iterations with the probability of crossover equal to 75% and mutation rate equal to 0.01% were done. The process of fitness changes during iterations is presented at figure 3.

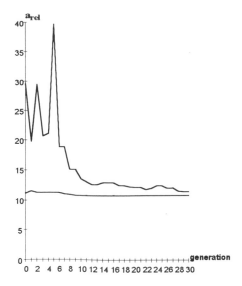

Figure3. The process of fitness changes during iterations (upper curve is the worst individual, lower one is the best individual)

In the initial population maximum value of goal function was equal to 29.65755, minimum was equal to 11.12535. During first 7 generations the maximum goal function value had big alterations. After this the tendency for maximum value decrease appeared. Minimum had the reduction tendency from the beginning. The best individual appeared in the ninth generation. Then it disappeared in the 11th generation and appeared again in the 12th one. The 30th generation has the worst individual with 11.39483 and the best one with 10.73364 value of goal function. There are 26 individuals in the population with such value of goal function. The fitness diagram on the 30th generation is presented at figure 4.

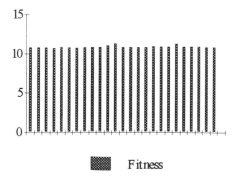

Figure 4. Fitness diagram of the thirtieth generation

The best and the worst individuals in this generation are presented in Tables 5 and 6.

Table 5. The best individual of the thirties population

λ	λ_f	V_{cr}	χ	c_0	η	H_{cr}	mat	fitness
10.5	8.5	750	35	0.10	2.0	10	D16	10.73364

Table 6. The worst individual of the thirties population

λ	λ_f	V_{cr}	χ	c_0	η	H_{cr}	mat	fitness
9	6.5	750	35	0.10	2.0	8.5	D16	11.39483

Some additional tests [4] showed that for population size equal to 50 and string's length of 20 bits the program reaches the stable optimum after 12-15 iterations.

7. Conclusions and further works

The paper demonstrates the ability of the GA to be implemented for aircraft structural optimization. The process of conceptual design based of simulation of airframe model evolution is presented.

The future works are planned in three directions.

1. Modification of POLINA package [9] with the purpose to increase the efficiency of the analysis and optimization codes.

2. Analyzing of other adaptive tools and looking for and creation of the more effective methods, algorithms and software for structural optimization.

3. Creation of the Net Shell for Designing Control (NSDC) [2,10] which provides a Chief Designer with the possibilities to present the design process as the set of subprocesses and tasks distributed through departments, teams, specialists and their desks; to prescribe the technology, data bases, data input, processing by particular software and output at each desk; to organize these data flow; to define time-table and responsibility, etc.

Acknowledgments

The research leading to this paper was supported by "ODA integrated systems" company.

References

1. Zarubin V.A., Multidisciplinary large-scale structural optimization and the place of genetic algorithms in it, 1996, *Proceedings of 1st Int. Conf. on Evolutionary Computation and Its Application "EvCA'96"*, Presidium of the Russian Academy if Science, Moscow

2. Zarubin V.A., Genetic algorithm in the role of a shell for structural evolution simulation at the conceptual deign stage, 1996, *Proc. of the 1st on-line workshop on soft computing, WSC1*, Nagoya University, Nagoya

3. Zarubin V.A., Structural sensitivity analysis and optimization in the RIPAK package. Part I, 1994, *Structural Optimization, Vol.8, No. 2/3*, Springer-Verlag International

4. Chernov A.V., Filatov E.F., Teplykh A.V., Genetic algorithm usage in aircraft preliminary design (in Russian), 1997, *Graduate project*, Samara State Aerospace University, Samara

5. Badyagin A.A., Eger E.M., Mishin V.F., Fomin N.A., *Aircraft design* (in Russian), 1984, Mashinostroenie, Moscow

6. Katyrev I.Ya., Neimark M.S., Sheinin V.M., *Designing of civil aircraft* (in Russian), 1991, Mashinostroenie, Moscow

7. Holland J.H., *Adaptation in natural and artificial systems*, 1975, Univ. of Michigan

8. Goldberg D.E., *Genetic algorithms in search, optimization and machine learning*, 1989, Addison-Wesly

9. POLINA package user's guide, 1995, *ODA Integrated Systems*, Samara

10. Zarubin V.A., Malgin A.S. et al., 1997, Multidisciplinary structural optimization on the basis of discrete modelling and informatuion technologies. Methods, alogrithms, software, 1997, *Proc. of the 1997 Joint ASME/ASCE/ SES/ summer meeting*, Northwestern University, Evanston

Multi-objective Optimisation and Preliminary Airframe Design

Dragan Cvetković[*] and Ian Parmee[†]
Plymouth Engineering Design Centre
University of Plymouth, UK

Eric Webb
British Aerospace
Warton, UK

Abstract

In this paper we explore established methods for optimising multi-objective functions whilst addressing the problem of preliminary design. Methods from the literature are investigated and new ones introduced. All methods are evaluated within a collaborative project with British Aerospace for whole system airframe design and the basic problems and difficulties of preliminary design methodology are discussed.

Our Genetic Algorithm is expanded to integrate different methods for optimising multi–objective functions. First, methods based on scalarisation and the utilisation of weights are addressed. Methods based on Pareto order are then analysed and two different sorting techniques (dominated/non–dominated and Pareto rank) investigated. Finally, several variants of sub–population based algorithms are presented.

All presented methods are also analysed in the context of whole system design, discussing their advantages and disadvantages.

1 Introduction

When dealing with industrial design problems, one immediately realises that there are significant differences between so called 'textbook optimisations problems' and 'real world applications'. In both cases, when attempting multi–objective optimisation, we have a function to optimise:

Definition 1 Let $n > 0$, $k > 0$, $\mathcal{D} = X_1 \times X_2 \times \ldots \times X_n \subseteq \mathbf{R}^n$, and $\mathcal{R} = \mathcal{Y}_1 \times \mathcal{Y}_2 \times \ldots \times \mathcal{Y}_k \subseteq \mathbf{R}^k$. Let further $f_i : \mathcal{D} \mapsto \mathcal{Y}_i$ for $1 \leq i \leq k$ and finally $\mathbf{F} : \mathcal{D} \mapsto \mathcal{R}$, so that $\mathbf{F}(x) = (f_1(x), \ldots, f_k(x))$.

Our goal is to optimise function $\mathbf{F}(x)$ under additional constraints i.e.

$$\max_{x} \mathbf{F}(x) \tag{1}$$

$$g_1(x, p) \leq 0, \ldots, g_l(x, p) \leq 0 \tag{2}$$

[*]e–mail: D.Cvetkovic@plymouth.ac.uk
[†]e–mail: I.Parmee@plymouth.ac.uk

where $p = (p_1, \ldots, p_u)$ are additional (real–valued) parameters. The problem itself is not easy because we have many optimisation criteria, some of which contradict one another. This problem is well known and a number of non–genetic [1, 2] and genetic algorithm approaches exists [3, 4, 5]. An additional problem is that not all objectives are equally important which necessitates the use of weights [6] and preferences [7].

However, initial discussion with the industrial partner makes the problem even harder due to the following additional (meta–)objectives:

- We have objectives and we have constraints. The difference between them is very fuzzy and some of them will move from objectives to constraints or vice versa. Some constraints are hard, some not; some will change or disappear whilst others may be introduced as the problem knowledge base expands.

- In many cases the variable ranges are also fuzzy and there is a requirement for exploration outside of the default regions.

- The output should contain both optimal solutions and suggestions of extending ranges and/or inclusion/removals of constraints.

- A set of results is required which the engineer can analyse off-line.

- The end–user (designer) should not be confused with the number of parameters and possibilities the (optimisation) program offers, we must avoid cognitive overload.

- The engineer wishes to interact with the search process by sampling results after N functions evaluations, and adapting parameters and/or constraints.

The problems of conceptual design relate to the fuzzy nature of initial design concepts and the many different variants that engineer wishes to try. Computers should be able to help him exploration of those variants whilst also suggesting some others as well. This problem has been investigated in Plymouth Engineering Design Centre (PEDC) before [8, 9]. The schema of computer aided design is simply presented in Figure 1.1: at the beginning, the computer system and the designer might have totally different concepts about the object. Ideally they should agree (after some 'negotiations') and have the same idea at the end of the process.

1.1 The BAe model

Our project is based on a collaboration with British Aerospace (BAe) and utilises their CAPS system. CAPS (Computer aided project studies) is an integrated computer software suite, developed at BAe for use by engineers and designers during the earliest investigation stages of a new aircraft project.

The scope of disciplines covered by CAPS is wide: preliminary geometric definition, aerodynamic analysis, mass estimation, performance analysis, cost estimation etc.

In a typical job the user programs CAPS to search for design solutions that meet performance requirements whilst satisfying a number of constraints.

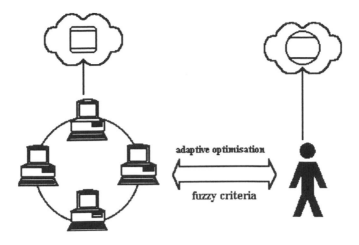

Figure 1.1: The initial phase of computer aided whole system design

BAe has developed a MINICAPS model [10] based upon the full CAPS representation for use by the PEDC. At the moment there are 8 input and 9 output parameters
(i.e. $n = 8$ and $k = 9$ in Definition 1). All constraints so far are variable domain. Input and output parameters of the function are presented in Figure 1.2. In the further
text we will usually denote input variables by $x_1, \ldots, x_8$ and the objectives $y_1, \ldots, y_9$.

The project commenced in April 1997, and this paper mainly describes the initial
phase and a comparison of the integration of well–known multi–objective techniques
with MINICAPS.

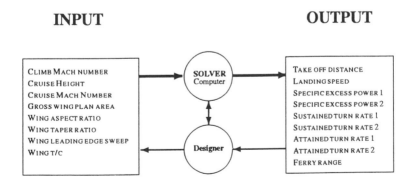

Figure 1.2: Input and output parameters of the BAe function and designer interaction.

2 Optimisation methods

2.1 Weighted sums

Weighted sums is a method of scalarisation of vector functions [1, p.32], [2].

Definition 2 For a function $\mathbf{F}(x) = (f_1(x), \ldots, f_k(x))$ from Definition 1 and a vector $w = (w_1, \ldots, w_k)$, so that $\sum_{i=1}^{k} w_i = 1$, define

$$\mathbf{F_w}(x) = \sum_{i=1}^{k} w_i \cdot f_i(x) \qquad (3)$$

Usually one assigns weights according to the importance of objectives: more important objective will get a higher weight, less important objectives will get lower weights. However, since not all the objectives have the same range of values, they must further be normalised.

One method for choosing weights is by calculating $w_i^* = 1/\max_x f_i(x)$. The vector $(\max_x f_i(x))_{i=1,k}$ is called the *positive ideal solution* [6, p.90]. If we are to minimise an objective y, there are two approaches:

1. Use identity $\min y = -\max(-y)$. However, the genetic algorithm has a problem with negative fitness values — that problem can be avoided by maximising $M - \max(-y)$ for a large enough M. More precisely, if objective y is positive and bounded i.e. $y \in [m, M]$, then $0 \le (M - y)/(M - m) \le 1$ and value $1/(M - m)$ can be used for the weight.

2. Assuming that $y > 0$, use identity $\min y = 1/\max(1/y)$ and so maximise $1/y$. In this case we have no problems estimating weights, but we can't always know in advance if the objective y is non–zero in all cases.

Example 1 The BAe function described above is $\mathbf{F} : \mathbf{R}^8 \mapsto \mathbf{R}^9$. Objectives 1 and 2 should be minimised and the rest of the objectives maximised. If we optimise for one objective only (i.e. just ignore the other 8 output values), we get the optima and weights w_i^* presented in Table 2.1.

i	1	2	3	4	5	6
$\text{opt}_x f_i$	217.800	56.8903	148.176	82.9057	16.6629	10.5922
w_i^*	-0.00108	-0.01186	0.00675	0.01206	0.06001	0.09441

i	7	8	9
$\text{opt}_x f_i$	27.8898	18.6173	10263.7
w_i^*	0.03586	0.05371	0.00010

Table 2.1: Results obtained by optimising BAe function one objective at the time

Using results from Table 2.1, optimising the function

$$\mathbf{F_{w^*}}(x) = \sum_{i=1}^{9} w_i^* \cdot f_i(x)$$

gives us the following as an optimum solution:

$$\mathbf{F}_{w^*}(0.86, 9033.6, 0.9, 56.04, 2.42, 0.1, 58.46, 0.03) = 4.027$$
$$\mathbf{F}(0.86, 9033.6, 0.9, 56.04, 2.42, 0.1, 58.46, 0.03) =$$
$$(259.97, 66.386, 114.75, 37.443, 12.578, 8.4048, 20.771, 18.617, 5761.2)$$

There are several drawbacks with this method. It is very computationally expensive, since for n objectives, we have to perform at least n GA runs (one for each objective) *before* we can start to optimise our function. This might make sense if we are not going to change (add or delete) our objectives and constraints: in that case we can just reuse those values. However, in the conceptual design phase, the probability of objectives and constraints variation is high, and thus the fitness landscape will change therefore, necessitating the re–calculation of our weights again and again. So, this method is not very feasible for our purpose. We will however use it in the further text in order to demonstrate different techniques and results obtained using them.

2.2 Lexicographic method

Lexicographic method for multi-objective optimisation is based on lexicographic ordering [11]:

Definition 3 We will say that point $x \in \mathbf{R}^n$ is *less in lexicographic order* then the point $y \in \mathbf{R}^n$, written $x \preceq y$ if either $x = y$ or the first nonzero component of $x - y$ is negative.

Obviously, since the sorting of individuals is done using relation $\preceq$, the first objective is the most important one, the second objective is a bit less important etc. so one has to be very aware of the importance of the particular objectives. According to [11]:

> Lexicographic optimization arises in those practical situations where optimal policies are determined by making decisions *successively*. First the most important objectives are met. Among the solutions which meet this objective a smaller set is then chosen to satisfy an objective second in importance etc.

So this method is quite opposite to Pareto sorting method, presented below, that treats all objectives equally.

However, there are some obvious drawbacks using this method, especially in preliminary system design where the designer doesn't always know the importance of particular objective. What is usually done in that case is to use *random lexicographic order*: the order of objectives we sort on is chosen randomly and varies from generation to generation. We have implemented this method within a GA and some results are presented in Table 2.3 on page 9.

2.3 Pareto optimisation

Consider a vector function $\mathbf{F} : \mathcal{D} \mapsto \mathcal{R}$ from Definition 1 again.

Definition 4 We will say that a point $x \in \mathcal{D}$ *Pareto–dominates* a point $y \in \mathcal{D}$ with respect to function $\mathbf{F}$, denoted $y \leq_{\mathbf{F}}^{P} x$, if $\wedge_{i=1}^{k}(f_i(y) \leq f_i(x))$ and at least one of inequalities is strict.

We say that point $x_P \in \mathcal{D}$ is *Pareto–optimal* or *non–dominated* (for a given function $\mathbf{F}$) if there is no point $y \in \mathcal{D}$ that Pareto–dominates x i.e. $(\neg \exists y \in \mathcal{D})(x_P \leq_{\mathbf{F}}^{P} y)$

Set $\mathcal{F} \subseteq \mathcal{D}$ is called the *Pareto front* with respect to function $\mathbf{F}$ if every element $x \in \mathcal{F}$ is Pareto optimal with respect to function $\mathbf{F}$. In other words, the Pareto front is the maximal set of non–dominated elements.

In the following text we will drop indices $\mathbf{F}$ and P and will just write $\leq$ instead of $\leq_{\mathbf{F}}^{P}$.

Example 2 In Figure 2.1 we can see an example of a partial order. All elements at the top of the Figure (i.e. points A, B, C, D and E (!)) are non–dominated and therefore in the Pareto front.

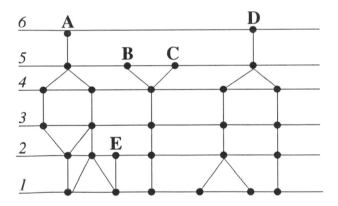

Figure 2.1: Example of partial order and ranking.

Example 2 also shows that not all elements from the Pareto front are automatically of a good fitness value (e.g. point E). Point E can be in Pareto front if one of the objective values is very high (higher that any other) but all others are low.

We have used Pareto method in Genetic Algorithms in the selections phase in the following way:

Pareto tournament Use standard tournament selection but the chosen individual will be the best according to Pareto ordering;

Pareto sort Sort first according to Pareto ordering and then, if both individuals are dominant (or non dominant), sort according to the fitness value;

Pareto truncation Choose parents only among the non–dominated parents.

It is interesting to observe how the number of non-dominated elements in the population increases. Figure 2.2, presents a diagram of number of non–dominated elements in population versus generation number. Population size used is 100. We have used standard tournament of size 2 as a selection method. The results are averaged over 50 runs and standard deviation is presented using error bars. We were optimising on y_3, y_4 and y_9 simultaneously. What we can see immediately is that after some 150 generations more then 50% of the populations is formed by non–dominated individuals and that the selection pressure gradually decreases with the number of generations.

In the case of conceptual design, this also means that the Pareto front, the set the designer is usually interested in, will be much too large for a human to handle. Even with a small population of 100 individuals, after 400 generations, the size of the Pareto front increases to 80 i.e. 80 solutions (more or less different) to deal with.

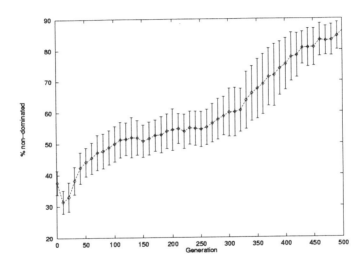

Figure 2.2: Non–dominated percent of the population for tournament selection of size 2, average over 50 runs. Average with standard deviation as error bars.

2.3.1 *Pareto ranking*

Instead of just a dominant/non-dominant scheme for Pareto sorting that doesn't distinguish elements very well (as we can see from the Figure 2.2 where at the end of the run more then 80% are non–dominated) and because of the arguments presented above, we can use a finer method: Pareto ranking (compare [12, 13]).

Definition 5 *Pareto rank r* in a set $X = \{x_1, \ldots, x_n\}$ is assigned in the following way:

$$(\forall x \in \mathbf{X}) r(x) \leftarrow 1$$
$$(\forall x \in \{x_1, \ldots, x_n\})(\forall y \in \{x_1, \ldots, x_n\} \setminus \{x\}) \text{ // sequentially!}$$
$$\quad \text{If } (r(x) = r(y) \wedge x > y) \; r(x) \leftarrow r(x) + 1$$
$$\quad \text{If } (r(x) = r(y) \wedge x < y) \; r(y) \leftarrow r(y) + 1$$

Using the above definition/algorithm, points A and D from Figure 2.1 would have rank 6, B and C rank 5 and point E rank 2.

2.4 Optimisation results

Figure 2.3 shows the Pareto front obtained by performing Pareto optimisation on the BAe function optimising y_3, y_4 and y_9.

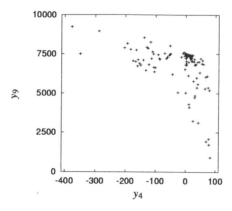

Figure 2.3: Pareto front of y_4 versus y_9 for BAe function.

2.4.1 *What to optimise?*

We have 9 output parameters to optimise, some of them are very complementary. For instance, if we want to optimise on y_3, y_4 and y_9, we have the situation[1] as presented in Table 2.2 (one run on GA with Pareto sorting).

Table 2.2 shows that optimising y_4 (specific excess power (SEP) for supersonic case) affects ferry range (y_9) a great deal and vice versa. If we want to maximise ferry range, we obtain catastrophic values for SEP, if we maximise SEP, we get ferry range 0. Since the goal of the project is to design of an aircraft that performs well in all situations, we must consider both sub-sonic and supersonic configurations.

If we use weights and the method of positive ideal solution for assigning weights, we get the results shown in Table 2.3. Because of the interaction between objectives,

---

[1]If we optimise on y, we use additional constraint $y \geq 0.001$

Optimise	y_3	y_4	y_9
y_3	148.1757	76.9852	0.0000
y_4	145.2489	82.9057	0.0000
y_9	-1.7305	-492.6476	10263.6426
y_3, y_4	145.3017	82.8891	2310.7244
y_3, y_9	23.2492	-380.7826	10096.2207
y_4, y_9	115.2269	0.0001	7886.8315
y_3, y_4, y_9	115.7906	0.0547	7873.7720

Table 2.2: Optimising different combinations of objectives gives different results

we have put additional constraints that every objective that we optimise on must be greater then 0.001.

Those results clearly show that choosing different optimisation methods and different objectives can give totally different results that are, globally, not necessary better or worse, just different.

What?	How?	y_3	$\sigma(y_3)$	y_4	$\sigma(y_4)$	y_9	$\sigma(y_9)$
y_3, y_4, y_9	PAR	119.72	4.27	0.68	3.65	7747.15	134.69
	LEX	115.61	5.17	0.42	1.33	7826.03	100.81
	WEI	146.84	ε	76.93	ε	6070.39	0.64
y_3, y_9	PAR	5.62	6.27	-459.52	28.27	10220.1	43.41
	LEX	0.34	0.64	-482.91	2.72	10249.8	10.47
	WEI	139.56	0.08	52.26	0.19	6959.5	5.93
y_4, y_9	PAR	117.33	4.41	0.29	1.19	7811.8	106.24
	LEX	113.68	5.35	0.17	0.80	7846.14	84.12
	WEI	145.72	ε	76.16	0.002	6198.94	0.29
y_3, y_4	PAR	145.30	ε	82.89	ε	1838.35	1734.38
	LEX	145.30	ε	82.89	ε	1527.58	1641.24
	WEI	145.29	ε	82.90	ε	1637.69	1686.49
y_9	PAR	-1.74	0.02	-491.83	1.25	10262.8	0.96
	LEX	-1.74	0.01	-491.98	0.79	10263.4	0.45
	WEI	-1.74	0.01	-492.19	0.81	10263.4	0.31
$y_1, \ldots, y_9$	PAR	59.67	ε	9.81	ε	4787.75	0.02
	LEX	59.67	ε	9.81	ε	4787.73	0.03
	WEI	114.66	0.16	37.19	0.21	5769.41	15.22

Table 2.3: Pareto (PAR), random lexicographic (LEX) and weighted sum (WEI) optimisations, average over 50 runs. Here $0 < \varepsilon < 0.005$ and $\sigma(y)$ is the standard deviation of y.

2.5 VEGA

VEGA stands for Vector Evaluated Genetic Algorithm and it was developed by Schaffer [14, 3]. The basic algorithm is presented in Figure 2.4.

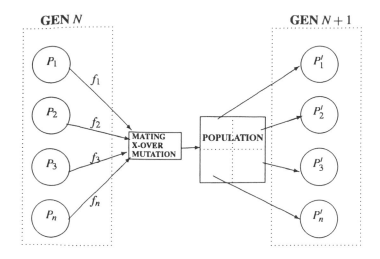

GEN N **GEN** $N+1$

Figure 2.4: Schema of Schaffer's (basic) VEGA algorithm

The whole population P is divided into several (equal to the number of objectives we are optimising) sub–populations. Operators used are the same as in a standard GA, only the selection step is different: from each subpopulation (of size m) choose individuals for the mating pool taking into account only that one objective. Then we perform mating, crossover and mutation in the standard way and split the new obtained population into sub–populations randomly. More formally:

Let $\mathcal{D}$ be a domain and $\mathbf{F}(x) = (f_1(x), \ldots, f_k(x))$ the function with domain $\mathcal{D}$. We define the following operators:

$$\text{CROSS} : \mathcal{D}^2 \mapsto \mathcal{D} \quad \text{– crossover}$$
$$\text{MUT} : \mathcal{D} \mapsto \mathcal{D} \quad \text{– mutation}$$
$$\text{SEL}_f : \mathcal{D}^m \mapsto \mathcal{D}^m \quad \text{– selection according to objective } f$$

Let $P(t) = (P_1, \ldots, P_k)$ be the population in generation t. Then

$$P(t+1) = \text{MUT}(\text{CROSS}(\cup_{i=1}^{k} \text{SEL}_{f_i}(P_i))^2) \qquad (4)$$

Later analysis of the algorithm, shows that a problem with VEGA is that it tends to average the solution. According to [15, p.191] (which further refers to [16]) "analysis of VEGA shows that the effect is the same as if fitness were a linear combination of attributes". Several approaches have been proposed to eliminate this disadvantage [17].

A similar approach to VEGA has been developed by Fourman [18]: selection is performed by comparing pairs of individuals, each pair according to one objective selected at random.

Actually, concerning Schaffer's VEGA, there are additional features that one can introduce with it. In the original, there is no real concept of subpopulations: we shuffle the populations together in every generation. So, the basic three ideas are

- Wait t_s generations before mixing them together;

- Don't mix them together, but every t_m generations copy n_b individuals from populations to other. In that case, if k is the number of subpopulations (i.e. number of objectives) and N population size (so, N/k is the size of each sub-population), we have constraint $n_b \leq N/k^2$.

- Similar to the previous idea, but don't copy, migrate individuals.

2.5.1 Results

We have implemented those ideas and some results and a comparison with the Pareto method are presented in Table 2.4. All results are averaged over 200 runs. All GA parameters are the same in all runs except for population size in the case of Schaffer's VEGA. We used uniform crossover with probability 1 and exponential mutation with probability 1/8 applied on a real–coded chromosomes.

method	y_3	$\sigma(y_3)$	y_4	$\sigma(y_4)$	y_9	$\sigma(y_9)$	$\sum y_i$
Fourman	115.30	8.37	12.23	10.90	7593.0	143.6	7720.48
Pareto	120.07	4.57	0.44	2.41	7731.7	145.3	7852.17
Schaffer	145.67	4.55	71.8	13.0	5886.4	230.3	6103.93
Subpop (mix)	116.76	8.90	20.7	13.6	7355.5	159.8	7492.96
Subpop (copy)	114.94	11.14	20.3	16.0	7116.1	260.0	7251.32

Table 2.4: Results on maximising y_3, y_4 and y_9 using different optimisation methods.

3   Conclusion and Future Work

In this paper we have analysed several different methods for optimising multi–objective functions. The results clearly show that there is a different problem here that we do not have in the case of single objective functions: the problem of relative importance and interaction of objectives. That also makes a problem of evaluating solutions much more difficult as it is not simple to say: this one is the best solution!

All optimisation methods presented above need a quantitative value for the relative importance of objectives and thus present an additional difficult problem. So, the requirement is for methods that can work with a qualitative in addition to a quantitative characterisation of importance ('this one is definitely more important than the other', 'this one is not so important' etc.). Preference methods [19, 20], as well as various fuzzy set methods [21, 22, 23] and combinations thereof [24] should be included here.

Also, human interaction should not be forgotten. Methods for exploring, expanding and sampling the search space in all directions are needed. In the future work it is intended to introduce some agent based systems which model, to some extent, the human–based solution evaluation process.

References

[1] Ching-Lai Hwang and Abu Syed Md. Masud. *Multiple Objective Decision Making – Methods and Applications*. Number 164 in Lecture Notes in Economics and Mathematical Systems. Springer Verlag, Berlin, 1979.

[2] Andrzej Osyczka. *Multicriterion Optimization in Engineering with FORTRAN Programs*. Ellis Horwood Series in Engineering Science. Ellis Horwood, 1984.

[3] J. David Schaffer. Multiple objective optimization with vector evaluated genetic algorithms. In Grefenstette [25], pages 93–100.

[4] Jeffrey Horn and Nicholas Nafpliotis. Multiobjective optimization using the niched Pareto genetic algorithm. Technical Report IlliGAL Report No. 93005, Illinois Genetic Algorithm Laboratory, 1993.

[5] Carlos M. Fonseca and Peter J. Fleming. An overview of evolutionary algorithms in multiobjective optimization. *Evolutionary Computation*, 3(1):1–16, 1995.

[6] Mitsuo Gen and Runwei Cheng. *Genetic Algorithms & Engineering Design*. Wiley Series in Engineering Design and Automation. J. Wiley & Sons, 1997.

[7] Carlos M. Fonseca and Peter J. Fleming. Multiple objective optimization and multiple constraint handling with evolutionary algorithms I: A unified formulation. Technical Report 564, University of Sheffield, UK, January 1995.

[8] Ian Parmee and Mike J. Denham. The integration of adaptive search techniques with current engineering design practice. In Ian Parmee, editor, *Adaptive Computing in Engineering Design and Control '94*, pages 1–14. Plymouth Engineering Design Centre, 1994.

[9] Ian Parmee. Strategies for the integration of evolutionary/adaptive search with the engineering design process. In D. Dasgupta and Z. Michalewicz, editors, *Evolutionary Algorithms in Engineering Applications*, pages 453–477. Springer Verlag, 1997.

[10] Eric Webb. MINICAPS – a simplified version of CAPS for use as a research tool. Unclassified Report BAe-WOA-RP-GEN-11313, British Aerospace, July 1997.

[11] Aharon Ben-Tal. Characterisation of pareto and lexicographic optimal solutions. In Fandel and Gal [20], pages 1–11.

[12] N. Srinivas and Kalyanmoy Deb. Multiobjective optimization using nondominant sorting in genetic algorithms. *Evolutionary Computation*, 2(3):221–248, 1995.

[13] Manuel Valenzuela-Rendón and Eduardo Uresti-Chare. A non–generational genetic algorithm for multiobjective optimization. In Thomas Bäck, editor, *Proceedings of the Seventh International Conference on Genetic Algorithms*, pages 658–665. Morgan Kaufmann, jul 1997.

[14] J. David Schaffer. *Some Experiments in Machine Learning using Vector Evaluated Genetic Algorithm*. PhD thesis, Vanderbilt University, Nashville, 1984. TCGA file No. 00314.

[15] Jon T. Richardson, Mark R. Palmer, Gunar Liepins, and Mike Hilliard. Some guidelines for genetic algorithms with penalty functions. In David Schaffer, editor, *Proceedings of the Third Internation Conference on Genetic Algorithm*, pages 191–197, 1989.

[16] M. Hilliard, G. Liepins, M. Palmer, and G. Rangarajan. The computer as a partner in algorithmic design: Automated discovery of parameters for multi-objective heuristics. In *Conference on the Impact of Recent Computer Advances in Operations Research*, 1988.

[17] I.C. Parmee, M. Johnson, and S. Burt. Techniques to aid global search in engineering design. In Frank D. Anger, Rita V. Rodrigez, and Moonis Ali, editors, *Industrial and Engineering Applications of Artificial Intelligence and Expert Systems IAE/AIE '94*, Austin, Texas, 1994. Gordon and Breach Science Publishers.

[18] M. P. Fourman. Compaction of symbolic layout using genetic algorithms. In Grefenstette [25], pages 141–153.

[19] Dragan Cvetković. The logic of preference and decision supporting systems. Master's thesis, Faculty of Mathematics, University of Belgrade, 1993. Also Technical Report MPI-I-93-260 of Max Planck Institute for Computer Science, Saarbrücken, Germany.

[20] G. Fandel and T. Gal, editors. *Proceedings of the Third Conference on Multiple Criteria Decision Making Theory and Application*, number 177 in Lecture Notes in Economics and Mathematical Systems, Hagen/Königswinter, West Germany, August 1979. Springer Verlag.

[21] Fabrice Wawek, Anne-Marie Jolly-Desodt, and Daniel Jolly. Modelisation of the criterias coming from the human operator for a fuzzy decision support system. In *4th European Congress on Intelligent Techniques and Soft Computing EUFIT '96*, pages 1349–1352, Aachen, 1996.

[22] Lotfi A. Zadeh. The concept of linguistic variable and its application to approximate reasoning. *Information Sciences*, 1975.

[23] Tamás Zétényi, editor. *Fuzzy Sets in Psychology*. Number 56 in Advances in Psychology. North–Holland, 1988.

[24] János Fodor and Marc Roubens. *Fuzzy Preference Modelling and Multicriteria Decision Support*, volume 14 of *System Theory, Knowledge Engineering and Problem Solving*. Kluwer Academic Publishers, 1994.

[25] John J. Grefenstette, editor. *Proceedings of the First Internation Conference on Genetic Algorithms*. Lawrance Erlbaum Associates, 1985.

Evolving robust strategies
for autonomous flight:
A challenge to optimal control theory

Philip W. Blythe
Department of Aeronautical Engineering
University of Sydney
NSW, 2006, Australia
Email: philby@aero.usyd.edu.au

Abstract: This article represents the second step in a series of papers posing a new challenge to the rigours of modern control theory, namely by developing an adaptive approach to create autonomous flight control systems. Based upon previous work which demonstrated that computational evolutionary adaptation is a generic extension of linear optimal control, here we explore the issues concerning the robustness of the technique in more rich and realistic domains. Robustness, like survivability, is a key driving initiative behind the acceptance of control systems in failure critical scenarios, like automatic flight control. It therefore seems worthwhile to seek ways of adjusting evolutionary optimisation methods to encourage robustness across all individuals that emerge from an evolutionary simulation. This can be achieved primarily by modifying the Evolutionary Strategy selection process to optimise across multiple simulation conditions, including noisy environments and control failure scenarios. By doing so, we have again empirically demonstrated a further prediction; evolutionary control designs are superior to their linear theoretic counterparts in terms of their robustness and likewise survivability in uncertain environments.

1 Introduction

Even though intelligent machines are not living in the true sense, like us, all they really want to do is to survive. Through technological progress, machines continually improve their abilities to manage in harsher environments, even though the manner in which this is accomplished is somewhat artificial. To the traditional control systems analyst, this ability is termed *robustness*, but to the scientist/engineer working under the philosophy of natural adaption, it equates to *survivability*. While the classical approaches to control design tend to start with an idealistic design, and subsequently patch its problem areas, robustness in animals is inherent – they simply would not exist if there were any glaring deficiencies in their ability to survive under difficult circumstances. For natural processes, survivability (or robustness) is the first priority from day one; optimal performance is secondary.

This article will investigate a number of adaptive techniques to improve the survivability of evolved flight control designs. While previous work has empirically demonstrated evolutionary processes could deliver the same or better levels of optimality and robustness than the central methods of linear control theory [1], this was largely because the fitness function was of a known form which naturally led to robust behaviour, provided the true optimum was found. As evolutionary methods open the problem scope to more complicated fitness criteria, the optimal solutions found by canonical adaptive strategies can no longer be assumed to be robust. Likewise for controller architectures which depart from the standard linear feedback form, an optimal solution to a specified cost cannot be blindly trusted in off-design conditions. To move ahead into these unknown territories, the adaptive strategies need to be modified to follow the same course as natural adaption; begin with minimalist and robust survival methods, and then gradually move towards an ecological optimum.

Ultimately, improving the survival abilities of individuals can only be done by including all of the real world dangers into the environment, or in the fitness function. For an operational autonomous aircraft there are many; failures of the powerplant, jamming of the actuators, loss of aerodynamic effectiveness through damage, to name but a few. Still, it is difficult to express these random factors in a cost formulation unless the fitness measure repeatedly evaluates the performance of the controller under each and every scenario – a computational impossibility. The alternative is to once again look to the naturalistic parallel, to see the way in which biological evolution has improved its failsafe capabilities, then emulating and incorporating these adaptions into the evolutionary design framework.

But before making additions to the optimisation scheme, a representative environment should be chosen which demands the functionality of autonomy, while including many of the operational hazards which a real aircraft must deal with. Figure 1 shows an example trajectory which the aircraft must follow, with the objective being to pass over all the waypoints within the minimum amount of time. Meanwhile it must withstand a number of unpleasantries; avoiding the elliptical constraints (punishable by death), maintaining stability in atmospheric turbulence, and being able to satisfactorily perform the mission when having experienced an actuator failure.

Even without considering any robustness improvements to the evolutionary strategies, controllers can be evolved to perform this operation for the more idealistic situations. Section 3 will develop some simple trajectory controllers using evolutionary simulation, as well as conventional methods, highlighting their respective strengths and weaknesses. Following this, we can then consider ways in which to improve the evolutionary optimisation method to produce safer, and more survivable designs. These

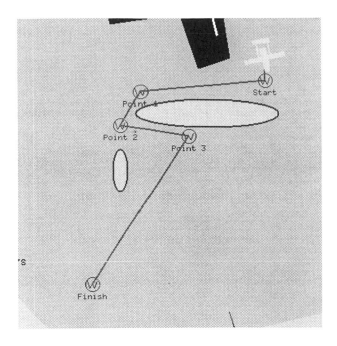

Figure 1: A representative environment: an autonomous mission to reach the specified waypoints in the minimum amount of time.

can be thought of as robustness improvements, yet "robust" design in the conventional sense has various connotations attached to it. But for the ecological approach, robustness qualities are purely a reflection of how the environment is constructed, and the machine's ability to survive in these representative surroundings. But firstly, we will quickly discuss some topical issues in modern control theory, and present an overview of the alternative ecologically adapted approach.

2   Deficiencies of modern control

The state of the art methods of optimal control design for aircraft, chemical plants, or power stations are based upon the analytic repertoire of linear optimal control. These rigorous mathematical methods grew out of frequency domain techniques devised by Bode [2], Harris [3], Hall [4] and later Nyquist [5] over 50 years ago. While analytic control theory has succeeded in producing acceptably stable control designs for over half a century, they are failing to adapt to modern day problem scenarios which demand more stringent levels of nonlinearity, optimal performance and also fault tolerance. The acceptance to the control community of computational techniques such as evolutionary adaptation becomes repeatedly hung upon one issue – robustness. With a mathematical basis, one can "prove" a controller's stability under certain conditions, but with a empirical simulation approach, you cannot. While

272

this aspect is fundamental to adaptive methods in general, we can counter this argument by demonstrating through simulation that analytic controllers provide no added benefits in terms of optimality or robustness across a wide range of environmental scenarios. And this should come as no surprise, since the approximations implicit in the theoretic models only aid to abstract them substantially from the true environment, which can really only be properly captured through advanced simulation.

To develop a *classical* control solution for the problem shown in figure 1, one would normally commit to a simplified a-priori analysis of the environment and aircraft plant, simplistic representations of turbulence and noise (if any), and linearisations of the complete nonlinear motion equations. From this simplified analytic model, a dynamically "stable" linear command tracking control law can be derived from linear optimal control theory, such as the LQR/LQT method, developed by Kalman in the late 50's [6][1]. Basically, the system is optimised to give the minimum deviations away from its pre-planned course over time, disregarding any other criterion of performance. These types of systems are easily upset by excessively noisy environments or undetected failures, and do nothing to directly encourage survival under difficult circumstances.

With the rapid availability of computational power, an alternative method presents itself; to evolve the feedback gains directly through evolutionary optimisation, by iteratively placing the candidate individuals into a flight simulation environment, and evaluating their environmental fitness. This basically conforms to the animat approach [8] characterised within the adaptive behaviour literature, where simulated animals and robots are evolved to carry out certain control tasks [9]. Although there is a number of routes to pursue under this theme, one must decide on both the brand of evolutionary optimisation to be used, and also on the control structure itself. Convergence of real valued parameters is most accurately done with a real valued optimisation technique, making Evolutionary Strategies the most attractive option (in terms of speed also [10]). As for the structure of the control law, the first-cut evolution of distributed network representations has been known to be problematic [11], while the simpler linear feedback systems have received little attention. However it seems that only through such humble beginnings can we directly compare traditional and adaptive methods on equal ground. From there, the evolutionary approach can be extended almost at infinitum to new environments, fitness criteria, and control structures, while the theoretical approaches remain intimately tied to their linear form, useful only as a solid baseline for empirical benchmarking.

3 Basic performance of evolutionary designs

The purpose of this experiment is to begin with some preliminary designs of tracking systems using the evolutionary method and comparing their performance and durability to conventional controllers. In these first basic implementations of a tracking

[1]More advanced control methods such as H_{inf} or μ synthesis are also extensions on Bellman's original dynamic programming approach [7].

controller, a set of feedback gains were evolved to move the aircraft towards its commanded flight path via its four control actuators; elevator (δ_e), aileron (δ_a), rudder (δ_r), and throttle (δ_t). The structure of the control law was a standard linear feedback system on the observable state vector ($\underset{\sim}{x}$), operating on deviations of the three velocity components of each axis (u, v, w), the associated rotation rates (p, q, r), and heading and bank angles (θ and ϕ), each away from their commanded values;

$$
\begin{bmatrix} \triangle\delta_e \\ \triangle\delta_a \\ \triangle\delta_r \\ \triangle\delta_t \end{bmatrix} = \begin{bmatrix} K_1 & K_2 & ... & K_8 \\ K_9 & K_{10} & ... & K_{16} \\ K_{17} & K_{18} & ... & K_{24} \\ K_{25} & K_{26} & ... & K_{32} \end{bmatrix} \begin{bmatrix} \triangle u \\ \triangle w \\ \triangle q \\ \triangle\theta \\ \triangle v \\ \triangle p \\ \triangle r \\ \triangle\phi \end{bmatrix} \tag{1}
$$

For these simulations, a rather simple and realistic fitness evaluation function is used[2], which represents an integral of the deviations of the normalised states away from their commanded values ($\underset{\sim}{\triangle x}$) and control movement penalties ($\underset{\sim}{\delta}$). i.e;

$$
Fit.Cost = \int_{t=0}^{t=t_{max}} [\underset{\sim}{\triangle x}Q + \underset{\sim}{\triangle\delta}R]dt \tag{2}
$$

One particular annoyance in creating fitness measures of this kind (in both conventional and evolutionary approaches), is finding the relative weighting[3] between goal achievement (state deviations, $\triangle xQ$), and energy consumption (control movements, $\triangle\delta R$). While theoretical control methods require a number of iterations from the designer to find an acceptable medium, the environment structure can be modelled to reflect a realistic set of constraints on control usage. Put simply, if an individual consumes too much energy, or is too violent in its movement, it will die for whatever reason. It was therefore found that the most well adapted designs were those which received no direct fitness penalties for control movement, with high gain solutions being eliminated by means of natural attrition [17].

For the decision domain of trajectory tracking, different wind conditions, turn radii and speed control are all examples of adjacent environment structures in which the controller must be able to maintain reasonable performance. The controller's abilities in these situations will reflect its overall robustness (in the classical sense of the word), and the representativeness of the environment in which it was designed. To this end, a somewhat different trajectory path was created as a test environment, while retaining a similar structure to the design setting. Figure 2 (c.f. figure 1) plots a comparison of

---

[2]Note that this is in contrast to the *squared*, or infinity norm deviation of states and controls used in optimal control theory – a rigid requirement that enables a form of least-squares regression to be used.

[3]This refers to the Q and R matrices.

the trajectories followed by the classical LQ controller and an evolved design, showing similar performance of both to this novel environment. On this problem, it took only a population of 50 individuals around 45 generations to converge on an accurate set of control gains (32 parameters), of which many variants existed throughout the population.

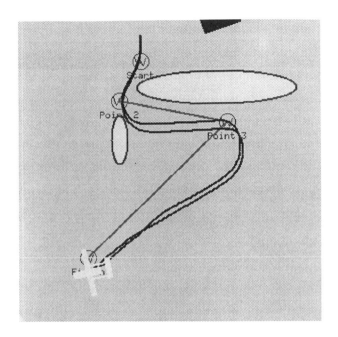

Figure 2: A comparison of off-design performance of the LQ and evolved controllers on a novel trajectory path.

Looking now towards more extreme changes in environmental representation, the fitness differences of the controllers can be measured in conditions of increased air disturbance (which is proportional to the wind strength). Of particular interest here are the effects of noise on controllers with high gain feedback. It is not uncommon that a large unsavory gust will drive a control system unstable, and so a pertinent indicator of safety and robustness is revealed by this kind of test environment. Having removed all penalties on control movement (tending to encourage high gain solutions), there is some concern over robustness in this scenario. The results of this study given in table 1 tell a different story – even with no control penalties, the robustness qualities of the evolved designs to large air disturbances (design level of a 20 knot wind, and off-design condition at 40 knots) are superior to the classical design methods.

As an interim conclusion, it seems that the robustness qualities of canonical evolutionary strategies are quite good, even without the robustness additions which will be detailed in the following sections. Provided that a representative design environment

Control method	20kt turb.	40kt turb.
LQR/LQT	2,007	2,979
evolved	1,949	2,046

Table 1: A tracking fitness cost comparison of evolutionary designs in off-design disturbance conditions on a novel trajectory, against conventional LQR/LQT optimal control methods.

is selected, then the extrapolative performance to novel and even harsher conditions is as good, if not better than the analytic methods. Another significant finding is the removal of control penalties from the cost function; by removing this constraint, more flexibility is given to the evolutionary method to find the appropriate control effectivities, rather than being ad-hocly selected by the designer. Still, some degree of caution must be exercised. By removing the constraints on the evolutionary process, the fidelity of the environmental model becomes crucial in determining the robustness characteristics of an evolved population. But for now, we will move on to more generic robustness improvements to the optimisation process itself, that encourage not only optimal performance, but also a safe operation across a range of ecological conditions.

4   Enhancing the robustness of evolution

In this section, we explore ideas in natural adaption which can help to promote robustness in evolutionary control design, delivering enhancements to the optimisation method. Evolutionary strategies, in their canonical form at least [12, 13], were not developed to be sturdy against non-deterministic environments, and so need to be appropriately modified. As evolutionary strategies are still a loose, but reasonable model of natural evolution, the methods which nature has adopted for improving robustness can also be implemented into these algorithms. Targeted primarily here are the different criteria of selection, and their influences on improving the population as a whole.

Though not having a direct bearing on the robustness of the final design, there are some other lessons from nature which can be added to the adaptive design framework to increase the efficiency of the optimisation method. When expanding the capabilities of artificial evolution to more complex control problems, this increase in complexity must be introduced in an analogous way to nature – gradually. This section will also explore a simple method for gradually tightening the cost objective, and in doing so, makes the optimisation converge more convincingly.

4.1   Aspects of robustness in natural evolution

Currently, the most significant flaw within the evolutionary simulation method is its inability to optimise in the presence of non-deterministic or stochastic environmental factors. For nature, dealing with this is an everyday occurrence. Each individual is not subjected to exactly the same fitness evaluation, and sometimes particular individuals receive a selective advantage – put more bluntly, the world is not a fair place. However, because natural evolution has operated over a long period of time, favourable environmental conditions do not always prevail. As a result, complacent genes which would otherwise only perform well in good environments have gradually been removed from the gene-pool. The only genes left are those which cope well in all sorts of adverse ecological conditions. Sad as it may sound, it is the benefits of unfortunate death which ultimately improve the robustness of the species over successive generations.

While this inhumane aspect of natural selection is good for promoting robustness of a species, other selection processes at work in societies can be a little more fair. In particular, the selection of a sexual partner operates on a much more level playing field, often with the female judging each candidate on similar grounds[4]. Still, elsewhere within the population, a different female may select a male based on another set of criteria. Altogether, the offspring produced by the entire population have been selected on a wide variety of different criteria. When this has operated over several generations, the different selection methods blend together, creating males which are a correspondingly optimal blend of all the desired characteristics.

As a final issue, not only does nature use natural selection (or death) as a means for removing genes which are not well adapted to their environment, but also as a method for regulating sophistication. Over the phylogenic history of most organisms, co-evolutionary pressures of predation from other species as well as competition for resources has meant that their criteria for survival has become increasingly harder. As the process of natural selection has continued to promote complexity, it appears that evolutionary development could not have occurred unless the pressures of survival started out very simple, and gradually led species to become more and more sophisticated. As with the evolutionary history of flight (for a good overview, see [14]), birds did not miraculously appear in the skies one fine sunny day, nor could they envisage the countless benefits of learning to fly. Instead, a gradual pathway of increased benefits lured them from their arboreal existence into controlled parachuting and gliding, and eventually into powered flight. .

4.2   Sexual selection criteria for evolutionary strategies

Canonical evolutionary strategies were designed to search functional fitness landscapes to find an optimal value, and do not provide any mechanisms to assist the optimisation in the presence of uncertainty. The evolution of simulated agents poses a very different problem, as it is important that the final designs can perform adequately a wide variety of changing environments. Although the robustness of (μ, λ) selection has been shown to cope better than $(\mu + \lambda)$ with gradual changes in environ-

[4]It is in the best interests of the choosier sex (usually female) to select fairly, as a bad choice only degrades her ability to survive.

ment [15], this only applies to environmental changes that occur between successive generations, and not with environment variabilities which are encountered by different individuals within a single generation. The inability to handle environmental uncertainty is therefore a manifestation of both (μ, λ) and $(\mu + \lambda)$ selection strategies. Both will always select for individuals who have received a more favourable treatment during their fitness evaluations, which in the long run, will not reward individuals who cope well under difficult conditions.

For the evolutionary process to encourage robust behaviour in the face of an unpredictable circumstances, a modification must be made to the selection strategy so that it does not penalise the unfortunate. This can be achieved by making each different selection process a fair one, but using a different fitness criterion for comparison on each selection trial[5]. Analogous to the natural process of sexual selection, these different fitness measures average out over successive generations. In a similar way also, if the differences in the fitness functions have a mean of zero, then overall, the evolution will remain stable; if they have a bias however, something akin to runaway sexual selection will occur, continually favouring one particular trait over others [16].

The effectiveness of this algorithmic modification will be examined later in section 5, where a controller is designed to satisfactorily function in the presence of actuator failures. But before this can be realistically simulated, environmental hazards must be properly modelled and integrated into the optimisation process. Not only does this create a more difficult and realistic design example, but the possibility of premature death has other benefits, which will now be discussed at length in the following section.

4.3   Improving survival affinity with artificial death

Implementing a realistic criterion of death into a simulated environment is quite a simple thing to do. For an aircraft, the most basic representation would include hitting the ground or an obstacle, or a structural failure due to excessive maneuvering. Even though the same circumstances equate to a particularly low fitness value, there are good reasons for immediately removing such candidates from the population. If a fair selection method is applied locally to a group of particularly poor candidates, then one of them will still make it through to the next generation, and continue its lineage. On the other hand, if the sufficiently bad offspring are punished by death, and other candidate offspring are created in their place, then there is no such propagation of unacceptably bad phenotypes. This concept has been implemented into the evolutionary design scheme, which dramatically improves the convergence of the algorithms over the first two or three generations.

This hard and fast rule for killing off hopeless offspring has another effect; it sets a minimum standard of "satisfactory" behaviour that is allowable when environmental

---

[5]As a point of clarification, the fitness criterion is composed of both the cost function and the stochastic processes of the environment. Here, the text refers to different fitness evaluations caused by the latter.

conditions are harsh. If this restriction is set too tightly, no individuals will be able to survive if something goes wrong. Consequently, it is in the best interests of the designer to implement a criterion of death which best represents a real world failure situation. If this is done properly, then the system will design a control method which takes advantage of this margin between optimal and satisfactory performance in times of dire need.

4.4   Grading a smooth course for incremental evolution

It is no mistake that evolution does not occur overnight. Even with the right ingredients – heat, light, carbon, etc – complexity does not spontaneously emerge out of vegetable soup. In our simulations of evolutionary adaptation, the same applies; evolution must be led through a series of incremental developments which slowly increases the capabilities of the evolved systems. Not only does this mean that the evolution should begin with the fewest evolvable traits, but the functionality demanded by the ecological cost should begin as simply as possible, and slowly increase in sophistication. If the changes are too abrupt or too difficult, the entire population can die, causing effective genetic extinction. Yet it is not the unique function of natural selection to promote increasing levels of sophistication; any selection mechanism which broadens its criteria over time can achieve the same end.

In extending the evolutionary paradigm to manoeuvre control, where the system must evolve to follow commanded states, there is an extensive set of manoeuvres which the controller must develop; some of which are harder than others. The gradual phasing in of more complicated movements can be accomplished by slowly increasing the lifespan of the individuals over successive generations. As part of the fitness evaluation, the simulation begins with gentle manoeuvres, and as the population becomes proficient at this task, their lifespans are increased whereupon they must tackle more complicated movements. The criterion used for judging proficiency is death; if a very high percentage (95%) of the population survive a generation, then the lifespan and the corresponding difficulty level is increased in their offspring.

5   Robustness methods applied to tracking control

Once again, this section returns to the application of evolutionary strategies for designing tracking control systems. While section 3 has gone to some lengths to show that the canonical evolutionary method produces rather robust control strategies, this is to some degree an artifact of chance. According to the arguments of natural adaption theory, a design need only be optimum for the environments to which it has been exposed. Unlike the classical analytic methods, this numerical technique cannot lie on the laurels of a safe operating history, so instead must guard itself against any operational hazards right from the beginning of the design process.

For the autonomous aircraft following a pre-planned course, there are a number of significant dangers which should be considered, some which have already been mentioned. But given that the control architecture is restricted to the linear feedback form given in equation 1, there are only three dangers to which the controller can develop some level of resilience; large atmospheric disturbances, jamming of the actuators, and changes in aircraft aerodynamic effectiveness. While robustness to turbulence has already been shown to be quite adequate in canonical evolutionary designs, coping well with discrete failures of actuators is not so automatic.

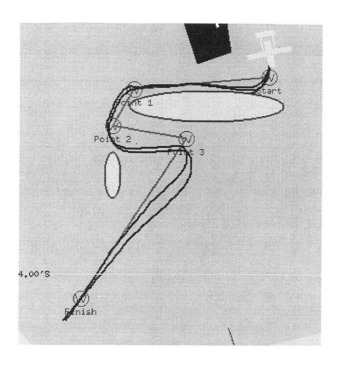

Figure 3: A plot of three trajectories followed by an evolved controller in different failure scenarios; the complete trajectories depict the paths of the fully operational and aileron jammed controllers, while the short path represents the controller with a rudder failure colliding with the obstacle.

To get some idea of the ability to which our evolved controllers from section 3 are able to cope with actuator failures, figure 3 shows three paths flown by a selected individual; one with all controls functioning, one with a failed aileron actuator, and the other with a failed rudder actuator. Through secondary yaw and roll effects of the aileron and rudder respectively, the controller is capable of steering the aircraft in a somewhat limited capacity. However in this instance, the results are not so good. Figure 3 shows a significantly less accurate track followed by an evolved controller which experienced a failed aileron, and the disastrous consequence of a rudder failure – flying straight into the side of the obstacle. This result is not surprising, since the evolved population did not experience any such scenarios in their development. Even more significantly, the evolutionary process was not made aware of the severity of the consequences if an actuator were to fail.

Using the extensions developed in section 4 for the evolutionary optimisation process, the design environment can be equipped with a probability that an actuator will fail – i.e. for each new offspring, there will be a chance that each specified actuator will not be operational. As each of the competing offspring will encounter the same environmental conditions, the individual who can manage the best in these adverse conditions will go forth into the next generation[6]. With the recombining population of individuals who experience both nominal and failure stricken environments, the final population will inherit a compromise of failsafe and optimal performance.

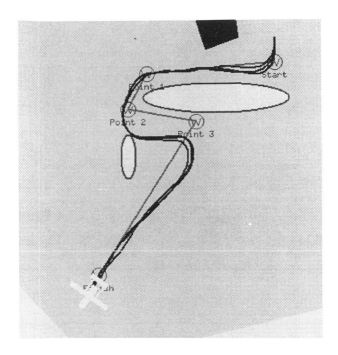

Figure 4: A final trajectory plot of an evolved design incorporating failure criteria. The three paths show that the fully functional, aileron jammed and rudder jammed scenarios all follow similar courses, successfully avoiding the obstacles.

Playing a supportive role in this situation is the implementation of artificial death. By penalising a collision with an obstacle with death, any strategy which does not meet this criterion is immediately removed from the optimisation process. In doing

---

[6]The probability of failure needed by each actuator to have a reasonable effect will be dependent upon population size and the number of generations used in the simulation. As a reasonable guide, there should be an average of at least one failure per generation to assure that each scenario is considered all the way through the optimisation process. This is especially important in the earlier generations, when most of the genetic removal from the gene pool takes place.

so, it not only steers the optimal individuals away from adopting such a strategy, but heavily influences individuals facing actuator failures to take evasive action using whatever resources are available. Figure 4 shows the recorded trajectories of a controller which has evolved with failure probabilities in both its aileron and rudder control surfaces. All three paths are now considerably closer together (c.f. figure 3), with the controllers able to avoid the obstacles with either the aileron or rudder rendered inoperative.

While this improvement in off-design performance was expected from the modifications to the evolutionary process, there was another improvement which was not predicted. Table 2 shows that with the incremental addition of faulty controls in the design, the fitness cost incurred by the fully functional system is likewise reduced. Albeit sensible in hindsight that the most robust solution is also the most optimal, it indicates however that the former designs were finding sub-optimal solutions, or even becoming trapped in local minima. While not being able to compare to a true theoretical optimum, it is difficult to say which is the case. For the moment however, what *is* clear is that evolutionary design methods can benefit tremendously from incorporation of robustness criteria into the optimisation procedure.

Design	No fault	Ail. fault	Rud. fault
Fault Free	6,317	13,105	crash
With Faults	4,930	8,158	7,343

Table 2: A fitness cost comparison of two evolved controllers; the first evolved with all actuators functioning, and the second designed with a probability of failure in both aileron and rudder. The costs tabulated represent trials respectively in the fully operational condition, in the presence of an aileron fault, and with an inoperative rudder.

These examples of increasing the robustness qualities by direct implementation into the design process clearly demonstrate the optimisation improvements outlined in the previous section. As well, it may also suggest a new paradigm for FDIR[7] – prevention. Instead of developing sophisticated systems for identifying faults, and then redesigning the control system appropriately, it may make more sense to have an over-redundant system which is designed to perform adequately in failure situations. This type of design would be much simpler to implement, and would supply much higher levels of security and performance than the mainstay of conventional approaches for fault reconfiguration.

---

[7]Fault detection, isolation, and reconfiguration.

6 Further Implications

The work presented in this article gives a quantitative description to the advantages of using evolutionary methods over conventional optimal control. Most particularly it is targeted towards improving the survivability of designs, partly via the optimisation technique, but mostly by creating a more ecologically realistic model of the world through computer simulation. Random events in environments can occur over any time span; from many times per second in high frequency turbulence, to discrete events that may affect only 1/50 individuals in a population. Ecologically adapted individuals will be intrinsically more attuned to these irregularities than generic analytic methods, simply on the virtues of it having adapted only to scenarios which are pertinent to its operation.

For the evolutionary approach to controller design, there are still more ways of increasing the reliability of individuals through more sophisticated control structures. For example, by replacing the linear feedback decision system with a bounded two layer neural network[8], a near tenfold decrease in fitness variance across the population has been recorded [17], while further reducing the overall fitness costs by over 35%. These results far exceed the abilities of classical linear control methods in terms of performance *and* reliability, suggesting not only that evolutionary adaptation is a sound approach, but a logical progression for the field of control theory.

Without reservation, this work is attempting to sow the ideas of natural adaption into the well established field of control theory. Through the two critical criteria of optimality and robustness, it now seems quite apparent that the evolutionary approach can deliver more appropriate control strategies for real world tasks than the most optimal linear techniques. While this result may seem obvious to some, researchers in adaptive behaviour and evolutionary robotics have failed to tackle this intermediate step, and instead have moved straight into the evolution of more complex neural control structures without adequate baselines. This has invariably led to difficulties in rating the performance of these more complex designs, as well as somewhat distancing the engineering control community. But moreover, very often in our excitement, we forget the golden principle of all bottom-up methodologies: start simple, and build towards complexity.

References

[1] P.W. Blythe. Evolutionary control design: Exploring the time domain. *AIAA Journal of Guidance, Control and Dynamics*, page (submitted), 1997.

[2] H. W. Bode. Relations between attenuation and phase in feedback amplifier design. *Bell System Technical Journal*, 19:421–454, 1940.

---

[8]Note that this can be achieved by simply mapping the linear control output through a $tanh$ transformation function.

[3] H. Jr. Harris. The analysis and design of servomechanisms. Technical Report Report 454, OSRD NDRC, Jan 1942.

[4] A.C. Hall. *The Analysis and Synthesis of Linear Servomechanisms*. Cambridge, Mass: Technology Press, 1943.

[5] H. Nyquist. The regeneration theory. *Transactions of the ASME*, 76:1151, 1954.

[6] R.E. Kalman. When is a linear control system optimal? *Journal of Basic Engineering*, page 51, 1964.

[7] R. Bellman. *Adaptive Control – a guided tour*. Princeton, NJ: Princeton University Press, 1961.

[8] S.W. Wilson. The animat approach to AI. In J.-A. Meyer and S.W. Wilson, editors, *First International Conference on Simulation of Adaptive Behavior: From Animals to Animats*, pages 15–21. Cambridge: MIT Press, 1991.

[9] D. Cliff, I. Harvey, and P. Husbands. Explorations in evolutionary robotics. *Adaptive Behavior*, 2(1):71–108, 1993.

[10] R. Salomon. Increasing adaptiviity through evolutionary strategies. In P. Maes, M.J. Mataric, J.-A. Meyer, J. Pollack, and S.W. Wilson, editors, *Fourth International Conference on Simulation of Adaptive Behavior: From Animals to Animats IV*, pages 411–420. Cambridge: MIT Press, 1996.

[11] R.K. Belew, J. McInerney, and N.N. Schraudolph. Evolving networks: Using the genetic algorithm with connectionist learning. In C.G. Langton, C. Taylor, J.D. Farmer, and S. Rasmussen, editors, *Workshop on artificial life: Artificial Life II*, pages 511–547. Redwood City: Addison-Wesley, 1992.

[12] H.P. Schwefel. Kybernetische evolution als strategie der experimentellen forschung in der strömungstechnik. Diploma thesis, 1965.

[13] I. Rechenberg. *Evolutionsstrategie: Optimierung technischer systeme nach prinzipien der biologischen evolution*. Stuttgart: Frommann-Holzboog Verlag, 1973.

[14] A. Feduccia. *The age of birds*. Harvard University Press, 1980.

[15] H.P. Schwefel. Collective phenomena in evolutionary systems. In *31st Annual Meeting of International Society for General System Research, Budapest*, pages 1025–1033, 1987.

[16] G.F. Miller and P.M. Todd. The role of mate choice in biocomputation: Sexual selection as a process of search, optimisation and diversification. In W. Banzof and F.H. Eeckman, editors, *Evolution and biocomputation: Computational models of evolution*, pages 169–204. Berlin: Springer-Verlag, 1995.

[17] P.W. Blythe. *Adaptive Strategies for Autonomous Flight Control*. PhD thesis, Dept Aeronautical Engineering, University of Sydney, 1997.

Chapter 6

Other Applications

Performance of Genetic Algorithms for Optimisation of Frame
Structures.
M.R. Ghasemi, E. Hinton

Global Optimisation in Optical Coating Design.
D.G. Li, A.C. Watson

Evolutionary Algorithms for the Design of Stack Filters Specified using
Selection Probabilities.
A.B.G. Doval, C.K. Mohan, M.K. Prasad

Drawing Graphs with Evolutionary Algorithms.
A.G.B. Tettamanzi

Benchmarking of Different Modifications of the Cascade Correlation
Algorithm.
D.I.Chudova, S.A. Dolenko, Yu. V. Orlov, D.Yu. Pavlov, I.G. Persiantsev

Determination of Gas Temperature in a CVD Reactor From Optical
Emission Spectra with the Help of Artificial Neural Networks and Group
Method of Data Handling (GMDH).
S.A. Dolenko, A.F. Pal, I.G. Persiantsev, A.O. Serov, A.V. Filippov

Multi-Domain Optimisation using Computer Experiments for Concurrent
Engineering.
R.A. Bates, R. Fontana, L. Pronzato, H.P. Wynn

Performance of Genetic Algorithms for Optimization of Frame Structures

M.R. Ghasemi, E. Hinton and S. Bulman
ADOPT Group, Department of Civil Engineering
University of Wales, Swansea
URL: http://www.swan.ac.uk/civeng/Research/adopt
email: e.hinton@swansea.ac.uk

Abstract. This paper describes work carried out at the University of Wales Swansea in the ADOPT Research group on research project on design optimization of engineering structures. Two design scenarios are presented for the optimization of 2D frame structures using specially developed genetic algorithms and related procedures. Examples are provided for each of these scenarios illustrating the procedures adopted.

1. Introduction

Background: Work is currently being carried out in the ADOPT (Adaptive Optimization) Research Group in the Department of Civil Engineering, University of Wales, Swansea on the 'FIDO project on Fully Integrated Design Optimization of Engineering Structures' [1]. This work is sponsored by the Engineering and Physical Sciences Research Council, EPSRC, and a consortium of industrial partners. This paper is a companion to a set of papers that deal with various aspects of the project and, in particular, describe the experiences of the ADOPT Group in developing and testing FIDO-TK, a computational tool-kit for design optimization of engineering structures. While the FIDO project also deals with optimal structural design using shell and continuum structural finite element (FE) models, the focus in this paper will be solely on structures modelled using 2D frame representations, highlighting the genetic algorithm (GA) aspects of FIDO-TK.

Two design scenarios: We will consider two separate structural design scenarios which depend on whether the structural designer has decided on the structural frame topology or not:

1. In the first design scenario, the structural designer has decided on the layout or topology of the 2D frame structure and wants to obtain the set of member types that lead to the minimum weight structure for a given displacement

constraint and set of loadings and boundary conditions. Here the frame members will be selected from a catalogue of standard steel I-sections and therefore the design variables which are the member types will be discrete in nature [2].

2. As an alternative design scenario, we will consider a more general problem of the following type. Given a structural domain in the form of a thin rectangular box and a set of boundary conditions and in-plane loadings, what is the stiffest frame structure that fits inside the box and has a specified volume fraction of the original box volume? Here, we will demonstrate the greater functionality of the FIDO-TK system. The initial topology of the frame will be obtained from the rectangular box domain using homogenization-based structural topology optimization procedures originally introduced by Bendsoe and Kikukchi [3]. This initial idealisation will be based on a 2D plane stress FE model. From the topological image thus obtained a frame idealisation will be extracted. Subsequently, the GA code will then be used to minimize the strain energy of the frame hence producing the stiffest frame with a constraint on the frame volume as used in the initial stage of the optimization. In this problem the frame will have members with rectangular cross-sections and as the width is defined by the width of the enclosing rectangular box, then the design variables will be the depths of the member cross-sections. Here, the variables will be continuous in nature although the GA method will treat them as a discrete subset of the infinite set of values between specified upper and lower bounds.

In both cases, we will use our own GA code to carry out the sizing optimization. Before running these two types of problems, we will benchmark the GA procedure adopted by tackling an analytical problem with a highly non-linear objective function to increase our confidence in our current GA implementation [4, 5].

Brief comment on FIDO-TK GA implementation: A full description of the GA procedure used in these studies is given in References [6-21] and will not be repeated here. Briefly, its main features are that it is a search procedure based on natural selection and unlike many mathematical programming algorithms does not require the evaluation of gradients of the objective function and constraints. Each variable is represented using a bitstring. Each bitstring is then merged to form a chromosome which represents a design. Later, the chromosomes can be decoded and the fitness function evaluated. The usual genetic operators of selection, crossover and mutation are included and a modified type of tournament selection is adopted.

GAs are designed for unconstrained optimization problems. In the present implementation in FIDO-TK, constrained problems are therefore converted into unconstrained ones using an adaptive type of function penalization details of which are described in References [12] and [14]. Also in the present work, a refinement technique of the feasible search space called 'rebirthing' is implemented. This technique is particularly useful when continuous variables are employed. At the start of the process, a small bitstring length may represent a wide range search space. When the convergence rate of optimization falls below a certain specified value the range is reduced to a smaller one by appropriately reducing the range of each design variable. See Reference [14] for further details.

2. Keane's highly non-linear objective function

Problem definition: Many test problems have been introduced to test optimization algorithms [4, 5, 21]. In a related paper [15], the performance of the GA procedure used here is studied for 14 test problems of increasing difficulty. Here, we focus on Keane's 'bump' problem [4, 5] which is defined as follows:

$$\max \quad \left| \sum_{i=1}^{n} cos^4(x_i) - 2 \prod_{i=1}^{n} cos^2(x_i) \right| / \sqrt{\sum_{i=1}^{n} ix_i^2} \qquad (1)$$

for

$$0 \leq x_i \leq 10 \qquad i = 1, ..., n \quad (n = 50) \qquad (2)$$

subject to

$$\prod_{i=1}^{n} x_i \geq 0.75 \quad \text{and} \quad \sum_{i=1}^{n} x_i \leq 15n/2 \qquad (3)$$

where the x_i are the variables and n is the number of dimensions. This function gives a highly 'bumpy' surface, where the true global optimum occurs with the product constraint active. The only way of illustrating the complexity of this function is by presenting a 2-dimensional representation of the function where $n = 2$; see Figure 1.

Solutions of Keane and others: Keane [4,5] attempted to obtain a solution to his problem using a parallel GA with 12-bit binary encoding, crossover, inversion, mutation, niche forming and a modified Fiacco-McCormick constraint penalty function. For n=20 he obtained values of approximately 0.76 after 20,000 evaluations and for n=50, values close to 0.76 after 50,000 evaluations. Good solutions to the problem, in both cases, $n = 20$ and $n = 50$, were obtained by program GENOCOP III and may be found in Reference [21].

Solutions obtained using FIDO-TK: Here, the problem was also solved with and without use of rebirthing, for both $n = 20$ and $n = 50$. Figure 2 shows the convergence history of the problem with $n = 20$. Without rebirthing

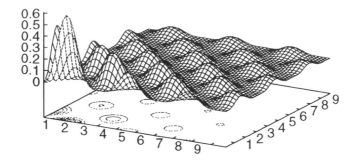

Figure 1   *A 2-dimensional representation of the bump problem without the constraints.*

technique, the solution converged after 15,800 function evaluations with an optimum value of 0.736. However, with rebirthing an optimum solution of 0.796 was obtained after 31,800 evaluations.

The optimum values $\mathbf{x}^*$ for the design variables for $n = 20$ are:

$$\mathbf{x}^* = [3.174, 3.081, 0.312, 3.060, 3.049, 3.002, 3.012, 2.932, 0.480, 0.835,$$
$$0.511, 0.458, 0.528, 0.459, 0.360, 0.417, 0.425, 0.399, 0.377, 0.456]^T$$

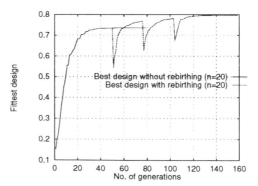

Figure 2   *Keane's bump problem: Convergence of the fittest design with no. of generations with n=20.*

The problem was also attempted for $n = 50$. Figure 3 illustrates the convergence history for the problem. Using the same GA parameters as for $n = 20$,

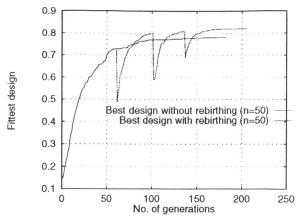

Figure 3   *Keane's bump problem: Convergence of the fittest design with no. of generations with n=50*

the best solution obtained without rebirthing was 0.780 which was achieved after 36,400 evaluations, while the problem with the option of rebirthing resulted in a better solution of 0.820 after 41,000 evaluations. For $n = 50$, the optimum values $\mathbf{x}^*$ for the design variables with use of rebirthing are:

$$
\begin{aligned}
\mathbf{x}^* = [&6.317, 6.346, 3.185, 3.267, 3.024, 3.154, 2.995, 3.118, 3.123, 3.108, \\
&3.132, 3.005, 3.029, 2.982, 3.066, 0.037, 5.727, 2.968, 3.068, 3.037, \\
&0.458, 0.450, 0.569, 0.547, 0.532, 0.465, 0.413, 0.447, 0.482, 0.454, \\
&0.505, 0.420, 0.468, 0.470, 0.531, 0.508, 0.461, 0.453, 0.426, 0.531, \\
&0.533, 0.422, 0.439, 0.466, 0.525, 0.410, 0.439, 0.481, 0.461, 0.454]^T
\end{aligned}
$$

These results as well as those obtained for the other test problems [15] provided confidence that our GA implementation was satisfactory.

3. Grierson's two-storey steel framework

Problem definition: A more practical problem is now considered involving the weight minimization of the two-storey frame illustrated in Figure 4. This problem has been adapted from one introduced by Grierson and Pal [2]. The overall height and width dimensions of the framework are given in Figure 4 and all exterior columns and floor and roof girders for the two stories are required to exist for all designs. The section types of the exterior columns and floor girders are variables, x_1 to x_6 and may have values selected from the 16 values given in Table 1. Moreover, the interior columns indicated by dashed lines in Figure 4 may or may not exist in the design of the framework. The section types of any existing interior column are variables, x_7 to x_{20} and may have values selected from the 16 values given in Table 2. They imply topology as well as size of members because the first catalogue value implies no member. The height of the first storey is a design variable y_1 which may have the discrete values given in Table 3. The framework has 18 free nodes and 54 degrees of freedom. The lateral displacement value at node 27 and the vertical displacement value at node 14 are constrained to be less than or equal to 12.5 mm and 10 mm respectively.

Table 1   Catalogue of steel sections for section type variables x_1 to x_6. See Figure 4.

Catalogue number	Section (AISC)	A $(10^3 \ mm^2)$	I $(10^6 \ mm^4)$
0	W460 × 52	6.63	212
1	W460 × 60	7.59	255
2	W460 × 61	7.76	259
3	W460 × 67	8.68	300
4	W460 × 68	8.73	297
5	W460 × 74	9.45	333
6	W460 × 82	10.40	370
7	W460 × 89	11.40	410
8	W460 × 97	12.30	445
9	W460 × 106	13.50	488
10	W460 × 113	14.40	556
11	W460 × 128	16.40	637
12	W460 × 144	18.40	726
13	W460 × 158	20.10	796
14	W460 × 177	22.60	910
15	W460 × 193	24.60	1020

Assumptions: Note that in Grierson's paper certain vital information was accidentally omitted and in the present problem, certain assumptions regarding lateral loading at nodes 10 and 15 and the value of the elastic modulus have been made. An elastic modulus of $E = 170 \times 10^3 N/mm^2$ has been assumed and the lateral loads are as indicated in Figure 4.

Solutions obtained using FIDO-TK: This problem was solved using FIDO-TK for populations of 50 and 150 and the results are compared with those of Grierson and Pal [2] in Table 4. Very good agreement is obtained between the three sets of results with slightly lower weights obtained using the present GA method. However, this may be because of the adaption of the problem

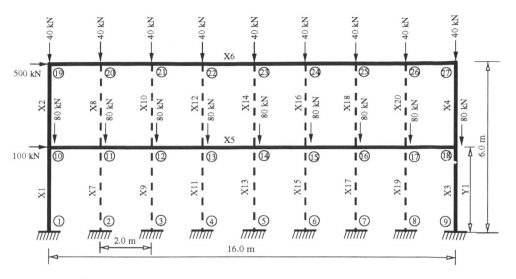

Figure 4   Two-storey steel framework problem adapted from work of Grierson et al [2].

Table 2   Catalogue of steel sections for section type variables x_7 to x_{20}. See Figure 4.

Catalogue number	Section (AISC)	A (10^3 mm^2)	I (10^6 mm^4)
0	nil	0.00	0
1	W460 × 60	7.59	255
2	W460 × 61	7.76	259
3	W460 × 67	8.68	300
4	W460 × 68	8.73	297
5	W460 × 74	9.45	333
6	W460 × 82	10.40	370
7	W460 × 89	11.40	410
8	W460 × 97	12.30	445
9	W460 × 106	13.50	488
10	W460 × 113	14.40	556
11	W460 × 128	16.40	637
12	W460 × 144	18.40	726
13	W460 × 158	20.10	796
14	W460 × 177	22.60	910
15	W460 × 193	24.60	1020

Table 3   Catalogue of discrete values for geometrical variable y_1. See Figure 4.

Catalogue No.	0	1	2	3	4	5	6	7
Storey height (mm)	2700	2800	2900	3000	3100	3200	3300	3400

Table 4   Optimal design variables for Grierson's framework.

Design variable	Grierson [ ] (10^3 mm^2) [†]	GA, 150 pop. (10^3 mm^2) [‡]	GA, 50 pop. (10^3 mm^2) [·]
x_1	6.63	6.63	6.63
x_2	6.63	6.63	7.59
x_3	9.45	7.59	7.76
x_{4-6}	6.63	6.63	6.63
x_7	7.59	7.59	7.59
x_8	0.00	7.59	7.59
x_9	9.45	9.45	8.68
x_{10}	0.00	0.00	0.00
x_{11}	7.59	7.59	8.68
x_{12-15}	0.00	0.00	0.00
x_{16}	7.59	0.00	0.00
x_{17}	0.00	0.00	0.00
x_{18}	8.73	7.59	7.59
x_{19}	0.00	0.00	0.00
x_{20}	7.59	7.59	7.59

[†]Grierson's solution: optimal weight=3481.0 kg, y=2700 mm, displacement at node 27=12.46 mm and displacement at node 14=-10.00 mm.
[‡]Present GA with 150 pop.: optimal weight=3417.7 kg, y=2900 mm displacement at node 27=12.38 mm and displacement at node 14=-9.89 mm.
[·] Present GA with 50 pop.: optimal weight=3451.3 kg, y=2800 mm displacement at node 27=12.47 mm and displacement at node 14=-9.64 mm.

definition as discussed earlier. See also Figure 5.

4. Stiffest frame problem

Problem definition: Let us now turn our attention to a situation in which the designer has not decided on ther structural topology. Imagine the following 2D frame design scenario: 'Given the loads and the support conditions what

Figure 5   *Optimum topology of the frame structure us-*
ing populations of 50 and 150 designs per gen-
eration.

is the stiffest structure that carries the loads to the supports, fits inside the given design space and satisfies a further prescribed volume constraint?' See, for example, the problem defined in Figure 6(a). Initially, the designer must decide on a layout or topology for the structure. Here, the objective is to maximize the overall stiffness of the structure whilst simultaneously reducing the volume to 15% of the starting volume.

The result of such a topology optimization is shown in Figure 6(c). By taking the layout image from the topology optimization a cursor is used to identify the joint positions and link the joints by appropriate straight lines representing the longitudinal axes of the frame members.

GAs are now used to minimze the strain energy of this 2D frame with the constraint that the volume must remain unchanged compared with the value specified in the initial topology optimization. As the frame members are assumed to be of rectangular cross-section and as their widths are defined by the width of the enclosing rectangular box-shaped design domain, the design variables are the depths of the member cross-sections. Here, the variables are 'pseudo-continuous' as the GA method treats them as a discrete subset of the infinite set of values between specified upper and lower bounds.

At this stage, it is worth considering the following point, if GAs are used to minimize the strain energy of the 2D frame, little change should be expected in the optimal strain energy obtained compared with the value given by the preliminary topology optimization. However, it is difficult to determine with precision the member depths from the structural image produced by the preliminary topology optimization. Furthermore, a different model based on 2D plane stress assumptions has been used in the preliminary topology optimization and it is necessary to confirm the optimal member cross-sections. Conse-

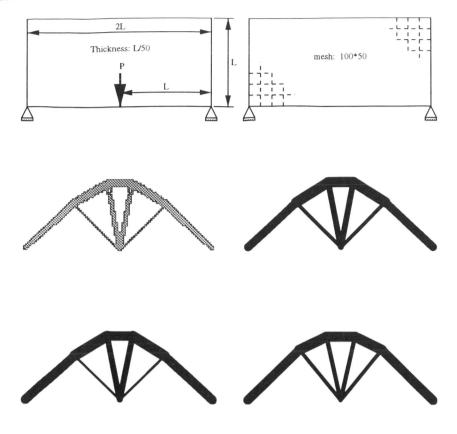

Figure 6 *Fully Integrated Design Optimization. (a) top left: initial design domain and boundary conditions, (b) top right: FE model, (c) mid-left: topology optimization results, (d) mid-right: post-processed frame structure, (e) bottom left: sizing optimization results, (f) bottom right: shape and sizing optimization results.*

quently, there is a definite need to carry out the subsequent GA optimization procedure.

In the present case, as expected, the use of GA sizing optimization results in only a 1.92% reduction in the value of the strain energy obtained compared to that of the preliminary topology optimization. Even, when the upper joints are allowed to move symmetrically (as design variables) within the rectangular box-shaped design domain, the reduction is only 4.81%. It is also worth noting that since section sizes are now known, further checks on member tensile and compressive stressses as well as buckling can be carried out if required.

5. Final Remarks

This paper has illustrated the use if FIDO-TK in the optimization of engineering structures. This computational toolkit integrates commercial and research software and provides the engineer with a interactive environment to control, monitor and manipulate the optimization process.

The modified GA procedure used in FIDO-TK has been used to study a highly non-linear bump function where semi-continuous variables were employed. The algorithm was also studied for the minimum weight optimization of a skeletal building structure with size, shape and topology design variables, where some practical steel sections were used. The use of rebirthing allows better optimal solutions to be obtained for problems involving continuous design variables provided that the solution at the first rebirthing is close to the optimal solution. A further example to find the stiffest 2D frame that fits within a given design domain with a given volume was also presented thereby illustrating the versatility of FIDO-TK.

Acknowledgements

Sponsorship from the following grants is gratefully acknowledged: Human Capital and Mobility grants: 'Advanced Finite Element Solution Techniques in Innovative Computer Architecture' (CHRX-CT93-0390) and 'Use of Image Processing Techniques to Directly Create Finite Element Models of Real Natural and Artificial Products and Systems', (CHRX-CT93-0386); ESPRIT IV award: 'A Parallel Automatic Optimization Scheme Applied to Die Design', EPSRC awards: 'Exploiting Parallelisms in Large Scale Finite Element Adaptive Stress Analysis and Structural Optimization'(GR/K22839); and 'Fully Integrated Design Optimization of Engineering Structures'; British Council sponsored links: with Stuttgart (Germany), Delft (Holland) and Rio de Janeiro (Brazil). The authors would also like to thank past and present members of the ADOPT group at Swansea. The authors also acknowledge helpful discussions with Prof. Ramm's group at the University of Stuttgart, Institut fur Baustatik, and here especially Prof. K. Bletzinger (now of University of Karlruhe) and Dipl.-Ing. K. Maute with whom some benchmarks have been developed within a British Council / DAAD sponsored ARC link.

References

1. Hinton E, Sienz J, Bulman S, Lee S J, 1997. Fully integrated design optimization of engineering structures using adoptive finite element models. In: Gutowski W, and Mroz Z (eds.), 1997. *Proceeding of Second World Congress of Structural and Multi-disciplinary Optimization, Vol. 1.* Institute of Fundamental Technological Research, Polish Academy of Sciences, pp 173-178.

2. Grierson D E, Pal W H, 1993. Optimal sizing, geometrical and topological design using a genetic algorithm. *Structural Optimization*, 6:151-159.

3. Bendsoe M P, Kikuchi N, 1988. Generating optimal topologies in structural design using a homogenization method. *Comp. Meths. Mech. Eng.*, 71:197-224.

4. Keane A J. http://www.soton-ac.uk/ ajk/welcome.html

5. Keane A J, 1994. Experiences with optimizers in structural design. In: I. Parmee I (ed.), 1994. *Proceedings of the Conference on Adaptive Computing in Engineering Design and Control 94*, pp 14-27.

6. Holland J H, 1992. *Adaptation in natural and artificial systems: 2nd edition.* MIT Press, Cambridge.

7. Goldberg D E, 1989. *Genetic Algorithms in Search, Optimization and Machine Learning.* Addison-Wesley, Englewood Cliffs.

8. Greffnstette J J, Fitzpatrick J M, 1985. Genetic search with approximate function evaluations. In: Greffenstette J (ed.) *Proceedings of First International Conference on Genetic Algorithms and their Applications*, Lawrence Erlbaum Associates, Hillsdale, New Jersey, pp 112-120.

9. Jenkins W M, 1991. Structural optimization with the genetic algorithm. *The Structural Engineer*, 69(24):418-422.

10. Hajela P, 1992. Genetic algorithms in automated structural synthesis. In: Topping B H V (ed.), *Optimization and Artificial Intelligence in Civil and Structural Engineering, volume 1*, Kluwer Academic Publishers, 639-653.

11. Adeli H, Cheng N T, 1994. Augmented lagrangian genetic algorithm for structural optimization. *Aerospace Engineering*, 7(1):104-118.

12. Ghasemi M R, Hinton E, 1996. Truss optimization using genetic algorithms. In: Topping B H V (ed.), *Advances in Computational Structures Technology*, Civil Comp Press, Edinburgh, pp 59-75.

13. Ghasemi M R, 1996. *Structural Optimization of Trusses and Axisymmetric Shells using Gradient-based Methods and Genetic Algorithms*. Ph.D. thesis, Dept. of Civ. Eng., University of Wales Swansea.

14. Ghasemi M R, Hinton E, 1997. Optimization of trusses using genetic algorithms for discrete and continuous variables. *Computers and Structures*, submitted, for publication.

15. Ghasemi M R, Hinton E, 1998. Testing genetic algorithms in FIDO-TK using set of difficult benchmarks. *Research Report, Dept. of Civil Engineering University College of Swansea.*

16. Anon. Why we need lots of optimization algorithms; http://cwp.mines.colorado. edu/html_ reports/coool.

17. Richardson J T, Palmer, Liepins G, Hilliard M, 1989. Some guidelines for genetic algorithms with penalty functions. In: *Proceeding of the Third International Conference on gentic Algorithms*, pp 191-197.

18. Homaifar A, Qi C X, Lai S H, 1984. Constrained optimization via genetic algorithms. *Simulation Journal*, 62(4):242-250.

19. Michalewicz Z, Janicow C, 1991. Genocop: A genetic algorithm for numerical optimization problems with constraints. *Communications of ACM*, 1991.

20. Anon. Genetic and evolutionary algorithm toolbox for use with matlab, http:// www.systemtechnik.tu-ilmenau.de/ pohlheim/GA-Toolbox/index.html.

21. Michalewicz Z, 1992. *Genetic Algorithms + Data Structures = Evolution Programs*. Springer, Department of Computer Science, University of North Carolina, Charlotte, NC 28223, USA, third, revised and extended edition edition, 1992.

Global optimisation in optical coating design

D.G. Li and A.C. Watson

School of Computer, Information and Mathematical Sciences

Faculty of Science, Technology and Engineering

Edith Cowan University

2 Bradford Street, Mount Lawley, W.A. 6050, Australia

Email: d.li@cowan.edu.au

Abstract. Many advanced local and global optimisation techniques, such as Gradient, Simplex, Flip-flop, Needle , Genetic and Simulated annealing, have been successfully applied to optical thin-film design. Any optimisation algorithm applied to a particular design problem should firstly address the issue of choosing a reasonable starting design, which is always a big obstacle to an inexperienced designer. To find the true global optimised solution for a thin film design problem, we need to solve an array of interlinked multi-dimensional simultaneous equations. For more than just a few layers, until recently this has been a very difficult task, requiring the use of a supercomputer and highly skilled programming. By using orthogonal Latin Square theory and an experimental design methodology into a search space reduction process, a Windows based program has been written that can operate on even a desktop personal computer. It can find the global optimum design for 23 layers design using any dispersive and lossy material within a period of several hours. Additionally this methodology (DGL-Optimisation, DGL is the short for D.G. Li) allows the use of target spectra such as s & p polarisation, with reflection and transmission simultaneously.

1. Introduction

The majority of problems faced by designers and engineers can be described as some form of local optimisation, trading off the improvement in one aspect against the worsening of another. In the thin film design problem, we need find how to stack optical thin film layers on top of one other to obtain the optimum spectral response - changing one layer thickness may improve the performance of the stack at one wavelength, but worsen it at another.

For a particular target spectral response (reflectance and / or transmittance), the variables not only include the refractive index of each layer and its thickness, but also weighting factors for each wavelength, the material loss and dispersion, the ray's polarisation and angles, and for non-parallel layers, it might even include the layer's x and y coordinates.

To find the solution that best matches the target is a very difficult task (especially when there is more than 1 condition to be satisfied, such as both reflection and transmission s and p polarisation), using dispersive and lossy materials. The merit function plot would appear as a multi-peak, multi-variable plot. Because there are an enormous number of inter-related possible layer combinations, the best film design cannot be found by any simple analytical process. The methods currently used in thin-film design software, such as the damped least squares, simplex, needle, flip-flop optimisations and annealing algorithm, etc., all depend on a starting condition either selected by the user or generated internally by the program. These starting conditions may be completely hidden from the user. Changing the initial conditions will give a different result, and the user has no way of knowing how much improvement could be effected.

With the DGL-global optimisation algorithm embodied in OpTeFilm software, the user knows that the design found is optimised within the criteria set - there is no need to try other starting conditions for the same layer structure, because there are no starting guesses. The algorithm inexorably must find the optimum solution that exists within the boundary conditions. This efficiency has powerful economic consequences. For example, previous designs needing excessive numbers of layers can now be fabricated with fewer layers, lowering cost, to get the same performance and better yields. A manufacturer can improve yields on marginal designs by using a design with a greater margin of error, as well as offering previously unavailable products.

2. The merit function

As in other methods, a merit function (M.F.) is calculated for each design the program finds. The M.F. is calculated by the sum over all the wavelengths (or angles) of a difference function between the value of the parameters calculated by the current design compared to the target. The optimisation process then becomes one of searching for the design that has the best merit function. Thus as an example, for a target having a reflection spectra for both s and p polarisation, the merit function would be calculated as :

$$MF = \sum_{\lambda} \{|Rp - Rpo| + |Rs - Rso|\}$$

where:

Calculated Merit Function - current design	**MF**	
Polarisation State	s & p	
Target spectra - Specified by the user	R_{so}	R_{po}
Transmittance or reflectance - current design	R_s	R_p

Therefore by finding the lowest value of MF we will have the best average design.

This formula can be made more sophisticated by introducing weighting factors to increase the importance of user specified wavelengths, as well as using other forms of the difference between the target and current design such as a square. A square form has the effect of weighting differences which are greater, thus flattening the deviation between the target and design.

3. A graphical analogy of optimisation methods

As discussed earlier, a single number (the M.F.) can be used to describe how close to the target a current design is located. By plotting a multi-dimensional graph with merit function as one of the axes, we can visualise the process. We require as many orthogonal axes as there are variables plus one for the M.F. Thus, for a 2 layer thin film problem, we require a 3 dimensional plot.

To see the process used in a simplified form, image a 2 dimensional array along the x and y axes, (which corresponds to a 2 layer problem). Let the value along the x axis represent a thickness of layer 1, and along the y axis the thickness of layer 2. In the z direction we plot the Merit Function value.

The task of the program is then to find the x and y values that generate the lowest value of the M.F. In this description, we shall invert the M.F. since a peak is easier to see than a valley. Therefore we choose the form of M.F. whose value increases as the design performance improves, i.e. we want to maximise the M.F. (One form of M. F. can be transformed into the other by taking a reciprocal of it). The game is to change each layer thickness in order to maximise the Merit Function. To help understand the optimisation process, consider the analogy of a man wandering in a cratered terrain, with a Global Positioning Satellite (GPS) receiver, which displays his absolute x, y, and z coordinates. Height (z) is the Merit function, and x, y represent the layer thicknesses of each layer in a 2 layer design. His task is to find the highest point (largest M.F.) bounded by the user defined maximum and minimum values of x and y . See Fig 1a.

3.1 Conventional methods

If he just walks upwards until he can go no further uphill, he will have found the local maximum. The 'best guess approach' is based on a starting design (position) that may be based on many years of experience. There are several mathematical methods available such as the gradient method, which alter several layer thicknesses simultaneously, see what happens to the merit function, and move the design in the direction of a maximum of the Merit Function. These find the 'Local Optimum, and the end design is completely dependent on the starting guess.

A derivation of this is the Simplex Method which, by the use of triangulation and in some variations, random numbers, is able to find a better 'near local' maximum quicker, exploring other designs by 'jumping away' from the nearest peak.

For multi-layer designs, using only two different materials, the flip-flop[1,2] method uses a method based on a large number of alternating sub-layers. Briefly, the merit function of the stack as each sub layer is flipped from low to high or vice versa is calculated. Stack sequences which improve the merit function when flipped, are preserved, if it was worsened, then the previous state is preserved. The program flips each sub-layer in turn, and the process is repeated when the first round is completed. By this process the number of layers is progressively reduced as adjacent high layers are coalesced. Similarly for low index layers. The process continues until the number of layers is below that specified by the user or no further improvement can be made.

The needle[3] optimisation method uses a completely different technique to optimise the design, by adding layers one by one, optimising the combination each time, and then adding another layer, and so on. This has the apparent advantage of keeping the layer number to a minimum for an acceptable design. It does have the advantage of being capable of producing complex designs with a minimum of user interaction, however the design found is unlikely to be the global optimum. The final design however is very dependent on the initial starting thickness, and many more layers than necessary are often required for a given performance.

3.2 The DGL method

The DGL-optimisation operates by a process of searching for all regions in the layer thickness space where a height greater than a specific level is located. This is akin to creating a contour map by slicing parameter space at a constant value of merit function.

In our 2 layer analogy, this is the equivalent of a plane parallel to the x-y plane at height z. See Fig 1b. This plane intersects the topography and identifies the entire region within which the peak is known to lie. By raising the slicing plane repeatedly, the region within which the peak must lie, is made smaller and smaller until only the highest peak remains - See Fig. 1c. Its coordinates correspond to the layer thicknesses of the optimum design. In practice, the surface is a mathematical construct of as many orthogonal dimensions as there are layers!

No starting guess is necessary (or even possible), and the operator only has to define the basic parameters, such as the number of layers, the max. and min. layer thicknesses for each layer (i.e. the boundary conditions), materials, and target spectrum. After the program is started, the operator can observe the values of max. and min. (within which the global optimum resides) for the various layers approaching each other. At the end of the run, there will be no layer those max. and min. value is greater than the specified value (as little as 1 nm). This stop value can be thought of as being the dimension of the peak - if one wishes one can make this as small as one would wish, in practice it just takes longer but with no practical benefit.

By using DGL global optimisation functions, a designer can be assured that he has found the best design physically possible, independent of his so-called best guess.

The mathematical procedures used in this form of global optimisation are possible to apply to a variety of other previously unsolved problems relating to the resultant of dependent variables, including experimental design and manufacturing variations. In the visible region, one of the global optimisation functions of OpTeFilm will even find all different designs having the same color, and compare them for manufacturing variations.

There are many other approaches people have adopted, but until now (with the exception of scanning), they all depend either on a starting design, some form of local optimisation or some random variation. Each method will usually give rise to different solutions. For designs using a large number of layers, these are still the only methods possible. In contrast, the DGL optimisation described here is a methodical global method.

4. Principles of the DGL global optimisation and the Latin Square

The true magnitude of the problem can be seen by considering a scanning approach, i.e. measuring the merit function value for every possible combination of layer thicknesses. The scanning method is guaranteed to find the global optimum, provided one does enough calculations. For example, let us consider the number of merit function calculations necessary to scan a 5 layer design at 2nm intervals over 10 nm to 350 nm (170 measurements per layer). The total number of possible combinations is 170^5, which would take years for even a super-computer!

The DGL optimisation uses a mathematical method based on orthogonal sets of numbers. By slicing the multi-dimensional parameter space with a horizontal plane of the Merit Function, with each parameter independent of the others. A peak is always be surrounded by a slope. By finding all regions in which the merit function has values above that of the plane, one can narrow the search region. After finding the boundary of all the isolated regions where this occurs, the plane is raised again, and the process repeated.

A Latin Square is an array of numbers widely used in experimental designs, By utilising orthogonal Latin Squares, one can form an array of several dimensions which are orthogonal to each other, and therefore allow the calculation of a resultant using many interdependent variables. This allows us to zoom in on the area of interest. Each iteration reduces the boundary of 1 or more of the variables in parameter space until the boundary dimension is less than that specified by the stop criteria. Combining Latin Square sampling with function domain contraction techniques, results in an optimisation with two desirable properties. Firstly, the number of function evaluations can be greatly reduced, and secondly, there is a guarantee of finding the global optimum solution. There is a text available on Latin Square theory and its applications[7].

5. Mathematical form of DGL optimisation

Consider a multi-dimensional continuous function f(x) with multiple global minima and local minima on subset G of R^n

(i) We define the local minima as follows

For a given point $x^* \in G$, if there exists a δ-neighborhood of x^*, $O(x^*, \delta)$, such that for

$x \in O(x^*, \delta)$,

and $f(x^*) \leq f(x)$ (1)

then x^* is called a local minimal point of f(x).

(ii) Definition of global minima

If for every $x \in G$ the inequality (1) is correct, then x^* is called a global minimum of f(x) on G, and the global minima of f(x) on G form a global minimum set.

(iii) How to find the global minima

Now for a given constant C_0 such that the level set

$H_0 = \{x | f(x) < C_0, x \in G\}$ is non-empty,

if $\mu(H_0) = 0$, where μ is the Lebesque measure of H_0,

then C_0 is the minimum of f(x) and H_0 is the global minimum set.

Otherwise, assume that $\mu(H_0) > 0$ and C_1 is the mean value of f(x) on H_0.

Then $C_1 = 1/\mu(H_0) \int_{H_0} f(x)d\mu$ (2)

and

$C_0 \geq C1 \geq f(x^*)$ (3)

We then gradually construct the level set H_k and mean value C_{k+1} of $f(\mathbf{x})$ on H_k as follows:

$$H_k = \{\mathbf{x} \mid f(\mathbf{x}) < C_k, \mathbf{x} \in G\} \qquad (4)$$

and

$$C_{k+1} = 1/\mu(H_k) \int_{Hk} f(\mathbf{x})d\mu \qquad (5)$$

With the assistance of Latin Square sampling, a decreasing sequence of mean values $\{C_k\}$ and a sequence of level sets $\{H_k\}$ are obtained.

Let
$$\lim_{k \to \infty} C_k = C^* \qquad (6)$$

and

$$\lim_{k \to \infty} H_k = H^* \qquad (7)$$

It can be proven that C^* is the minimum of $f(\mathbf{x})$ on G, and H^* is the global minimum set.

There are several strategies to avoid missing the global optimum when seeking the minimum solution. Among these, the most important step is to design a suitable orthogonal Latin Square with which the function within domains can be repeatedly sampled. The algorithm is automatically constrained to stay within the function domain and will not request function evaluations outside this domain.

There are two stopping criteria possible; either when the target M.F. value is reached, or when the maximum domain length is smaller than the user selected value. OpTeFilm uses the latter stop criteria, corresponding to the variation possible for each layer thickness - which can be as little as 1 nm. This means that the global minimum has been found for a particular layer thickness range of each layer, with a variation of less than 1 nm for each layer, strictly speaking then, the global optimum is not defined at a point but as lying within a region.

6. Application of DGL global optimisation to practical filter design

In fig 1a, 1b & 1c, we see a representation of a two layer problem. Using a search analogy, to find the peak, the region in which the highest peak must lie is narrowed each time the plane is raised. This occurs until there is only one peak left. Its coordinates correspond to the layer thicknesses of the optimum design.

The examples in this paper are intended only for the comparison of solutions obtained with the DGL method with those obtained with other thin film synthesis programs. Dispersion of the refractive indices and residual absorption within the dielectric layers have not been allowed for since they do not materially affect the solution. Any refractive indices lying within reasonable upper and lower limits were accepted.

OpTeFilm is the worlds first true global optimisation program to run under windows on a PC. The parameters that the user defines must include the materials (with specified refractive index and loss over the wavelength range), the number and order in which the layers are to be placed, the incident angle etc. The target can be selected form a variety of pre selected parameters, including s and p polarisation. The advantage of this methodology is that a global optimisation can be made on parameters that may or may not be related. For example a merit function can be calculated including s and p polarisation of transmission and just p reflection. A weighting factor can also be applied to specific target wavelengths.

Figure 2 shows a typical design process for an edge filter. Starting from a theoretical design, the structure can be optimised locally by the gradient method. A Simplex optimisation improves the design considerably. This would often be the endpoint. Using a global optimisation however, with the same number of layers and no starting guess, yields the best performance. Note must be made that a different performance can arise if the order of the materials is reversed. i.e. the high index next to the substrate is substituted by a low index material..

Figure 3 shows a comparison between the best design found in a competition[4,5] for the best anti-reflection germanium I.R. filter. The global optimisation clearly yields a superior performance.

Figure 4 is a comparison of designs for a 50% beamsplitter. Here the comparison is between the Gradient[8] and the DGL optimisation.

The application of this technique will undoubtedly have implications well beyond thin film design. Already some of these principles have been applied to a variety of difficult mathematical problems involving image recognition, hydrology and lens design, as well as the theoretical solving of mathematical problems by evolutionary optimisation[6].

7. Conclusion

By using a Latin Square and other mathematical techniques, it is possible to create a global optimisation program for thin film design on the desktop. A global optimisation program such as OpTeFilm allows the solution of multilayer designs using mixed parameters such as separate p and s transmission spectra, weighted targets, using lossy and dispersive materials without any starting design. Using the criteria selected by the user, the function will methodically proceed to the optimum design. The primary advantages of this technique are that mixed targets with dispersive and lossy materials can be used, the global optimum is always found, excellent designs can be found with little prior knowledge, and the new merit functions can be created according to whatever combination of parameters are required.

ACKNOWLEDGEMENTS

The authors are indebted to Burt Nathan, of Raylight Pty. Ltd. who markets OpTeFilm software, (P.O. Box 1089, Aldinga 5173, S.A. Australia. Fax: +61-8-885574071, Email: raylight@adelaide.dialix.com.au, and Wed: http://www.webvert.com.au/ray/opte), for his assistance at various stages of preparing this paper and for many stimulating discussions.

References

1. J.A. Dobrowolski. "Comparison of the Fourier transform and flip-flop thin-film synthesis methods," Applied Optics **25**, 1966 (1986)

2. J.A. Dobrowolski and R.A. Kemp. "Flip-flop thin film design program with enhanced capabilities," Applied Optics **31** 3807 (1992)

3. A.V. Tikhonravov. "Some theoretical aspects of thin-film optics and their applications," Applied Optics **32** 5417 (1993)

4. J.A. Aguilera et al. "Antireflection coatings for germanium IR optics: a comparison of numerical design methods," Applied Optics **27**, 2832 (1988)

5. P. Baumeister. "Starting designs for the computer optimization of optical coatings," Applied Optics **34**, 4835 (1995)

6. H. Bersini et al. "Results of the first intenational contest on evolutionary optimisation," IEEE journal ... (1996)

7. W.G. Cochran and G.M. Cox, "Experimental designs", John Wiley & Sons, Inc, (1957)

8. J.A. Dobrowolski and R.A. Kemp, "Refinement of optical multilayer systems with different optimization procedures", Applied optics **29** 2876 (1990)

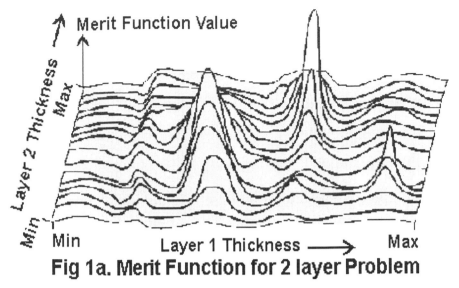

Fig 1a. Merit Function for 2 layer Problem

In this description of a 2 layer design, the boundary conditions are defined by the layer's maximum and minimum thicknesses. We seek to find the layer thicknesses that give rise to the maximum value of the Merit Function.

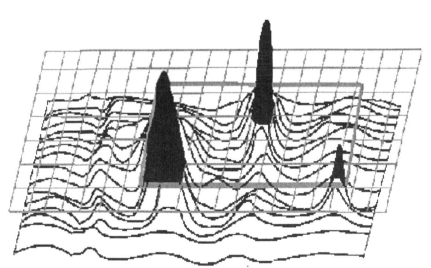

Fig 1b. Regions of Merit Function > Plane value

A plane is constructed of 'constant Merit Function' and the boundary of regions having a higher merit function than that of the plane are identified.

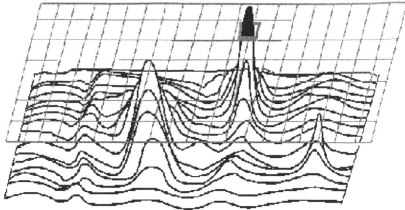

Fiq 1c. Merit Function Peak reqion

As the plane is raised, the region within which the peak of the Merit Function exists, is narrowed. The process is repeated until the layer thicknesses which give rise to the highest peak are uniquely identified.

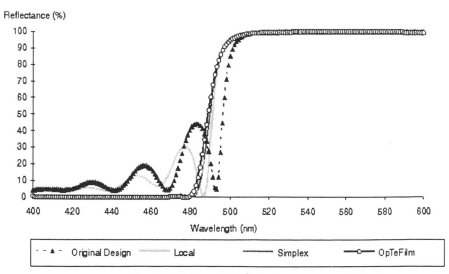

Fig 2a. Edge Filter Performance - Detail

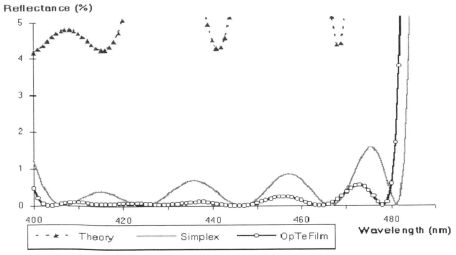

Fig 2b. Edge Filter Performance - Detail

A comparison of the results of using different optimisation processes. The global optimum design has a significantly lower average reflectance. All designs use the same number of layers.

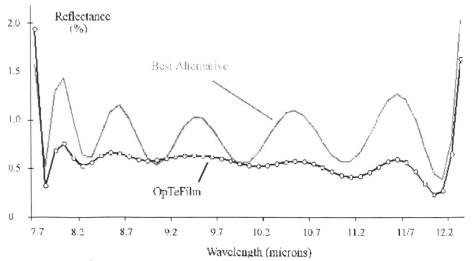

Fig 3. Germanium I.R. Filter - Detail

As above, the starting criteria selected for the OpTeFilm design consisted of the same as that of the published design, viz. Number of layers and their order, materials, etc.

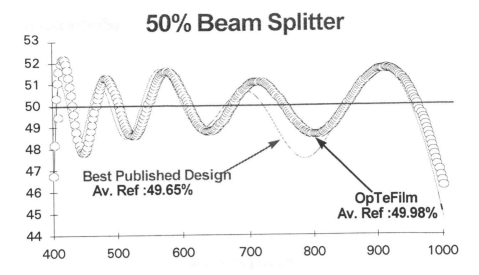

Fig. 4. 50% Beamsplitter - The starting criteria selected for the OpTeFilm design consisted of the same as that of published design obtained by Gradient optimisation, viz. Number of layers and their order, materials, etc.

Evolutionary Algorithm for the Design of Stack Filters Specified using Selection Probabilities

A. Beatriz Garmendia Doval
Chilukuri K. Mohan
Dept. of Electrical Eng. & Computer Sci.
2-120 Center for Science & Tech.
Syracuse University
Syracuse, NY 13244-4100, U.S.A.
abgarmen/mohan@top.cis.syr.edu,

Mohit K. Prasad

Room 55E-212
1247 S. Cedar Crest Blvd.
Lucent Technologies, Bell Labs.
Allentown, PA 18103, U.S.A.
mkprasad@lucent.com

Abstract: Evolutionary algorithms are applied to the design of a class of nonlinear discrete-time filters: the positive Boolean function defining a stack filter is derived from its properties (specified in terms of *selection probabilities*). For window size 9, best results for this computationally intractable problem were obtained using a tree representation for each stack filter.

1 Introduction

Many commercially important digital nonlinear filters fall into the category of stack filters [1, 2]. Examples include the median filter, rank-order filters, max-median filter, midrange estimators, and the weighted median filter. This paper shows that stack filters, specified using selection probabilities [3], can be successfully designed by the application of variants of the genetic algorithm.

A stack filter operates on a window of finite width. Raw measurements for each component in the window are "stacked" into columns of bits, with the 1s below the 0s; the number of 1s in a column represents the magnitude of the raw measurements. Columns are juxtaposed, resulting in a bit-array. Various operations can be applied over the rows and columns of this bit-array. A median filter, for instance, can be described as counting the number of row-sums that exceed a threshold value that equals half the number of rows.

Example 1 For a window size three, if the measurements are $(3 \quad 1 \quad 1)$, one gets the following bit-array:

$$\begin{bmatrix} 1 & 0 & 0 \\ 1 & 0 & 0 \\ 1 & 1 & 1 \end{bmatrix}$$

The sum of elements in each column corresponds to one of the raw measurements $(3 \quad 1 \quad 1)$. Applying the median filter to this bit-array, comparing row-sums with a threshold of 3/2, we obtain:

$$
\begin{array}{ccccccccc}
1 & + & 0 & + & 0 & = & 1 & \longrightarrow & 0 \\
1 & + & 0 & + & 0 & = & 1 & \longrightarrow & 0 \\
1 & + & 1 & + & 1 & = & 3 & \longrightarrow & 1
\end{array}
$$

(exceeds threshold)

Adding the elements in the last column (the number of row-sums that exceed the threshold), the result obtained is 1.

Prasad and Lee [3] present a new technique for the statistical analysis of stack filters. *Selection probabilities* are defined, based on the observation that the output of the stack filter is always chosen from the samples in the input window. The *rank selection probability* (RSP) is the probability that the output equals a sample with a certain *rank*, i.e., its position when the elements in the window are sorted in increasing order of value. The *sample selection probability* (SSP) is the probability that the output equals a sample with a certain position within the filter window. These measures have been used to characterize robustness and detail-preserving abilities of stack filters [3]. The properties desired of a stack filter can hence be formulated in terms of the desired selection probabilities.

Stack filters can be defined using positive Boolean polynomials in variables that correspond to components of the input windows. In these polynomials, addition represents disjunction (OR) and multiplication represents conjunction (AND). The number of possible terms in a polynomial is less than 2^N, where N is the input window size. The coefficient of each such term is 1 or 0, indicating that the term is present or absent in the polynomial.

We cast the stack filter design problem into the following form. find a positive Boolean polynomial that represents a stack filter with the desired (given) selection probabilities. The search space is huge (exceeding 2^{35} for window size 7), precluding exhaustive systematic search.

This paper describes a new successful approach for such filter design problems. Section 2 describes the application of a genetic algorithm using bit-string representations for each stack filter. Section 3 describes innovations that result in much better results for problems of window size 9. The last section compares our results with previous results [6, 7].

2   Genetic Algorithm for Stack Filter Design

A prototypical Genetic Algorithm (GA) [4, 5], works on large collections of individuals ("populations"). A "generation" consists of the population at a given instant during the evolutionary process. The quality of an individual is measured using a "fitness" function. A new generation of individuals is derived from the previous, applying "crossover" and "mutation" operators. Crossover generates offspring individuals by combining components of two parent individuals. Mutation randomly modifies components of an individual. Individuals with better fitness are allocated greater number of offspring, providing "selection pressure" for the evolutionary process.

For stack filter design, the fitness function we use is a positive linear function of the sum of the absolute differences between desired and actual RSP and SSP components, with maximum fitness value of 100 for the goal filter. The desired RSP and SSP components are given as inputs to the program. For each individual in the population, RSP and SSP calculations are as follows [3].

For each j, let h_j, g_j be Boolean functions such that the filter is described by the Boolean function $f(X) = x_j h_j(X_j) + g_j(X_j)$, where $X_j = (x_1, ..., x_{j-1}, x_{j+1}, ..., x_N)$. The *Combination matrix (C-matrix)* entries are:

$$C_{i,j} = \sum_{X_j:\ \omega_H(X_j)=N-i} h_j(X_j)\overline{g_j(X_j)},\ 1 \le i,\ j \le N$$

where the summation is arithmetic, over all binary vectors X_j whose Hamming weight $\omega_H(X_j)$ (the number of 1s in X_j) equals $N - i$. Then,

$$RSP[i] = \sum_{j=1}^{N} \frac{(N-i)!(i-1)!C_{i,j}}{N!},\ \text{and}\ SSP[j] = \sum_{i=1}^{N} \frac{(N-i)!(i-1)!C_{i,j}}{N!}.$$

A bit-string of length $2^N - 1$ was used to represent each polynomial, with one bit used to indicate the presence or absence of each term in the polynomial. Some terms in a polynomial render others redundant. For example, when the term $x_1 x_3$ is present in one individual, it "masks" all other terms $x_1 x_3 * T$, since the Boolean function $x_1 x_3 + x_1 x_3 * T$ is equivalent to $x_1 x_3$. A *normalization* procedure removes those terms ($x_1 x_3 * T$) from the individual to avoid this redundancy. For instance, if 1s were present in bit-positions indicating $x_1 x_3$ as well as $x_1 x_2 x_3$, the latter 1 would be changed to 0.

We used "one-point crossover," in which parent bit-strings are broken at some randomly chosen point, and the pieces are recombined to generate offspring that inherit one piece from each parent:

$$\begin{aligned}
(b_1 b_2 \ldots b_i b_{i+1} \ldots b_N) + &\quad (b_1' b_2' \ldots b_i' b_{i+1}' \ldots b_N') \quad \rightarrow \\
&\quad (b_1 b_2 \ldots b_i b_{i+1}' \ldots b_N') \quad + (b_1' b_2' \ldots b_i' b_{i+1} \ldots b_N)
\end{aligned}$$

Following crossover, mutation randomly reverses each bit with low probability.

The selection method used was "linear ranking," in which individuals in a population are ranked according to their fitness, and the number of offspring allocated to each individual linearly depends on this rank, with the highest number to the fittest and the lowest number to the least fit. This method is relatively insensitive to the precise choice of the fitness function.

A population size of twice the individual length was chosen, e.g., 62 for window-length 5, since each individual contains $2^5 - 1$ bits.

Figure 1 summarizes results obtained using three versions of the GA:

Prog. 1: High mutation rate (30%); best few individuals are preserved for next generation.

Prog. 2: Low mutation rate (3%); about half the best individuals are preserved for next generation; unnecessary crossover steps are prevented by comparing parents; duplicates in the population are replaced by randomly generated individuals.

Prog. 3: Low mutation rate (3%); low relative reproductive rates for better individuals; duplicate individuals permitted.

Stack filter	Prog. 1	Prog. 2	Prog. 3
$x_2x_4 + x_1x_3x_5$	95	12	10
$x_3 + x_2x_5 + x_1x_4$	7	3	6
$x_3x_4x_5 + x_2x_4 + x_2x_3x_5 + x_1x_4x_5$ $+x_1x_3x_4 + x_1x_2x_5 + x_1$	167	57	28
$x_3x_4 + x_2x_4x_5 + x_2x_3 + x_1x_3x_5 + x_1x_2x_4$	39	283	25
$x_4x_5 + x_3x_5 + x_2x_5 + x_2x_3x_4$	93	45	21
$x_3x_4 + x_2x_4 + x_2x_3$	7	48	7
$x_3x_5 + x_3x_4 + x_2x_3 + x_1x_3 + x_1x_2x_4x_5$	72	36	16

Figure 1: Number of generations required, using bit-string representations

These algorithms were fast and worked well on all inputs (attempted) with window size 5. But when those algorithms were applied to inputs with window size 9, the perfect filter was found only in cases that represent filters with a small number of terms, with most terms containing few variables.

3 A Tree-Based Evolutionary Algorithm

Performance was significantly improved by modifying the GA as follows:

1. Improving efficiency of fitness computation procedure

2. Adopting a new representation for filters

3. Using new operators

4. Using a steady state algorithm

5. Using random restarts to overcome premature convergence

The rest of this section describes these innovations, and consequent results.

3.1 Fitness computation

Terminology: A *true vector* for the stack filter f is a binary vector (subset of variables) for which the output of f is 1. A true vector is a *minimal vector* if none of its proper subsets corresponds to a true vector.

The algorithm in Section 2 for computing C-matrix entries has a complexity $O(N2^{N-1})$, too high when N is large. Prasad [8] has suggested a more efficient algorithm, outlined below.

- Step 1: Obtain all true vectors of the Boolean function.

- Step 2: Construct sets corresponding to the true vectors such that a 1 in the jth position implies x_j is a member of the set.

- Step 3: For each set, list its subsets corresponding to the minimal vectors (as in column 3 of Figure 2).

- Step 4: For each true vector, construct the intersection of sets corresponding to the minimal vectors.

- Step 5: For each true vector X, the rank is computed to be $N+1-\omega_H(X)$.

- Step 6: Generate the C-matrix as follows:

  (a) Initialize each element $C_{i,j}$ of the C-matrix to zero.

  (b) For each true vector, if i is its rank, for each x_j in its intersection set, increment $C_{i,j}$ by 1.

True vector	Set	Minimal sets	Intersection set	Rank
$(1,0,0,0)$	$\{x_1\}$	$\{x_1\}$	$\{x_1\}$	4
$(0,1,0,0)$	$\{x_2\}$	$\{x_2\}$	$\{x_2\}$	4
$(0,0,1,1)$	$\{x_3,x_4\}$	$\{x_3,x_4\}$	$\{x_3,x_4\}$	3
$(1,1,0,0)$	$\{x_1,x_2\}$	$\{x_1\},\{x_2\}$	$\emptyset$	–
$(1,0,1,0)$	$\{x_1,x_3\}$	$\{x_1\}$	$\{x_1\}$	3
$(1,0,0,1)$	$\{x_1,x_4\}$	$\{x_1\}$	$\{x_1\}$	3
$(1,1,1,0)$	$\{x_1,x_2,x_3\}$	$\{x_1\},\{x_2\}$	$\emptyset$	–
$(1,1,0,1)$	$\{x_1,x_2,x_4\}$	$\{x_1\},\{x_2\}$	$\emptyset$	–
$(1,0,1,1)$	$\{x_1,x_3,x_4\}$	$\{x_1\},\{x_3,x_4\}$	$\emptyset$	–
$(0,1,1,0)$	$\{x_2,x_3\}$	$\{x_2\}$	$\{x_2\}$	3
$(0,1,0,1)$	$\{x_2,x_4\}$	$\{x_2\}$	$\{x_2\}$	3
$(0,1,1,1)$	$\{x_2,x_3,x_4\}$	$\{x_2\},\{x_3,x_4\}$	$\emptyset$	–
$(1,1,1,1)$	$\{x_1,x_2,x_3,x_4\}$	$\{x_1\},\{x_2\},\{x_3,x_4\}$	$\emptyset$	–

Figure 2: For the filter $x_1 + x_2 + x_3x_4$, the above table yields the C-matrix

$$\begin{pmatrix} 0 & 0 & 0 & 0 \\ 0 & 0 & 0 & 0 \\ 2 & 2 & 1 & 1 \\ 1 & 1 & 0 & 0 \end{pmatrix}$$

To further speed up computation, only the true vectors relevant for the C-matrix need to be constructed. In the example given, in the first column of Figure 2, all true vectors are listed. The fourth true vector, $x_1 x_2$ has the null set as its intersection set. Once this is known, we can avoid the creation of all the true vectors that contain both variables, because each of them will also have the null set as its intersection set. Taking this into account and also being careful not to process a true vector more than once, we developed a new algorithm of complexity $O(K)$, where K is the sum of the cardinality of the intersection sets; $K = 8$ in the example shown.

3.2 Tree representation

The normalized representation described in Section 2 is biased against terms with many variables. Smaller terms are more likely to be represented in an individual than bigger terms. This explains why good results were not obtained with the GA in Section 2, for cases with window size 9 that require many terms formed by three or more variables. Avoiding normalization did not help, since the effect of the larger terms is completely masked out until the smaller term is deleted. Removing the smaller term results in a significantly different filter, hence it does not help to randomly delete either the smaller or bigger term.

Our final solution to this problem was to change the representation. Instead of using the bit-string representation used in most genetic algorithms, we adopted the tree representation used with success in genetic programming. Each level in the tree represents one of the variables in the window. Each node can have one of the following values: '+', '1', '0', and 'L', with the following meanings:

+ : This node has descendants that include terms that contain the variable corresponding to this level (left subtree), as well as terms that do not contain the variable (right subtree).

1 : All the terms represented by the descendants of this node contain the variable corresponding to this level.

0 : None of the terms represented by the descendants of this node contain the variable corresponding to this level.

L : Leaf node, the path to which (from the root node) describes a term.

For instance, Figure 3 shows the tree representation of the polynomial $x_1 x_2 + x_1 x_4 + x_2 x_3 x_5 + x_4 x_5$.

3.3 New operators

- The new 'crossover' operation creates two new individuals through the exchange of subtrees in the same relative position between the parents.

- The mutation operation replaces one randomly chosen subtree by another randomly generated subtree.

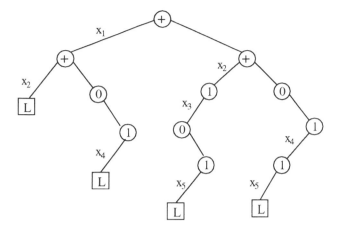

Figure 3: Tree representation of $x_1x_2 + x_1x_4 + x_2x_3x_5 + x_4x_5$

- We also introduced a new operator, called *expansion*, that tentatively replaces a term T by the terms corresponding to $T * (x_1 + x_2 + ... + x_N)$. The replacement is accepted if this change increases fitness.

- After generating an individual, a fixed number of hill-climbing steps (10) is applied to the individual. In each step, the current individual is compared with a mutation of the same, and the mutant replaces the current one if this increases fitness. Hill-climbing is used to find the best filter without requiring too many generations. In one case, hill-climbing was found to be necessary to find the best match. Without it, a suboptimal solution (99.6% of optimal fitness) was found.

3.4 Steady state algorithm

The algorithm used was "steady state," i.e., only a small number of individuals are replaced at a time, instead of replacing the entire population as in canonical "generational" GAs.

In each iteration, two individuals are chosen with probability depending on fitness. From these, two new individuals are created (by crossover or mutation). Each new individual is tested against the worst individual in the population. In case the new individual is better, it replaces the worst individual. For comparison, a "generation" in a canonical (generational) algorithm requires roughly as much computation as $|P|/2$ iterations of the steady state algorithm, where $|P|$ is the number of individuals in the population.

3.5 Random restarts

When the population appears to have converged, so that subsequent evolution is extremely slow, we retain a small elite (5%) and create new individuals

Algorithm Stack_Filter_Design;

Initialize population randomly;
while (no. of iterations < *max_iter*) and (best fitness is unacceptable), **do**

1. Use linear ranking selection to obtain *parent₁* and *parent₂* from the current population;

2. Obtain two offspring by applying one of the following operators:

   - With probability 0.7, apply crossover to *parent₁* and *parent₂*;
   - With probability 0.26, apply expansion operator to *parent₁* and *parent₂*;
   - With probability 0.04, apply mutation to *parent₁* and *parent₂*;

3. Perform a fixed number of hill-climbing steps on each offspring;

4. Replace the least fitness individuals in the current generation by the hill-climbed offspring, in case the latter have higher fitness.

5. In the current population, if best fitness ≈ least fitness, then replace 95% of the population by randomly generated individuals, retaining 5% of the best members of the population.

end_while.

Figure 4: New evolutionary algorithm for evolving stack filters

randomly to replace the rest of the population. The population is considered to have converged when the highest and least fitness (of individuals in a generation) differ very little (0.15% fitness).

3.6 Results

Figure 5 gives the average CPU time for the inputs of window size 5. Even in the worst case, the desired solution was obtained in about 250 operator applications, the equivalent of about 8 generations of a generational algorithm. For inputs of window size 9, Figure 6 gives the average equivalent number of generations and the average CPU time on a Sun ULTRA1 workstation.

The results obtained are much better than for the bit-string representation. In most cases, the best filter (desired goal) was found on average in less than one hour, often in just a few minutes. Even in the worst case, the best individual's fitness exceeded 99.0 within one hour of computation.

Polynomial description (stack filter)	Run time (sec.)
$x_2x_4 + x_1x_3x_5$	0.8
$x_3 + x_2x_5 + x_1x_4$	0.1
$x_3x_4x_5 + x_2x_4 + x_2x_3x_5 + x_1x_4x_5$	4.7
$\quad\quad + x_1x_3x_4 + x_1x_2x_5 + x_1x_2x_3$	
$x_3x_4 + x_2x_4x_5 + x_2x_3 + x_1x_3x_5 + x_1x_2x_4$	2.0
$x_4x_5 + x_3x_5 + x_2x_5 + x_2x_3x_4$	1.3
$x_3x_4 + x_2x_4 + x_2x_3$	0.6
$x_3x_5 + x_3x_4 + x_2x_3 + x_1x_3 + x_1x_2x_4x_5$	0.5

Figure 5: CPU time for window size 5, averaged over 10 runs.

	Equivalent no. of generations	Run time (minutes)
input1	2	2.6
input2	1	1.1
input3	4	9.1
input4	18	85.6
input5	6	7.2
input6	13	20.7

Figure 6: Computational expense for window size 9, averaged over 5 runs, using a SUN ULTRA1 workstation; note that equivalent number of generations = $\lceil$number of iterations/(population size/2)$\rceil$.

4 Discussion

We have presented a new method for designing stack filters. The desired properties of a stack filter can first be specified in terms of selection probabilities. An evolutionary algorithm is then used to discover a stack filter whose selection probabilities are as close to the desired values as possible. Current implementations have been successful on problems with window sizes 5 and 9. Further work involves extending this method to larger window sizes.

In recent work, Chu [6] and Oakley [7] apply GAs to filter design, using signals corrupted with noise as input data. The filter's output is compared with the original uncorrupted signal to measure filter performance. However, a large amount of input data would be required to accurately specify each filter and distinguish it from other filters. Only certain kinds of noise can be simulated when raw signals are used to design filters in this manner. Quote Chu: "An alternative, perhaps more realistic, performance measure has to be formulated to avoid the use of known signals." This paper accomplishes that task: selection probabilities are a logical way to describe filter properties.

The early results of Chu [6] and Oakley [7] provide a yardstick for evaluating our results:

1. Both Chu and Oakley experimented with filters with window size 7 whereas our algorithm has been successful for window size 9; the number of possible bit-strings (for the simple GA) increases from $2^{2^7} \approx 3.4 \times 10^{38}$ to $2^{2^9} \approx 1.2 \times 10^{154}$. Thus, our results are obtained for much larger search spaces, although the actual search space size is smaller (not known exactly) than the number of bit-strings.

2. In Chu's work, the fitness of a filter was evaluated in terms of its performance on specific signals; it was not possible to describe the exact optimal solution. Similarly, Oakley's experiments yielded non-zero error values. By contrast, each of our experiments had a specified goal in mind, reached with zero error; significantly less time was needed to find suboptimal solutions with fitness values extremely close to the optimal solution.

3. Computation times are not clearly mentioned in Chu's work; experimental results, shown for 80 generations of a typical run, indicate that solution quality did not improve substantially with time, after the spurt in fitness found in the first ten generations. Oakley's results indicate that computation times of the order of a day were required for windows of size 7. By contrast, we obtained optimal solutions for window size 9 requiring computational times of the order of ten minutes.

References

[1] P.D.Wendt, E.J.Coyle, and N.C.Gallagher, Jr., *Stack Filters,* IEEE Trans. Acoustics, Speech and Signal Processing, **36**(4):898-911, Aug. 1986.

[2] E.J.Coyle, *The theory and VLSI implementation of stack filters*, in VLSI Signal Processing II, IEEE Press, NY, 1986.

[3] M.K.Prasad and Y.H.Lee, *Stack Filters and Selection Probabilities*, IEEE Trans. on Signal Processing, **42**(10):2628-2643, Oct. 1994.

[4] J.H.Holland, Adaptation in Natural and Artificial Systems, Univ. of Mich. Press, Ann Arbor (MI), 1975.

[5] D.E.Goldberg, Genetic Algorithms in Search, Optimization, and Machine Learning, Addison-Wesley, 1989.

[6] C.H.Chu, *A Genetic Algorithm Approach to the Configuration of Stack Filters*, Proc. Third Int'l. Conf. Genetic Algorithms, pp. 219-224, 1989.

[7] H.Oakley, *Two Scientific Applications of Genetic Programming: Stack Filters and Non-Linear Equation Fitting in Chaotic Data*, in Adv. in Genetic Prog. (ed. K.E.Kinnear, Jr.), pp. 369-389, MIT Press, 1994.

[8] M.K.Prasad, *A Structured Design Approach for Non-Linear Filters*, submitted for publication.

Drawing Graphs with Evolutionary Algorithms

Andrea G. B. Tettamanzi[1,2]

[1] Università degli Studi di Milano, Dipartimento di Scienze dell'Informazione
Via Comelico 39, I-20135 Milano, Italy
E-mail: tettaman@dsi.unimi.it
[2] Genetica—Advanced Software Architectures S.r.l.
Viale Monte Nero 68, I-20135 Milano, Italy
E-mail: genetica@tin.it

Abstract. This paper illustrates an evolutionary algorithm for drawing graphs according to a number of esthetic criteria. Tests are carried out on three graphs of increasing difficulty and a comparison is made on the performance of three different recombination operators. The results are then briefly discussed.

1 Introduction

A number of data presentation problems involve the drawing of a graph on a two-dimensional surface, like a sheet of paper or a computer screen. Examples include circuit schematics, communication and public transportation networks, social relationships and software engineering diagrams. In almost all data presentation applications, the usefulness of a graph drawing depends on its readability, i.e. the capability of conveying the meaning of the diagram quickly and clearly. Readability issues are expressed by means of *esthetics,* which can be formulated as optimization criteria for the drawing algorithm [3].

An account of esthetic criteria that have been proposed and various heuristic methods for satisfying them can be found in [20]. An extensive annotated bibliography on algorithms for drawing graphs is given in [9] and [3]. The methods proposed vary according to the class of graphs for which they are intended and the esthetic criteria they take into account. For most reasonable esthetic requirements, however, it turns out that solving this problem exactly is prohibitively expensive for large graphs.

Evolutionary algorithms are a broad class of optimization methods inspired by Biology, that build on the key concept of Darwinian evolution [4, 6]. It is assumed that the reader is already familiar with the main concepts and issues relevant to evolutionary algorithms; good reference books are [19, 12, 5, 17, 16, 1]; [13, 10, 18] are more of a historical interest.

2 The Problem

Given a partially connected graph $G = (V, E)$, we want to determine the coordinates for all vertices in V on a plane so as to satisfy a certain number of esthetic requirements.

The esthetic criteria that were employed in this work are the following:

- there should be as few edge crossings as possible, ideally none;
- the length of each individual edge should be as close as possible to a parameter L;
- the angles between edges incident into the same vertex should be as uniform as possible (ideally the same).

A criterion that is usually considered requires that the vertices be evenly distributed on the available space. In fact, this criterion is entailed by the second and third criteria stated above, since all the edges should be approximately the same length and the edges departing from a vertex should spread as much apart from one another as possible.

2.1 Complexity

In general, the optimization problems associated with most esthetics are NP-hard [14, 15]. For instance, it has been proven that even just minimizing the number of edge crossings is an NP-hard problem [11]. Therefore, the problem described above, which requires in addition to satisfy two other criteria, is also NP-hard, thus providing a valid motivation for resorting to an evolutionary approach.

2.2 Related work

The use of evolutionary algorithms for drawing *directed* graphs, a problem of great practical importance in relation with interactive software tools that use diagrams such as transition or structure diagrams in the form of directed graphs (cf. for instance [8]), has already begun to be explored by Michalewicz [17]. Surprisingly, however, that application disappeared from subsequent editions of his book. Michalewicz's work considers only two criteria, namely that arcs pointing upward should be avoided and there should be as few arc crossings as possible.

Direction of edges is not addressed in the work described here, which deals with other types of graphs arising, for example, when we want to represent on paper telecommunication networks, where links are always bidirectional.

3 The Algorithm

The overall flow of the proposed evolutionary algorithm is the following, which operates on an array *individual*[*popSize*] of individuals (i.e. the population); even though this is not explicitly demonstrated in the pseudo-code for sake of readability, crossover and mutation are never applied to the best individual in the population.

```
SeedPopulation(popSize)
generation := 0
while true do
      for i := 1 to popSize do
          EvaluateFitness(i)
      end for
      Selection
      for i := 1 to popSize step 2 do
          Crossover(i, i + 1, p_cross, σ²_cross)
      end for
      for i := 1 to popSize do
          Mutation(i, p_mut, σ²_mut)
      end for
      generation := generation + 1
end while
```

The various elements of the algorithm are illustrated in the following subsections.

3.1 Encoding

How a graph is drawn on a plane, i.e. a candidate solution to our problem, is completely determined by the (x, y) coordinates assigned to each vertex.

Therefore, a genotype consists in a vector $((x_1, y_1), \ldots, (x_{\|V\|}, y_{\|V\|}))$, where (x_i, y_i) are the coordinates of the ith vertex of the graph, encoded as two integer numbers in $\{-4096, \ldots, 4095\}$, which can be considered the basic *genes* of an individual. This gives a virtual page of $8192 \times 8192 = 67,108,864$ pixels on which the graph can be drawn.

It is worth noticing that this is not a bit-string representation and thus genetic operators always act on pairs of coordinates (or vertex positions) according to their meaning.

3.2 Initialization

Initial vertex positions for each individual are randomly generated independently and according to the same uniform probability over the whole drawing page.

Generation of initial graphs could also be carried out with the help of greedy algorithms, which indeed have been tried, although they are not discussed in this paper.

3.3 Crossover

Three different types of crossover have been experimented with:

- *uniform* crossover, whereby the vertex positions that make up an individual's genotype have the same probability of being inherited from either parent;

- *single point* crossover, where the crossover point cannot split a vertex position into two;
- *convex hull* crossover, where the coordinates of each vertex in the offspring are a linear combination of the coordinates of the same vertex in its parents: suppose that (x_1, y_1) and (x_2, y_2) are the coordinates of a vertex in the two parents, then the same vertex in the offspring will have coordinates (X, Y), where X and Y are two independent, normally distributed random variables with mean respectively $\frac{x_1+x_2}{2}$ and $\frac{y_1+y_2}{2}$ and with the same variance σ_{cross}^2, which is a parameter of the algorithm. This kind of recombination operator is loosely inspired by the *intermediate recombination* operator widely used in evolution strategies [18, 19], which is just the averaging operator.

Whatever the type that is being used, crossover is applied with a given probability, p_{cross} to each couple of individuals in the population.

3.4 Mutation

Mutation perturbs individuals by adding independent Gaussian noise to each of their vertices. If (x, y) is the position of a vertex before mutation, the position of the same vertex after mutation will be given by (X, Y), where X and Y are two independent normally distributed random variables, with mean respectively x and y and with the same variance σ_{mut}^2, which is a parameter of the algorithm, and therefore it remains constant throughout evolution (unless the human operator changes it interactively, a possibility that is given by the software package implementing the evolutionary algorithm).

This mutation operator is very similar to the convex hull crossover described above, the main difference between the two operators being that in the case of crossover both parents participate in setting the mean of the new vertex position distribution; also, we have always used the mutation operator described above with a much bigger standard deviation than the convex hull crossover.

Self-adaptation of the mutation variance [1, 2], one of the crucial properties of evolution strategies, is likely to give good results in this setting and is certainly going to be implemented in the future.

3.5 Fitness

Three factors contribute to determining an individual's fitness, one for each esthetic criterion:

- the number of edge crossings, χ;
- the mean relative square error σ of edge lengths defined as

$$\sigma = \frac{1}{\|E\|} \sum_{e \in E} \left(\frac{\|e\| - L}{L} \right)^2, \tag{1}$$

where $\|e\|$ is the length of edge e;

– the cumulative square deviation Δ of edge angles from their ideal values, defined as

$$\Delta = \sum_{v \in V} \sum_{k=1}^{N_v} \left(\psi_k(v) - \frac{2\pi}{N_v} \right)^2, \tag{2}$$

where N_v is the number of edges incident into vertex v and the $\psi_k(v)$, $k = 1, \dots, N_v$, are the angles between adjacent vertices.

An individual's fitness, f, to be maximized, is then calculated as follows:

$$f = a\frac{1}{\sigma + 1} + b\frac{1}{\chi + 1} + c\frac{1}{\Delta + 1}, \tag{3}$$

where a, b and c are constants that control the relative importance of the three criteria and compensate for their different numerical magnitudes. Their values have been empirically determined as $a = 0.02$, $b = 0.8$ and $c = 0.18$ for the experiments described below; however, the reader should be aware that by modifying these constants the drawings produced by the algorithm can widely vary.

It can be easily verified that only in very few special cases can the fitness defined in Equation 3 approach the theoretical maximum of one. This is so because the three criteria employed conflict with each other and cannot in general be all completely satisfied at the same time.

3.6 Selection

Elitist fitness proportionate selection, using the roulette-wheel algorithm, was implemented to begin with, using a simple fitness scaling whereby the scaled fitness $\hat{f}$ is

$$\hat{f} = f - f_{\text{worst}}, \tag{4}$$

where f_{worst} is the fitness of the worst individual in the current population.

Linear ranking selection was tried as well, but the results were comparable to those obtained with fitness proportionate selection; therefore it was not taken into account in the experiments illustrated below.

Overall, the algorithm is elitist, in the sense that the best individual in the population is always passed on unchanged to the next generation, without undergoing crossover or mutation.

4 Experimental Results

The algorithm described above was implemented and run on several Pentium 120 workstations, under the Windows 95/NT operating systems.

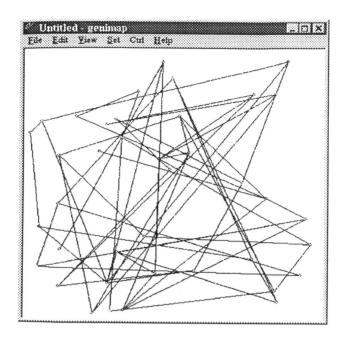

Fig. 1. An example of random layout for Graph G_1, having fitness 0.006193.

4.1 The Test Suite

Preliminary experiments were performed using three test graphs, that will be denoted G_1, G_2 and G_3.

- G_1 represents the road plan of a neighborhood of Milan and contains $\|V_1\| = 39$ vertices and $\|E_1\| = 58$ edges.
- G_2 is a less connected graph of a more irregular nature, having $\|V_2\| = 51$ vertices and $\|E_2\| = 58$ edges.
- G_3 is a graph describing a co-authorship relation among a number of researchers in the field of evolutionary computing. This graph is much bigger than the other two, having $\|V_3\| = 160$ vertices and $\|E_3\| = 165$ edges.

Both G_1 and G_2 were designed by hand so as to be planar and regular. Graph G_3, apart from being much bigger, results to be non-planar. Furthermore it has sets of non-connected vertices.

Figure 1 shows an example of how a randomly generated layout (i.e. a member of the initial population) for the simplest graph G_1 looks like.

The desired length for edges was set to

$$L = \frac{8192}{\sqrt{\|V\|}},$$

which is the average distance between closest vertices when they are uniformly distributed over the whole drawing plane.

4.2 Test Runs

After a first phase during which common optimal values for crossover and mutation rates were roughly established ($p_{\text{cross}} = 0.3$ and $p_{\text{mut}} = 0.15$) for a population size of 100 individuals, four sets of evolutionary algorithm runs were performed for each test graph. The characteristics of each set of runs are the following:

1. single-point crossover;
2. uniform crossover;
3. convex hull crossover with standard deviation $\sigma_{\text{cross}} = 100$;
4. convex hull crossover with standard deviation $\sigma_{\text{cross}} = 200$.

In all four runs, σ_{mut} was set equal to L. All runs were carried on for 10,000 generations, corresponding to an average execution time of about one hour.

4.3 Results

Throughout all the experiments performed, the set of runs featuring convex hull crossover proved significantly better than the other two, getting very close to finding a planar solution for graphs G_1 and G_2.

Tables 1 to 3 compare the performance of the different crossovers on the three test graphs.

Gen. no.	convex 100	convex 200	single-point	uniform
0	0.006133	0.006033	0.007258	0.006193
1000	0.042044	0.036540	0.077847	0.176708
2000	0.069960	0.052057	0.089722	0.217646
3000	0.070431	0.061988	0.091262	0.218769
4000	0.071041	0.063126	0.091565	0.219746
5000	0.071357	0.069572	0.091880	0.220178
6000	0.071468	0.069710	0.092377	0.220233
7000	0.071740	0.070211	0.092476	0.220305
8000	0.074128	0.070427	0.092660	0.220572
9000	0.074448	0.070489	0.092745	0.221003
10000	0.074633	0.070568	0.092809	0.221208

Table 1. Fitness of the best individual during four sample evolutions for Graph G_1 across 10,000 generations.

Gen. no.	convex 100	convex 200	single-point	uniform
0	0.005065	0.005285	0.005448	0.005459
1000	0.104270	0.148000	0.215165	0.283912
2000	0.212809	0.216858	0.216344	0.285629
3000	0.215021	0.284077	0.218658	0.285913
4000	0.416083	0.284694	0.219015	0.286115
5000	0.417016	0.417343	0.219184	0.286291
6000	0.418009	0.417981	0.219327	0.286588
7000	0.418725	0.418200	0.219455	0.286774
8000	0.418951	0.418361	0.219686	0.286876
9000	0.419220	0.418427	0.219897	0.286983
10000	0.419366	0.418585	0.219917	0.287002

Table 2. Fitness of the best individual during four sample evolutions for Graph G_2 across 10,000 generations.

Gen. no.	convex 100	convex 200	single-point	uniform
0	0.001045	0.001125	0.001087	0.001102
1000	0.017012	0.014273	0.017142	0.018259
2000	0.017805	0.015395	0.027558	0.040031
3000	0.018426	0.015815	0.041464	0.079277
4000	0.018536	0.016085	0.059769	0.093941
5000	0.018536	0.016444	0.066907	0.103466
6000	0.018870	0.016444	0.071187	0.115019
7000	0.019059	0.016602	0.071678	0.115365
8000	0.019059	0.016602	0.077160	0.115678
9000	0.019216	0.016882	0.104661	0.115932
10000	0.019404	0.017786	0.104922	0.149165

Table 3. Fitness of the best individual during four sample evolutions for Graph G_3 across 10,000 generations.

A quick inspection of the test data in Tables 1 to 3 points out that uniform crossover dominates single-point crossover. Convex hull crossover with a smaller standard deviation seems to dominate the one with a greater standard deviation. However, no simple ordering appears to hold between uniform (or single-point) crossover and convex hull crossover.

Uniform crossover exhibited a consistently good performance on all the three test graphs, and the best one on Graphs G_1 and G_3. On Graph G_2 convex hull crossover performed remarkably better than both single-point and uniform crossover, but its performance on the other two graphs was disappointing. This fact suggests that more study should be devoted to the tuning of standard deviation in convex hull crossover as a function of graph characteristics. If such a function exists, it is not a simple one, as is the case with mutation, where it was soon clear that standard deviation should be directly proportional to parameter L.

Figures 2 to 4 show the best results, in terms of fitness, obtained by the evolutionary algorithm applied to Graphs G_1, G_2 and G_3. However, compare Figure 3 with the less fit but more pleasing layout found in Figure 5.

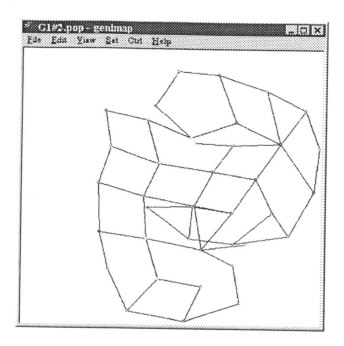

Fig. 2. A layout of Graph G_1 found by the algorithm with uniform crossover after 10,000 generations, having fitness 0.221208.

5   Discussion

There is a troubling aspect with convex hull crossover used in combination with mutation: when genetic diversity in the population is low, i.e. when it is highly likely that the same vertex is placed in the same position by both parents selected for sexual reproduction, that recombination operator turns into a kind of mutation, and it would thus be expected to contribute to maintaining a certain level of diversity, which in turn would avoid premature convergence and allow the algorithm to escape from local minima. Despite of this, however, runs using convex hull crossover got stuck in local minima in two of the three test cases. On the other hand, when genetic diversity in the population is high, convex hull

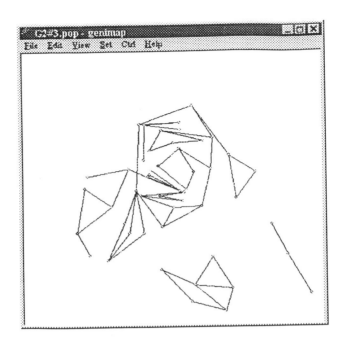

Fig. 3. A layout of Graph G_2 found by the algorithm with convex hull crossover with standard deviation 100 after 10,000 generations, having fitness 0.419366.

crossover is more likely to cram vertices in the center of the drawing plane; the algorithm then has a hard time spreading them back apart in the right way.

Even though one might think that using evolutionary algorithms for graph drawing represents a departure from classical algorithmic strategies found in computational geometry literature, a number of implicit similarities with well-known classical heuristics can be found.

For instance, including a term inversely proportional to σ as defined in Equation 1 in the fitness function recalls the so-called *spring embedder* force-directed method [7], whereby the drawing process is to simulate a mechanical system, where vertices are replaced by rings and edges are replaced by springs: the springs attract the rings if they are too far apart and repel them if they are too close.

The natural continuation of the work described here would consist in making comparisons with other stochastic algorithms based for example on taboo search or simulated annealing and with relaxation algorithms like the spring embedder mentioned above. For the latter, it would be easy to develop a hybrid evolutionary algorithm incorporating them either as Lamarckian mutation operators or as decoders of the genotype into the actual solution to be evaluated.

A further idea, since the settings of the parameters controlling the relative weights of the esthetic criteria are very arbitrary and the quality of a drawing is

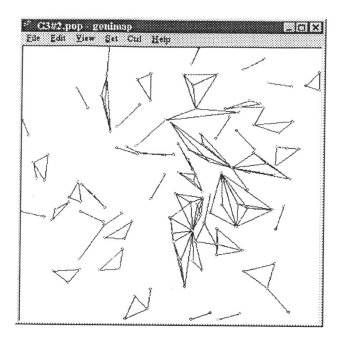

Fig. 4. A layout of Graph G_3 found by the algorithm with uniform crossover after 10,000 generation, having fitness 0.149165.

inherently subjective, would be to allow the algorithm to interact with a human operator, using its responses to optimally tune the parameters relevant to the fitness function.

Acknowledgments

This work was partially supported by M.U.R.S.T. 60% funds. The author thanks Prof. Gianni Degli Antoni for his support and useful discussions and Genetica— Advanced Software Architectures for the extensive use of their computing equipment.

References

1. T. Bäck. *Evolutionary algorithms in theory and practice.* Oxford University Press, Oxford, 1996.
2. T. Bäck, U. Hammel, and H.-P. Schwefel. Evolutionary computation: History and current state. *IEEE Transactions on Evolutionary Computation,* 1(1):3–17, 1997.
3. G. Di Battista, P. Eades, R. Tamassia, and I. G. Tollis. Algorithms for drawing graphs: An annotated bibliography. Technical report, Available on the Internet, URL: ftp://wilma.cs.brown.edu/pub/papers/compgeo/gdbiblio.ps.Z, 1989.

336

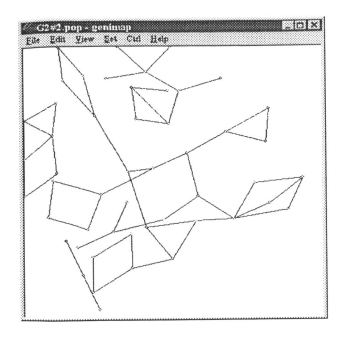

Fig. 5. A layout of Graph G_2 found by the algorithm with uniform crossover after 10,000 generations, having fitness 0.287002.

4. C. Darwin. *On the Origin of Species by Means of Natural Selection.* John Murray, 1859.

5. L. Davis. *Handbook of Genetic Algorithms.* VNR Computer Library. Van Nostrand Reinhold, New York, 1991.

6. R. Dawkins. *The blind Watchmaker.* Norton, 1987.

7. P. Eades. A heuristics for graph drawing. *Congressus Numerantium*, 42:149–160, 1984.

8. P. Eades and X. Lin. How to draw a directed graph. Technical report, Department of Computer Science, University of Queensland, Australia, 1989.

9. P. Eades and R. Tamassia. Algorithms for drawing graphs: An annotated bibliography. Technical Report CS-89-09, Department of Computer Science, Brown University, 1989.

10. L. J. Fogel, A. J. Owens, and M. J. Walsh. *Artificial Intelligence through Simulated Evolution.* John Wiley & Sons, New York, 1966.

11. M. R. Garey and D. S. Johnson. Crossing number is np-complete. *SIAM Journal on Algebraic and Discrete Methods*, 4(3):312–316, 1983.

12. D. E. Goldberg. *Genetic Algorithms in Search, Optimization & Machine Learning.* Addison-Wesley, Reading, MA, 1989.

13. J. H. Holland. *Adaptation in Natural and Artificial Systems.* The University of Michigan Press, Ann Arbor, 1975.

14. D. S. Johnson. The np-completeness column: An ongoing guide. *Journal of Algorithms*, 3(1):89–99, 1982.

15. D. S. Johnson. The np-completeness column: An ongoing guide. *Journal of Algorithms*, 5(2):147–160, 1984.
16. J. R. Koza. *Genetic Programming: on the programming of computers by means of natural selection*. The MIT Press, Cambridge, Massachussets, 1993.
17. Z. Michalewicz. *Genetic Algorithms + Data Structures = Evolution Programs*. Springer-Verlag, Berlin, 1992.
18. I. Rechenberg. *Evolutionsstrategie: Optimierung technischer Systeme nach Prinzipien der biologischen Evolution*. Fromman-Holzboog Verlag, Stuttgart, 1973.
19. H.-P. Schwefel. *Numerical optimization of computer models*. Wiley, Chichester; New York, 1981.
20. R. Tamassia, G. Di Battista, and C. Batini. Automatic graph drawing and readability of diagrams. *IEEE Transactions on Systems, Man and Cybernetics*, 18(1):61–79, 1988.

This article was processed using the LaTeX macro package with LLNCS style

Benchmarking of Different Modifications of the Cascade Correlation Algorithm[1]

D.I.Chudova[*], S.A.Dolenko, Yu.V.Orlov, D.Yu.Pavlov[*],
I.G.Persiantsev

Microelectronics Dept., Nuclear Physics Institute, Moscow
State University, Vorobjovy gory, Moscow, 119899, Russia

* Dept. of Computational Mathematics and Cybernetics, Moscow
State University, Vorobjovy gory, Moscow, 119899, Russia

E-mail: pers@ailab.npi.msu.su

Abstract. We apply different modifications of Fahlman and Lebier's cascade correlation algorithm to a real-world application in order to fulfil a comparison of these approaches in terms of generalisation ability and time elapsed for training. Investigation shows that the algorithm should be varied in view of different statements of training and retraining problems for optimal performance.

1. Description of Different Modifications of the CC Algorithm

In this paper we consider two neural network architectures synthesised and trained by Fahlman and Lebier's [1] Cascade Correlation (CC) algorithm. The first is a classic architecture described in [1] - every hidden unit added by the algorithm is connected to all previous hidden units and the input layer. In what follows we shall address this architecture as non-MLP architecture in view of the fact that resulting network will not have an architecture of the multilayer perceptron (MLP). Another architecture considered here is MLP architecture - CC algorithm is applied to the initial network which is a feed-forward network with one hidden layer and adds hidden neurons to that very layer thus creating a well-known MLP architecture. The main difference between architectures lies in that network with non-MLP architecture has extra connections between layers and is significantly worse studied.

Training a network with CC algorithm consists of two stages which are repeatedly applied one after another - training the net with error back-propagation (BP) algorithm [2] and adding a hidden neuron. The latter stage as described in [3] is very computationally intensive since it requires calculation of correlation between the residual error in the network output and the output of the hidden neuron being added ("candidate-neuron") and its gradient maximisation. In what

[1]This work was supported by the Russian Foundation for Basic Research, project no. 96-01-01964.

follows we shall address this algorithm as gradient CC. We propose a stochastic CC algorithm which requires less time to add a hidden unit and we compare it with the classic one. The main idea of the stochastic algorithm is to choose random weights, calculate the correlation, and save the network which gives the maximum value of correlation in the pre-specified number of attempts.

Thus, we compare two architectures - MLP and non-MLP in each of which we used either classic gradient correlation maximisation algorithm or a stochastic one. We also make a comparison of CC algorithm in different modifications with BP algorithm. The purpose of our investigation is to find out the conditions and problem statements under which this or that version of the algorithm is preferred with respect to the minimum test set error.

2. Statement of Problems to be Solved with CC Algorithm. Quality of Performance

Suppose we have a training set which is to be classified by a supervised neural network into the given number of classes. Test set is also given and is used to estimate the performance of the trained neural network classifier. The main task is to train the network and to obtain the minimum possible value of error on the representative test set. We shall call this task training a network. Another task which often arises is retraining the network which was trained before to classify patterns into a lesser number of classes. We shall call this task retraining the network.

We suppose that initial network architecture is a three layer perceptron with sigmoid transfer functions in the output and hidden layers. To estimate the performance of the network, we introduce mean squared error (MSE), wrong answers percentage (WAP), and time of training (TT) criteria. WAP criterion which is the most important in the tasks under consideration is calculated in the following way. A pattern is considered to be misclassified if either the maximum value over output neurons activation is less than 0.5 or in the opposite case if the maximum is achieved on a neuron corresponding to the class other than that of the example.

We draw your attention to the fact that usually the number of hidden neurons sufficient to perform the classification task is not known. Thus, BP algorithm has a significant drawback - one has to choose the number of hidden units in advance. CC algorithm is free from this demand - it builds the feature layer dynamically. In all experiments, BP algorithm is applied to a network for which CC algorithm has already detected the optimal number of neurons.

3. Statement of the Problem for Benchmarking

We used CC and BP algorithms for training and retraining tasks on a database of utterances of words used to animate the standard Windows calculator. The database includes Russian utterances of digits and commands, such as "Equal",

"Plus", "Copy", "Paste", "Exit", which were recorded, FFT pre-processed, and normalised in a standard way. Input dimension is equal to 1000, number of output neurons is defined by the number of classes. Each of the 36 classes contained utterances spoken by 19 speakers, 17 of these were used for training set and 2 for test set.

We point out that we wanted not to get the best performance on this problem but to compare different methods of its solution. For achieving best performance, the database should be much larger and more attention should be paid to image pre-processing. For this study, we have chosen this database as a suitable example of complex and linearly unseparable data.

It is also important what criterion has been used to stop training and to pass from one stage of training to another. We stop training if WAP error on test set hasn't improved for 50 epochs. We pass to neuron adding if MSE on test set has decreased in last 10 epochs less than q% of its value (q=5). If gradient method of correlation maximisation is used we stop it if correlation in the last iteration has increased less than p% of its value (p=4).

4. Numerical Results

We began our experiments with an MLP network with 1 hidden neuron and trained it with CC algorithm to fulfil the training task. Then we applied CC algorithm to train the net to recognise first 12 classes and then retrained the net three times adding 6 more classes out of 30 at a time. We also applied BP algorithm to an MLP net with number of hidden units equal to that obtained by CC.

Standard graphs of MSE versus number of epochs for gradient and stochastic CC algorithms are given in Fig. 4.1. Black dots placed on the graphs stand for the moments when a new hidden neuron was added to the net. It can be seen that adding a neuron leads to a sufficient decrease in MSE. Main characteristics of graphs for MLP and non-MLP architectures practically coincide.

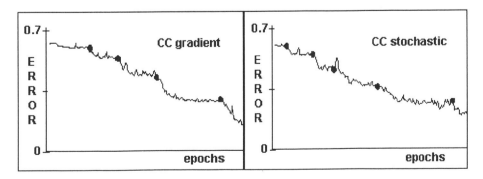

Figure 4.1. MSE versus number of epochs

All experiments were conducted on PC Pentium-120MHz, 16Mb RAM, with all training paradigms implemented in C as Dynamic Link Library NeuroWindows™ from Ward Systems Group, Inc.

Below (Tables 1-5) we present the results of training the net to solve the training and retraining tasks as described above. Each table of results has the following structure: the first column gives information on the task being performed (for instance, 12->18 means that the net previously trained on the 12 classes is retrained to perform classification of 6 extra classes), subsequent columns contain results on respectively training time (sec), number of training epochs, best MSE on test set, best WAP on test set (%), and number of hidden neurons (#) added to the initial architecture 1000-1-30.

Main conclusions that can be drawn from Tables 1-5 are the following:

1. If one solves a complicated classification task (for instance, classification into a relatively large number of classes) with gradient CC algorithm, then it is better to divide the task into a sequence of subtasks with increasing complexity (number of classes) and to solve retraining tasks until the whole task is solved. This approach leads to a smaller MSE and WAP error on test set and takes less time than solution of the whole task with the same algorithm. (See Table 1).

2. If compared with gradient CC algorithm, stochastic CC algorithm is superior on the time elapsed criterion. The results (MSE and WAP error) obtained for the training task (lines starting with 0-> in Table 2) with this algorithm are as good as these with gradient CC algorithm. However, nets trained with this algorithm aren't suitable for further retraining - MSE and WAP error after the completion of retraining task are higher than on the corresponding nets trained with gradient CC algorithm. Refer to Table 2.

3. As far as BP algorithm is concerned, we also tried to perform training and retraining tasks with it. In all experiments with BP algorithm, the number of neurons in the hidden layer was determined by CC algorithm in advance. For solution of the training task we chose the number of hidden neurons in the net sufficient for subsequent retraining of this net. Training the net with BP algorithm results in a larger MSE and WAP error than these obtained by gradual retraining with gradient CC algorithm (both MLP and non-MLP architectures). Retraining the net with the procedure described above gives even worse results and shouldn't be considered at all. Refer to Table 5.

4. Nets with non-MLP architecture show practically the same results as nets with MLP architecture, although the former can be superior on a more complicated task solution. Refer to Tables 3 and 4 for details.

5. It can be easily concluded from Tables 2 and 5 that if the classification problem to be solved is not very complicated, it is worth using stochastic CC algorithm rather than BP algorithm. This approach will not only save the time spent on trying different dimensions of hidden layer but it will also be efficient in terms of time of training, MSE, and WAP error.

Table 1. MLP Architecture with Gradient CC Algorithm.

Task	Time	Eps.	MSE	WAP	#
0->12	2517	205	0.38	4.1	5
12->18	874	63	0.27	5.5	6
18->24	3247	101	0.43	8.3	8
24->30	2469	96	0.50	20.0	9
0->18	6367	278	0.44	11.1	7
0->24	16295	270	0.47	16.6	9

Table 2. MLP Architecture with Stochastic CC Algorithm.

Task	Time	Eps.	MSE	WAP	#
0->12	421	172	0.23	4.1	4
12->18	467	67	0.37	8.3	5
18->24	845	102	0.52	16.6	6
24->30	2000	130	0.62	26.6	7
0->18	2032	222	0.31	5.5	7
0->24	2254	247	0.41	12.5	8

Table 3. Non-MLP Architecture with Gradient CC Algorithm.

Task	Time	Eps.	MSE	WAP	#
0->12	2310	236	0.29	4.1	6
12->18	1625	63	0.29	5.5	6
18->24	2502	128	0.42	12.5	7
24->30	4261	89	0.46	16.6	8
0->18	5474	211	0.41	8.3	7
0->24	10237	190	0.49	18.7	8

Table 4. Non-MLP Architecture with Stochastic CC Algorithm.

Task	Time	Eps.	MSE	WAP	#
0->12	721	204	0.26	8.3	5
12->18	832	150	0.35	5.5	6
18->24	866	110	0.43	10.4	7
24->30	1532	145	0.50	21.0	9
0->18	1517	203	0.33	8.3	6
0->24	2553	225	0.43	10.4	8

344

Table 5. MLP Architecture with BP Algorithm.

Task	Time	Eps.	MSE	WAP	#
0->12	198	58	0.26	4.1	8
12->18	603	89	0.33	13.8	8
18->24	327	67	0.48	16.6	8
0->18	442	64	0.29	5.5	8
0->24	347	70	0.46	14.5	8
0->30	625	85	0.51	20.0	9

5. Conclusions and Recommendations

This paper is devoted to benchmarking and comparison of different modifications of Cascade Correlation algorithm. It is shown that CC algorithm is applicable to network synthesis and incremental learning without destroying the information gained by the previously trained nets. We outline the CC algorithm based approaches to solution of two main tasks which arise in solution of classification problems: task of training and task of retraining a net.

We show that training task is better solved by either gradient CC algorithm with gradual retraining or stochastic CC algorithm with immediate training to a prespecified number of classes. However, the latter approach suffers from low potential to further retraining.

The retraining task can be performed in different network architectures - we state that MLP and non-MLP architectures allow one to obtain practically identical results on MSE and WAP error criteria. However, the non-MLP architecture might perform better on a more complex problem.

In any case, it is better to use CC algorithms than BP in which one will not only obtain a worse error on test set but also will have to use *ad hoc* approach to hidden layer construction.

References

1. Fahlman S.E., Lebiere C., 1990. The Cascade-Correlation Learning Architecture. In: Touretzky D.S. (ed), 1990. *Advances in Neural Information Processing System, Vol 2.* Morgan Kaufmann, pp 524-532.
2. Rumelhart D.E., Hinton G.E., Williams R.J., 1986. Learning Internal Representations by Error Propagation. In: *Parallel Distributed Processing, Vol. 1.* MIT Press, Ch.8.
3. Hoehfeld M., Fahlman S.E., 1992. Learning with Limited Numerical Precision Using the Cascade-Correlation Algorithm. *IEEE Transactions on Neural Networks* 3:No.4.

Determination of Gas Temperature in a CVD Reactor from Optical Emmission Spectra with the Help of Artificial Neural Networks and Group Method of Data Handling (GMDH)

S.A.Dolenko, A.V.Filippov*, A.F.Pal, I.G.Persiantsev, and A.O.Serov

Microelectronics Dept., Nuclear Physics Institute, Moscow State University, Vorobjovy gory, Moscow, 119899, Russia

* Troitsk Institute of Innovation and Fusion Research, Troitsk, Moscow Region, 142092, Russia

E-mail: dolenko@micnel.npi.msu.su

Abstract. Optical emission spectroscopy is widely used for monitoring of low-temperature plasmas in science and technology [1]. However, determination of temperature from optical emission spectra is an inverse problem that is often very difficult to solve steadily by conventional methods. This paper reports successful application of artificial neural networks (ANN) and group method of data handling (GMDH) for determination of gas temperature from model optical emission spectra analogous to those recorded in a DC-discharge CVD reactor used for diamond film deposition.

1. Introduction: Determination of Gas Temperature from Optical Emission Spectra

Low-temperature plasma in chemical vapour deposition (CVD) reactors has been an object of investigation for many researchers, both from academia and industry [2-4]. Knowledge of physico-chemical properties of plasma in CVD reactors is necessary to build adequate theoretical models of plasmochemical processes in such reactors, and to develop technology routines required to produce radicals necessary for deposition of films with required properties.

Gas kinetic temperature is one of the basic parameters of a reactive system affecting chemical compound of the system, because many gas-phase reaction rate coefficients are strongly dependent on gas kinetic temperature. Generally, gas kinetic temperature is equal to the rotational temperature of the electronically excited states of diatomic molecules, because virtually every gas kinetic collision changes the rotational quantum number, while collisions that change the electronic state or vibrational quantum number are much less frequent. Therefore, gas kinetic or translational temperature is derived by *in situ* optical emission spectroscopy method from measurement of relative intensities of rotational lines within a single vibrational band of some band system of diatomic molecules well studied in spectroscopy, such as N_2, CO, H_2 etc.

For temperature measurement, rotationally resolved or unresolved emission spectra may be used. In the case of resolved spectrum, the task of temperature determination is mathematically more simple; however, one needs an expensive spectrometer of high resolving power to obtain such spectra. Therefore, this method can hardly be applied in technology. In the case of unresolved spectrum, a spectrometer of relatively low resolving power is sufficient; however, the mathematics procedure of determination of gas kinetic temperature is more complicated.

The problem of determination of gas temperature from optical emission spectra is an incorrect inverse problem in its classical statement: knowing some function, one determines some parameters governing the behaviour of this function.

Usually such problem is solved by fitting theoretically calculated optical emission spectra to measured spectra with a least-squares routine. However, this approach encounters some difficulties. Presence of a large number of local minima hampers the procedure of minimisation of error. Under conditions of high temperature, when thermal equilibrium emission is considerable and when emission intensities of molecular band systems are affected by fluctuations, this method turns out to be inapplicable for determination of gas temperature in real time.

During the last decade, artificial neural networks (ANN) [5] proved to be a powerful technique for solution of different problems in image recognition, classification and forecasting [6]. The ability of neural networks to extract important features from raw input data, restoring problem parameters even in cases when the influence of these parameters on the measured signal is unknown or is distorted by noise, has been used to solve incorrect inverse problems in other areas of science and technology, e.g. [7]. It is also important that a trained network is able to produce an answer in fractions of a second, so it may be used for real-time measurements.

Group method of data handling (GMDH) is a powerful method of mathematical modelling, that is able to find the solution of modelling problem by constructing a sophisticated analytical expression that can describe the behaviour of the modelled system most closely [8]. GMDH has been successfully used in a wide area of different applications [9]. Considering the unknown underlying dependence between the determined parameter (temperature) and measured function (spectrum shape) as the object of modelling, one can apply GMDH for solution of inverse problems of the type discussed here. Such medium-sized problems with continuous output turn out to be most suitable for GMDH [10]. Note that once the analytical GMDH model has been found, application of this model is very quick and computatively cheap, so GMDH is ideally suitable for work in real time.

In this paper, we report first results of application of ANN and GMDH for determination of gas temperature from model optical emission spectra of molecular CO, analogous to those recorded in a DC-discharge CVD reactor used for diamond film deposition.

2. Object of investigation. Program for solution of direct problem.

Although by now ANN and GMDH were used by the authors to determine temperature from model spectra only, these model spectra were calculated in the spectral region and in the conditions corresponding to those of the existing CVD reactor. Moreover, the final goal of this investigation is use of ANN and GMDH for determination of temperature from actual optical emission spectra of the discharge in this reactor. That is why a short description of the experiments is given below.

Experiments were carried out with the set-up described in [11]. Direct current glow discharge was investigated under gas pressure of about 150 Torr and with gas flow rate of 20 m/s in the mixtures $CO+H_2$ with different component percentage with a small admixture of Kr in some cases. Typical discharge parameters were: discharge current I_d - 300 mA, with discharge radius being equal to 0.2 cm, voltage 680 V, discharge gap 1.4 cm. Emission spectra were recorded with scanning monochromator MS-80. Gas kinetic temperature is to be determined by ANN and GMDH from vibrational band 0-1 of CO $(B^1\Sigma \rightarrow A^1\Pi)$ Angstrom system (it has been already determined from spectra of this band by least squares fitting [11]). Measured and calculated spectra of this band are shown in Figure 2.1.

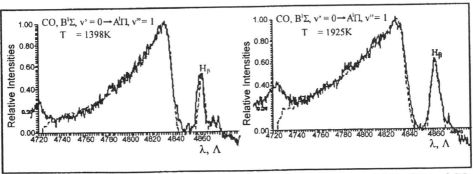

Figure 2.1. Measured (solid) and calculated (dashed) spectra of vibrational band 0-1 of CO $(B1\Sigma \rightarrow A1\Pi)$ Angstrom system:
Left: 5%$CO+H_2$, p=153 Torr, I_d=210 mA;
Right: 5%CO+4%$Kr+H_2$, p=153 Torr, I_d=275 mA.

Spectra modelling was performed by a special computer program in the following way. The intensities of separate rovibronic lines were determined using spectroscopic data for the Angstrom band system from [12-16], from the following expression:

$$I_{n'v'J' \to n''v''J''} = const\,(v_{J'J''})^4 S_e q_{v'v''} S_{J'J''}\, exp(-hcF'(J')/kT),$$ (1)

where $v_{J'J''}$ is the wave number of rovibronic transition **n'v'J'→n''v''J''**, **n** denotes the electronic state, **v'J'** and **v''J''** are vibrational and rotational quantum numbers of upper **n'** and lower **n''** electronic states, S_e is the electronic moment of the transition, $q_{v'v''}$ is the Franck-Condon factor, $S_{J'J''}$ is the Honl-London factor, **F'(J')** is the rotational energy of the upper state in reciprocal centimetres.

After calculation of the positions and intensities of separate rovibronic lines, the spectrum was obtained by convolution with apparatus function of the monochromator, which has been recorded with H_β line of the hydrogen atom. The resulting modelled spectrum was obtained with the step of 0.125 Å, with 1000 steps and 125 Å total range.

Sample model spectra for gas (rotational) temperatures from 500 to 2000 K are presented in Figure 2.2.

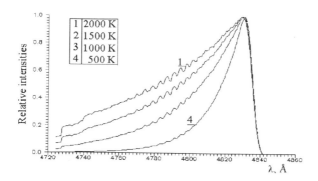

Figure 2.2. Calculated spectra of vibrational band 0-1 of CO $(B^1\Sigma \to A^1\Pi)$ Angstrom system.

3. Data preparation and design of neural network and GMDH algorithm

For use with ANN and GMDH, the program described above was used to model spectra of vibrational band 0-1 of CO $(B^1\Sigma \to A^1\Pi)$ Angstrom system in temperature range from 505 to 2500 K. For these calculations, trapeziform apparatus function having left and right wings with 5 Å width and plate with 1 Å width was used. Each model spectrum was averaged by three neighbouring points, making 333 input points in each pattern used for temperature determination with the help of ANN and GMDH.

To perform training and validation of the networks, the following sets of data were created.

- 200 input patterns for temperatures ranging from 510 to 2500 K with 10 K step were divided into training set (TRN) with 160 patterns and test set of 40 patterns by extracting each 5th pattern into test set.
- Production (validation) sets. Main production set, PR0, contained 200 patterns, for temperatures from 505 to 2495 K with 10 K step. All the other sets were obtained by adding random noise to PR0. For each of the following data sets, 10 different spectra were produced by adding noise with the same amplitude to each spectrum in the PR0 data set, thus making 2000 patterns in each of the following sets. Production sets PE1, PE3, PE5 and PE10 corresponded to noise levels of 1%, 3%, 5%, and 10%, respectively, with noise calculated as fraction of spectrum intensity in each point. Production sets PM1, PM3, PM5 and PM10 corresponded to noise levels of 1%, 3%, 5%, and 10%, respectively, with noise calculated as fraction of total spectrum intensity (normalised into unit).

Two types of noise addition were used to account for two types of noise in actual registration system: the registration error (depending on signal intensity in each point) and the darkness noise of the detector (independent of the signal but dependent on the condition of the detector, noise intensity equal in all measurement points).

All the work with ANN and GMDH was conducted with the help of software package *NeuroShell 2 rel.3.0* from *Ward Systems Group, Inc.* The following network architectures were used in this investigation [17]:

- 3-layer perceptron with standard connections – 8 or 16 hidden neurons, logistic activation function in the hidden layer, linear activation function in the output layer, small values of learning rate (about 0.01) and substantial values of momentum (about 0.9);
- General Regression Neural Network (GRNN) – search of the best smoothing factor on test set in an iterative manner; note that for curves with noise, the best smoothing factor for each noise level was chosen manually;
- Group Method of Data Handling (GMDH) – full cubic polynoms within each layer, Regularity criterion, Extended linear models included, random choice of 30% of possible models to test.

4. Results and discussion

The following statistics are used below to estimate quality of the obtained nets:

RSQ - R squared, the coefficient of multiple determination, a statistical indicator usually applied to multiple regression analysis. *NeuroShell 2* Manual [17] recommends it for use as the main indicator of net quality. It compares the accuracy of the model to the accuracy of a trivial benchmark model wherein the prediction is just the mean of all of the samples. A perfect fit results in an R squared value of 1, a very good fit near 1, and a very poor fit near 0. If the neural model predictions are worse than one could predict by just using the mean of pattern outputs, the R squared value is 0. The formula for RSQ is

$$R^2 = 1 - \frac{SSE}{SS_{YY}},$$

where $SSE = \sum (T - \widetilde{T})^2$;

$SS_{yy} = \sum (T - \overline{T})^2$;

T is the actual value, $\widetilde{T}$ is the predicted value of T, and $\overline{T}$ is the mean of T values.

SD - standard deviation, square root of the mean squared error, in degrees Kelvin (K):

$$SD = \sqrt{\frac{1}{N} \sum (T - \widetilde{T})^2} .$$

MAE - mean absolute error, in degrees Kelvin (K):

$$MAE = \frac{1}{N} \sum |T - \widetilde{T}|.$$

The results for the three tested architectures are presented below in Tables 4.1-4.3. Note that retraining the neural nets several times with different initial values of weights did not cause substantial change in the results.

Table 4.1. Results of temperature determination from model spectra, three-layer perceptron.

Data set	RSQ	SD, K	MAE, K
TRN	0.9999	4.1	2.8
TST	0.9999	4.2	2.8
PRO	1.0000	4.0	2.8
PE1	0.8184	246.0	181.1
PM1	0.7711	276.2	200.0
PE3	0.7522	287.4	225.6
PM3	0.7069	312.6	243.8
PE5	0.6334	349.6	281.4
PM5	0.5134	402.8	319.2
PE10	0	–	–
PM10	0	–	–

Table 4.2. Results of temperature determination from model spectra, GRNN.

Data set	RSQ	SD, K	MAE, K
TRN	0.9253	157.8	89.8
TST	0.9250	158.1	89.3
PRO	0.9252	157.9	89.8
PE1	0.9254	157.6	92.3
PE3	0.9237	159.5	102.4
PE5	0.9206	162.7	112.7
PE10	0.9035	179.3	141.3

Table 4.3. Results of temperature determination from model spectra, GMDH.

Data set	RSQ	SD, K	MAE, K
TRN	1.0000	0.05	0.04
TST	1.0000	0.03	0.02
PRO	1.0000	0.06	0.05
PE1	0.9997	9.2	6.5
PM1	0.9990	18.0	14.3
PE3	0.9954	39.2	28.3
PM3	0.9603	115.0	92.3
PE5	0.9808	79.9	57.3
PM5	0.7446	291.8	232.1
PE10	0.8476	225.4	162.0
PM10	0	–	–

The following facts are worth to be noted and analysed:

- The testing level of measurement precision, that can be derived from comparison with other methods of temperature measurements in CVD reactors, is about 50-70 K. If adaptive computing methods would be able to maintain such (or higher) precision of temperature measurements, then it could be considered a success and an evidence for using these methods for temperature determination.

- The three-layer perceptron network showed good results on all data sets without noise . At the same time, even addition of noise with amplitude equal to 1% of signal intensity in each channel (PE1) makes the performance of this network very poor. To make the network more tolerant to noise, patterns with noise should be used in the training set along with patterns without noise.

- GRNN nets demonstrated very weak dependence of performance on noise level. At the same time, the performance was unsatisfactory even on training and test set. In these conditions, large values of the smoothing factor required for the algorithm to be able to process all the patterns confirm the conclusion that for this problem GRNN algorithm turns out to be too "global-looking". Properties of GRNN that usually make it the preferable choice for work with sparse and incomplete data, make it unable to describe relatively strong influence of temperature on small details of spectra. At the same time, global change of spectrum shape with temperature easily observed in Fig. 2 fails to provide enough information to determine temperature with sufficient precision. All this can be explained by the fact that the spectra used are partially rotationally resolved (but the apparatus function is too wide to make it possible to resolve it completely). Change of relative intensities of separate rovibronic lines (expression (1) in Section 2) cause the little spikes on the spectra not only to change their intensities, but also to move their positions. Such relatively strong changes in spectrum shape may be beneficial for other methods, but not for GRNN.

- GRNN failed to make an estimation for most of patterns in PM data sets, considering the input noisy patterns to be too different from these in the training set. Other architectures, though (obviously) making larger errors on PM data sets compared with PE data sets, yet were capable to make an estimation of temperature with reasonable error.
- For use with real experimental spectra, GRNN would require estimations of intensity and type of noise in experiment – in order to select the correct value of the smoothing factor. This problem would require a special investigation.
- Contrary to GRNN, GMDH demonstrated not only extremely high precision in absence of noise, but also satisfactory results for noise levels up to 5%. Such remarkable performance on independent data, even on noisy data, may be related to use of Regularity criterion, and also to the ability of GMDH algorithm to select most important features of the spectra.

5. Conclusions and future work

Neural nets and GMDH were successfully applied for determination of gas temperature from model optical emission spectra analogous to those recorded in a DC-discharge CVD reactor used for diamond film deposition. GMDH demonstrated excellent results on independent data without noise and relatively high tolerance to noise in the input data, the three-layered perceptron gave high precision on independent data without noise, and GRNN algorithm in the conditions of this problem failed to achieve reasonable precision.

Future work should include:
- Investigation of opportunities of other ANN architectures for the studied problem;
- Introducing noise during ANN training, in order to increase tolerance of the resulting nets to noisy input data (especially for backpropagation networks);
- Testing different methods of spectra pre-processing (averaging over a larger number of points, spectra smoothing etc.)
- Application of the method to real life experimental spectra recorded on the described CVD reactor and comparison of the results to these obtained by least squares fitting [11].

References

1. Gottscho R.A., Miller T.A., 1984. Optical Techniques in Plasma Diagnostics. *Pure & Appl Chem* 56:189-208.
2. Reeve S.W., Weimer W.A., 1995. Plasma diagnostics of a direct-current arc jet diamond reactor. II. Optical emission spectroscopy. *J Vac Sci Technol A* 13:359-367
3. Pastol A., Catterine Y., 1990. Optical emission spectroscopy for diagnostic and monitoring of CH_4 plasmas used for a-C:H deposition. *J Phys D: Appl Phys* 23:799-805.
4. Chu H.N., Den Hartog E.A., Lefkow A.R. et al., 1991. Measurements of the gas kinetic temperature in a CH_4-H_2 discharge during the growth of diamond. *Phys Rev A* 44:3796-3803.
5. Lippmann R.P., 1987. An introduction to computing with neural nets. *IEEE ASSP Magazine* 3:No 4:4-22.
6. Orliov Yu.V., Persiantsev I.G., Rebrik S.P., Babichenko S.M., 1995. Application of neural networks to fluorescent diagnostic of organic pollution in a water. *Proc SPIE*, 2503:150-156.
7. Cennini P., Cittolin S., Revol J.P. et al., 1995. A neural network approach for the TPC signal processing. *Nucl Instrum and Methods in Phys Research A*, 356:507-513.
8. Ivakhnenko A.G., 1971. Polynomial theory of complex systems. *IEEE Trans Syst Man & Cybern* SMC-1:364-378.
9. Farlow S.J., ed., 1984. *Self-Organizing Method in Modeling: GMDH Type Algorithms.* (Statistics: Textbooks and Monographs, No 54).
10. Dolenko S.A., Orlov Yu.V., Persiantsev I.G., 1996. Practical implementation and use of group method of data handling (GMDH): prospects and problems. In: Parmee I.C. (ed.), 1996. *Adaptive Computing in Engineering Design and Control. Conference proceedings.* PEDC, Univ. of Plymouth, UK, pp 291-293.
11. Kashko D.V., Pal' A.F., Rakhimov A.T., Serov A.O., Suetin N.V., Filippov A.V., 1996. The deposition of diamond films in dc discharge with close circle gas circuit. *Proc. 7th Int. Symp. "Thin films in electronics"*, Moscow - Ioshkar-Ola, pp 19-32.
12. Huber K.P., Herzberg G., 1979. *Molecular Spectra and Molecular Structure IV. Constants of Diatomic Molecules.* Van Nostrand Reinhold Co., New York, pp160-169.
13. Field R.W., Wieke B.G., Simmons J.D., Tilford S.G., 1972. Analysis of perturbations in the $A^3\Pi$ and $A^1\Pi$ states of CO. *J Mol Spectr* 44:383-399.
14. Murthy N.S., Prahllad U.D., 1979. Integrated intensity measurements and relative band strengths of CO (Angstrom) bands. *Physika* 97C:385-387.
15. Robinson D., Nicholls R.W., 1958. Intensity measurements on the O_2^+ Second Negative, CO Angstrom and Third Positive and NO gamma and beta molecular band systems. *Proc Phys Soc* 71:957-964.
16. Kovacs I., 1969. *Rotational structure in the spectra of diatomic molecules.* Akademiai Kiada, Budapest.
17. *NeuroShell 2 User Manual.* Fourth edition - June 1996. Ward Systems Group, Inc., Frederick, MD, USA.

Multi–Domain Optimisation Using Computer Experiments for Concurrent Engineering

R A Bates[†], R Fontana[‡], L Pronzato[§] and H P Wynn[†]

†University of Warwick, Coventry CV4 7AL, UK.
‡Centro Ricerche FIAT, Orbassano, Torino, Italy.
§Laboratiore i3S, CNRS, Sophia Antipolis, France.

Abstract

It is a challenge to optimise several simulators in a multi-objective way in Computer-Aided Engineering (CAE). The technology of replacing each simulator with a fast statistical emulator based on a space-filling computer experiment is introduced. A case study based on a rudimentary model of a car suspension system is conducted in full and is one of a series in a collaborative EU project (No. BE96-3046) which has a particular emphasis on multi-domain problems: mechanical, thermal, electrical etc. In the case study the objectives from three separate simulator responses are combined in different portmanteau criteria; global optimisation is then used to find Pareto solutions.

1 Background

Industry is urgently engaged in improving its competitive position: shortening time to market, cost reduction, product quality improvement. Computer-aided engineering (CAE) occupies a strategic position through its support for design optimisation and reduction of costly physical experimentation. Using CAE for multi–criteria, global design optimisation paves the way for the construction of a truly Concurrent Engineering environment.

High computational effort is a barrier to successful and fast optimisation, vital for Concurrent Engineering. Typical practice involves running simulation software in a sequential, manual fashion with local, piecemeal optimisation, leading at best to local optima. These limitations are made more serious by the increased complexity of design problems, the interactions between different domains (mechanical, thermal, electronic etc.) and the large range of available simulators.

This paper describes work completed as part of an EU–funded research programme: $(CE)^2$ Computer Experiments for Concurrent Engineering (Project No. BE96-3046). The philosophy of $(CE)^2$ is to determine optimal engineering solutions to design problems by modelling the Computer-Aided Engineering

356

analysis of engineering designs with fast statistical models, *emulators*. The paper sketches the background and presents a preliminary case study which concerns the optimisation of a rudimentary car suspension system over three different road profiles.

2 The $(CE)^2$ project

The $(CE)^2$ project has, at its core, mathematical functions, *emulators*, that represent simulator code. An emulator is a highly adaptive but still simple mathematical model, which mimics the simulator over the whole design space. It is built by experimenting on the simulator using a well-chosen set of inputs. Different emulators can be combined for global optimisation, multiobjective optimisation, sensitivity analysis and fast graphical display. This substitution of the simulator represents a trade-off between accuracy and fast analysis, see Table 1.

	Speed	Accuracy	Optimisation
Simulator	slow	very good	local
Emulator	fast	good	global

Table 1: Comparison of simulators with emulators

The project will culminate in a decision support system that incorporates state-of-the-art multi-objective algorithms on a supporting foundation of statistical emulators, derived from the design and analysis of computer experiments conducted on the simulators. A major innovation of the project is the concentration on "multi–domain"optimisation, that is simultaneous optimisation of mechanical, electrical, thermal etc. simulators.

3 Emulation of computer simulators

The simulator to be emulated is considered as a deterministic model,

$$y = f(\mathbf{x}), \tag{1}$$

where y is the simulator *response* and $\mathbf{x} = x_1, \ldots, x_d$ is the vector of d design parameters, or *factors*. The response y is obtained by running the simulator at the factor values given by $\mathbf{x}$. The emulator model, $\hat{f}$, approximates the response,

$$\hat{y} = \hat{f}(\mathbf{x}). \tag{2}$$

The model is built by conducting an experiment which consists of running the simulator at a set of carefully selected *design points*, $\mathbf{s}_i = s_1, \ldots, s_d$. This set of n design points, $S = \{\mathbf{s}_1, \ldots, \mathbf{s}_n\}$ is called an *experimental design*. The experimental design, S, and the corresponding set of responses, $Y = y_1, \ldots, y_n$, contain the information used to build the emulator model.

The $(CE)^2$ project itself will use a library of alternative emulators. We discuss next the emulator arising from previous projects [1, 2, 3, 5, 7].

3.1 The DACE emulator

Consider the model

$$y(\mathbf{x}) = \beta + \mathbf{Z}(\mathbf{x}) \tag{3}$$

where $\mathbf{Z}(\mathbf{x})$ is a random function with zero mean and β is an unknown constant. At two sets of inputs, $\mathbf{x} = x_1, \ldots, x_d$ and $\mathbf{x}' = x'_1, \ldots, x'_d$, the covariance between $\mathbf{Z}(\mathbf{x})$ and $\mathbf{Z}(\mathbf{x}')$ is

$$Cov(\mathbf{Z}(\mathbf{x}), \mathbf{Z}(\mathbf{x}')) = \sigma^2 \mathbf{R}(\mathbf{x}, \mathbf{x}'). \tag{4}$$

The covariance function $\mathbf{R}(\mathbf{x}, \mathbf{x}')$ is chosen to reflect the characteristics of the data to be modelled. For $(CE)^2$ the choice is the family

$$\mathbf{R}(\mathbf{x}, \mathbf{x}') = \prod_{i=1}^{d} \exp(-\theta_i |x_i - x'_i|^{p_i}) \tag{5}$$

The parameters θ_i and p_i are unknown and need to be estimated along with the parameters β and σ^2. These estimates are made using maximum liklelihood estimation (MLE).

Let $Y = (y_1, \ldots, y_n)$ denote the observed responses at an experimental design of n input vectors, $S = \{\mathbf{s}_1, \ldots, \mathbf{s}_n\}$, and write

$$\mathbf{r_s} = [\mathbf{R}(\mathbf{x}, \mathbf{s}_k)] \qquad \mathbf{R}_s = \{\mathbf{R}(\mathbf{s}_k, \mathbf{s}_{k'})\} \tag{6}$$

which are an $n \times 1$ vector and an $n \times n$ matrix, respectively.

It can be shown that the best linear unbiased predictor of $Y(\mathbf{x})$ when $\mathbf{R}$ is known is

$$\hat{Y}(\mathbf{x}) = \hat{\beta} + \mathbf{r}'_s \mathbf{R}_s^{-1}(Y - \hat{\beta}\mathbf{1}) \tag{7}$$

where $\mathbf{1}$ is a vector of 1's. This is the equation that the DACE emulator uses for interpolation. With this correlation structure two points, x_i and x'_i, that are close together will have highly correlated g's. The predictor also has the exact interpolation property

$$\hat{Y}(\mathbf{s}_i) = Y(\mathbf{s}_i) \qquad i = 1, \ldots, n. \tag{8}$$

3.2 Cross validation

Cross validation is used to check the accuracy of the DACE emulator. This involves predicting at each design point in turn when that point is left out of the predictor (Equation 7). Let $\hat{y}_i(\mathbf{x})$ be the estimate of $y(\mathbf{x}_i)$ based on all the design points except $y(\mathbf{x}_i)$. The prediction error (the estimated root mean square error, ERMSE) is then calculated as

$$ERMSE = \sqrt{\frac{1}{n} \sum_{i=1}^{n} [\hat{y}_i(\mathbf{x}) - y(\mathbf{x}_i)]^2} \tag{9}$$

The estimate $\hat{y}_i(\mathbf{x})$ is computed using the values $\hat{\theta}$ and $\hat{p}$ estimated using the full experimental design S. This avoids the overhead of recomputing the model parameters at each prediction, and has been shown to not adversely affect the estimate of prediction error [7].

4 Experimental design

Latin Hypercube Sampling (LHS) designs [4] are good for filling space in high dimensions. They are also fast to compute because of their pseudo-random nature. These facts make them highly suited for use as experimental design plans for computer experiments in high dimensions, involving many input factors. Normalising the input factors so that they all lie in the range $[-0.5, +0.5]$, all possible combinations of d input factors will occur in the space $[-0.5, +0.5]^d$. For an experiment with n runs an LHS design is constructed by dividing the interval $[-0.5, +0.5]$ into n equally spaced values for each of the d factors and randomising them. Let $\mathbf{z} = [0, 1, ..., n-1]$, where n is the number of runs in the experimental plan. Then

$$s_j = \frac{\pi_j(\mathbf{z}) - 1/2}{n} \ , \quad j = 1, ..., d \tag{10}$$

is the jth column of the experimental design S, where $\pi_1, ..., \pi_d$ are independent random permutations of $\mathbf{z}$. This algorithm places the design points in the centre of the randomly selected sections of a grid.

5 Multiobjective optimisation

Except in very rare situations, it is impossible to minimize all the responses at the same time, and multiobjective optimization can be considered as the process of obtaining *reasonable compromises* from previously defined *preferences*. A well-admitted definition for "reasonable" is *non-dominated* in the sense of a binary relation between the responses, see for example [6, 8, 9].

A classically used preference is the *Pareto preference*: $\mathbf{f}^1$ is preferable to $\mathbf{f}^2$ if and only if $f_i^1 \leq f_i^2$, $i = 1, ..., m$, and $\mathbf{f}^1 \neq \mathbf{f}^2$. A non-dominated solution $\mathbf{f}^* \in \mathcal{F}$ is such that there is no $\mathbf{f}$ in $\mathcal{F}$ preferable to $\mathbf{f}^*$. The set of non-dominated solutions is the *Pareto set*, that we shall denote $\mathcal{F}^*$. The associated set in the $\mathcal{X}$ space is the Pareto-optimal solution set, to be denoted $\mathcal{X}^*$. A natural objective in this context is to obtain points in $\mathcal{X}^*$.

Minimizing a convex combination $F(\mathbf{f}(.)) = w^T \mathbf{f}(.) = \sum_{i=1}^m w_i f_i(.)$, with the vector $w = (w_1, ..., w_m)^T$ in the canonical simplex

$$\mathcal{C} = \{w \in I\!\!R^m \mid \sum_{i=1}^m w_i = 1, \ w_i \geq 0, i = 1 ..., m\},$$

is a simple method to get a point in $\mathcal{X}^*$. Note, however, that not all the points in $\mathcal{X}^*$ can be obtained in this way, by varying $w \in \mathcal{C}$, if the $f_i(.)$'s are not convex. Also note that, even if the $f_i(.)$'s are uniextremal (no local minimum exists), $w^T f(.)$ can have local minima. For complex multi-objective problems the use of global optimization methods is recommended.

5.1 Multi–domain optimisation

The use of emulator models in optimisation naturally leads one to consider multi–domain optimisation. A common design scenario is where several different responses of a particular design are considered, for example in the design of a turbine blade, one needs to conduct stress analysis and thermal analysis of the turbine blades. The advantage of the emulator approach is that each design response is modelled using a common emulator form. This provides, perhaps for the first time, global optimal solutions to complex multi–domain engineering design problems.

6 A preliminary case study

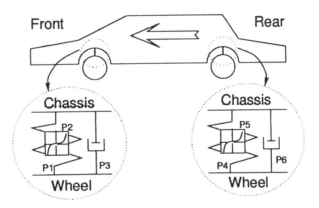

Figure 1: Rudimentary car suspension system.

The following case study is one of a series carried out by a team as part of the $(CE)^2$ project. It is based on an in-house simulation code at Centro Ricerche FIAT.

The car suspension system is modelled in a very basic form using three components: a spring, a rubber stop and a damper, characterised by three parameters: spring stiffness, rubber characteristic and damping coefficient. The effect of the rubber characteristic is to introduce a nonlinear response to the spring. As the car passes over the (uneven) road surface, energy is transmitted to the car chassis and the suspension parameters determine how much energy is absorbed by the suspension system and how much is transmitted through to the chassis. With reference to Figure 1, the design parameters and corresponding ranges are:

- $P1, P4 \in [1.6, 2.6]$ spring stiffness [Kg/mm] (front,rear)

- $P2, P5 \in [0.0, 40.0]$ rubber characteristic [mm] (front,rear)

- $P3, P6 \in [1000.0, 4000.0]$ damping coefficient [N/ms^{-1}] (front,rear)

Note that the rubber characteristic defines the point at which the rubber stop begins to influence the behaviour of the spring between the wheel axel and the chassis.

The suspension system is modelled with three different road profile simulators to explore the following conditions:

1. STEP: step of 80 mm (car speed of 60 km/h)

2. INSAC: long road undulation (car speed of 80 km/h)

3. PAVE: random road profile.

Each road profile simulator returns the following outputs:

1. *Stp*: spring maximum load,

2. *Isc*: half of the difference between the maximimum and the minimum of the vertical acceleration of a predefined point of the drivers seat, and

3. *Pve*: RMS of the signal corresponding to the vertical acceleration of a predefined point of the drivers seat.

The objective of the case study is to find a design solution which minimises *Stp, Isc* and *Pve* by experimenting on the three simulators, building emulator models of the simmulators and using these emulators to build an objective function for optimisation.

6.1 The computer experiment

A LHS design of $n = 24$ design points was generated for the 6 parameters. This design was used to obtain corresponding responses for *Stp, Isc* and *Pve*, from the three analytical models supplied by CRF. Using this data, a DACE model emulator [5] was fitted to each of the three responses.

For the *Stp* response the emulator fitted to the $n = 24$ point LHS design was not very accurate. This was caused by one of the response values being much higher than the other 23 values. Discussion with the CRF design engineer revealed that this point corresponded to a design in which the spring stiffness and the damping coefficient at the front of the car were very low and the rubber characteristic was high. Consequently most of the energy was being transmitted to the car body via the rubber stop, modelled as the nonlinear part of the spring response.

A second modelling experiment with $n = 96$ design points was carried out to improve the accuracy of emulating the *Stp* response. Table 2 shows the Cross Validation (CV) results associated with each emulator model, the range of the response and an estimate of the percentage error associated with predicting at a new design point. The CV results, plotted in Figure 2, show that the emulators are accurate with less than 5.8% error over the output range.

Response	n	CV ERMSE	Range	% Error
Stp	96	29.05	964.7	3.011
Isc	24	7.661×10^{-4}	5.959×10^{-2}	1.286
Pve	24	2.599×10^{-3}	4.497×10^{-2}	5.778

Table 2: Cross Validation results for the CRF case study.

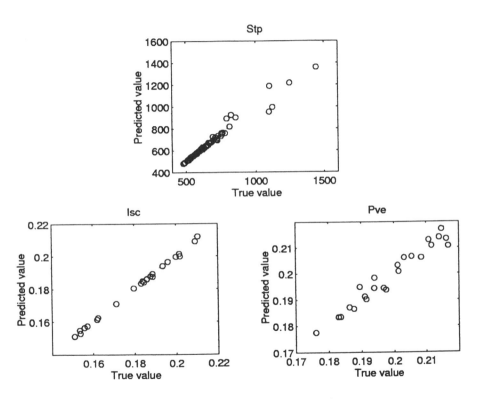

Figure 2: Cross Validation results

The main effects plots of Figure 3 describe the individual linear effect of each design parameter on the response, with the effects of the other parameters integrated out of the model, see [7].

6.2   Optimisation

The three emulator models were used to formulate the optimisation problem. The objective function for optimisation was to find values for $P1$ to $P6$ which

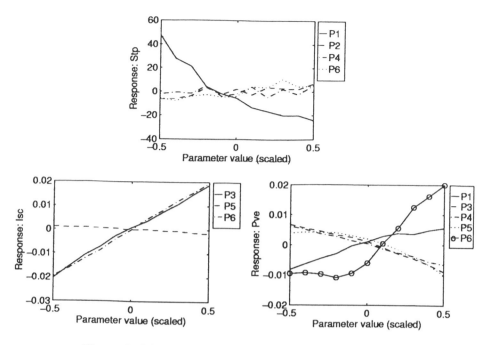

Figure 3: Main effects results for the three emulators.

minimise Stp, Isc and Pve. The objective function was defined as:

$$\min(w_1 Stp' + w_2 Isc' + w_3 Pve') \tag{11}$$

where $W = (w_1, w_2, w_3)$ is a vector of weighting factors, and for each response $Y \in \{Stp, Isc, Pve\}$,

$$Y' = \frac{Y - Y_{min}}{Y_{max} - Y_{min}} \tag{12}$$

and Y_{max}, Y_{min} are the highest and lowest values of Y obtained during the experiment.

Optimal solutions for 10 different sets of weights in the canonical simplex (5), described in Table 3, were calculated using a multi–objective optimiser. Figure 4 shows the results of the 10 different global optimisations.

The optimal design for $w_1 = w_2 = w_3 = 1/3$ is shown in Table 4 along with emulator responses, the corresponding simulator confirmation responses and the percentage error associated with emulating the optimal design.

7 Discussion

A brief description of the $(CE)^2$ research programme has been given. A case study involving the optimisation of a car suspension system has shown the capability of the $(CE)^2$ methodology in dealing with multiple design objectives

Solution no.	w_1	w_2	w_3
1	1/3	1/3	1/3
2	1	0	0
3	0	1	0
4	0	0	1
5	1/2	1/2	0
6	1/2	0	1/2
7	0	1/2	1/2
8	2/3	1/6	1/6
9	1/6	2/3	1/6
10	1/6	1/6	2/3

Table 3: Weights for multi-objective optimiser

Parameter	Value	Range	Response	Emulator	Simulator	% Error
$P1$	2.245	[1.6,2.6]	Stp	780.9	719.5	8.53
$P2$	7.372	[0,40]	Isc	0.1782	0.1596	2.87
$P3$	1276	[1000,4000]	Pve	0.1403	0.1506	0.63
$P4$	2.6	[1.6,2.6]				
$P5$	40	[0,40]				
$P6$	1000	[1000,4000]				

Table 4: The optimal design for $w_1 = w_2 = w_3 = 1/3$ with corresponding responses.

in different domains. The multi-objective optimisation of the car suspension system gave 10 pareto solutions in the canonical simplex. An analysis of these results (Figure 4) shows that weighting schemes 1, 9 and 10 provide the best among the ten trial solutions. The solutions for weighting schemes 2, 3 and 4 should, in theory, provide the best design solutions for Stp, Isc, and Pve. The results show that this is the case for 2 but that solutions 3 and 4 are not quite the best mono-objective solutions, however the solutions are almost the best and this is encouraging. The results of solution 1 are given and compared with the actual simulator results for the same design. The results show that the emulators are very accurate for the Isc and Pve responses (2.87% and 0.63% error respectively) and fairly accurate for the Stp response (8.53%). The case study results show the ability of the $(CE)^2$ methodology to deal with different simulations of an engieering model and provide a framework for the use of global multi-objective optimisation methods in design.

References

[1] R A Bates, R J Buck, E Riccomagno, and H P Wynn. Experimental design for large systems. *Journal of the Royal Statistical Society B*, 58:77–94, 1996.

[2] R A Bates, R J Buck, and H P Wynn. Generic circuit response interpolation for robust design. *IEE Proc Circ. Dev. and Syst.*, 142:131–134, 1995.

364

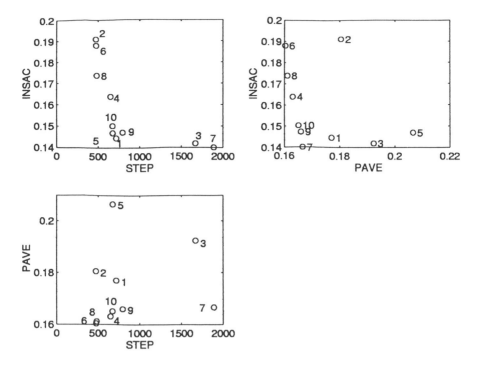

Figure 4: Multi-objective optmisation results - values rescaled to original values after optimisation.

[3] R A Bates and H P Wynn. Tolerancing and optimisation for model–based Robust Engineering Design. *Quality and Reliability Engineering International*, 12:119–127, 1996.

[4] M D McKay, W J Conover, and R J Beckman. A comparison of three methods for selecting values of input variables in the analysis of output from a computer code. *Technometrics*, 21:239–245, 1979.

[5] J Sacks, W J Welch, T J Mitchell, and H P Wynn. Design and analysis of computer experiments. *Statistical Science*, 4:409–435, November 1989.

[6] R E Steuer. *Multiple Criteria Optimization: Theory, Computations and Applications*. Wiley, New York, 1986.

[7] W. J. Welch, R. J. Buck, J. Sacks, H. P. Wynn, T. J. Mitchell, and M. D. Morris. Screening, predicting, and computer experiments. *Technometrics*, 34(1):15–25, 1992.

[8] P L Yu. *Multiple Criteria Decision Making: Concepts techniques and Extensions*. Plenum Press, New York, 1985.

[9] P L Yu. Multiple criteria decision making: five basic concepts. In G L Nemhauser et al., editor, *Handbooks in OR & MS vol.1*, pages 663–699. Elsevier, Amsterdam, 1989.

AUTHOR INDEX

Andre, D. 177
Averill, R.C. 121
Bates, R.A. 355
Bennett, F.H. 177
Bierwirth, C. 59
Blythe, P.W. 269
Bulman, S. 287
Chen, K. 221
Chudova, D.I. 339
Coello Coello, C.A. 151
Cvetkovic, D. 255
Dimopoulos, C. 69
Dolenko, S.A. 339, 345
Doval, A.B.G. 315
El-Beltagy, M.A. 111
Elby, D. 121
Filippov, A.V. 345
Fontana, R. 355
Gen, M. 95
Gero, J.S. 3, 207
Ghasemi, M.R. 287
Goodman, E.D. 121
Hajela, P. 12
Hinton, E. 287
Holden, C.M.E. 233
Husbands, P. 161
Keane, A.J. 111
Keane, M.A. 177
Kim, D.G. 161

Koza, J.R. 177
Li, D.G. 301
Li, Y. 95
Mattfeld, D.C. 59
Mohan, C.K. 315
Montano. L. 85
Orlov, Yu.U. 339
Pal, A.F. 345
Parmee, I.C. 27, 193, 221, 255
Pavlov, D.Yu. 339
Persiantsev, I.G. 339, 345
Prasad, M.K. 315
Pronzato, L. 355
Punch, W.F. 121
Schnier, T. 207
Schoenauer, M. 137
Sen. P. 45
Serov, A.O. 345
Sourd, F. 137
Tettamanzi, A.G.B. 325
Todd, D.S. 45
Troyani, N. 85
Watson, A.C. 301
Watson, A.H. 193
Webb, E. 255
Wright, W.A. 233
Wynn, H.P. 355
Zalzala, A.M.S. 69
Zarubin, V.A. 245